*Symmetry and Structure*

# *Symmetry and Structure*

S. F. A. KETTLE

*Professor of Inorganic and Theoretical Chemistry*
*University of East Anglia*

JOHN WILEY & SONS

Chichester • New York • Brisbane • Toronto • Singapore

*Library of Congress Cataloging in Publication Data:*

Kettle, S. F. A. (Sidney Francis Alan)
Symmetry and structure.

Includes index.
1. Chemical structure. 2. Symmetry (Physics)
3. Groups, Theory of. 4. Chemical bonds. 5. Molecules.
I. Title.
QD471.K47516 1985 541.2′2 84-17365

ISBN 0 471 90501 1 (Cloth)
ISBN 0 471 90705 7 (Paperback)

*British Library Cataloguing in Publication Data:*

Kettle, S. F. A.
Symmetry and structure.
1. Molecular structure 2. Symmetry (Physics)
I. Title
541.2′2 QD461

ISBN 0 471 90501 1 (Cloth)
ISBN 0 471 90705 7 (Paperback)

Typeset by Mathematical Composition Setters Ltd, 7 Ivy Street, Salisbury, Wiltshire
Printed by St. Edmundsbury Press, Suffolk

# *Contents*

# *Preface*

It has been claimed that group theory is little more than applied commonsense. If this is so, then it should be possible to present it in a way which avoids explicit use of formal mathematics and, in particular, matrix algebra. Matrix algebra is not a difficult topic and many students will have met it before entering University. To write a book simply to avoid its use would, then, seem a pointless exercise. Yet this is such a book; the reason for its existence lies deeper. In my experience, chemists prefer to think in terms of models and pictures rather than mathematics; they find it easier to describe a model mathematically than to start with a mathematical development and derive a picture from it. For a full understanding of group theory both picture and mathematics are needed and so it is usual to develop them together. Unfortunately, group theory is a sequential subject — each stage depends on preceding stages so that in a text there is a need for constant referral back to earlier pages. Because the mathematical treatment is more precise and comprehensive than the pictorial, the back references are most readily made to mathematical sections. My experience is that for most students the physical picture becomes more and more hazy as the mathematics takes over. This is the reason for the structure of the present book. The subject is presented pictorially — but accurately — so that the student is not referred back to mathematical equations but rather to pictorial explanations. If the text provides the pictures where, then, is the mathematics? The answer is 'in the Appendices'. These comprise a significant proportion of the book and, whilst I hope that they more or less stand on their own as a text on mathematical group theory, they are also integrated with the main text. In this way I hope to have gained something of the best of both worlds! In the text itself I have avoided mathematics by using chemical topics as vehicles for the group theory. For much of the book the vehicle is chemical bonding but in the later chapters other subjects are covered.

Throughout, I have tried to choose topic and group theory in a supportive manner; a topic is only included if it enables some development of the group theoretical theme. Conversely, the use of symmetry arguments must tell us something new about the particular topic. I have therefore been more concerned with establishing a firm basis than with its development. Thus, although both ligand field theory and the Woodward–Hoffman rules are included, neither is treated at any length because no new principles would have immediately emerged in an extension which are not covered elsewhere.

In writing the book I have assumed an elementary knowledge of valence theory — in particular, the shapes of atomic orbitals. However, whenever I have first used terms such as 'orthogonal' and 'normalized' I have given a reminder of their definition.

The content of the book is largely determined by that group theory which should be of utility to an undergraduate, although there are a few points at which I have included something simply because I find the topic fun and hope the reader does too! There is no discussion of space groups because for these it is the symmetry operations, and not the associated group theory, with which most students will be concerned. There is no section on the formation of hybrid orbitals because this would not have been in keeping with the presentation of the rather different approach to chemical bonding which forms a major theme in the book. However, this neglect should not be seen as a reflection on the concept of hybrid orbitals, which, in fact, has a secure basis (see D. B. Cook and P. W. Fowler, *Amer. J. Phys.*, **49**, 857, 1981) and appear at several points in the text.

Chapters 1 to 7 of the book contain basic material whereas Chapters 9, 10 and 11 generally cover more advanced topics. Chapter 8 forms a bridge between these two sections. Whilst I hope that the presentation in the last three chapters is one which the average reader will readily follow, I have taken the opportunity to include aspects which are omitted in many textbooks. Here, too, I have used a chemically relevant topic as a vehicle to introduce some new aspect of group theory.

Those who wish to approach some of the material in this book from a related but different viewpoint may find it helpful to listen to two audiotapes, called *Symmetry in Chemistry*, which I have recorded for the Educational Techniques Group Trust of the Royal Society of Chemistry, London.

I am grateful to those institutions at which I have been able to write or revise parts of the manuscript — the University of Massachusetts (Amherst), Northwestern University, The Université de Paris-Sud and Universita di Torino.

In writing this book I have been helped by the constructive criticism of many colleagues at the University of East Anglia, by Drs. H. Fritzer and A. Hutcheon in particular as well as several anonymous reviewers. To Mrs. J. Johnson and Mrs. M. Livock I owe a debt of gratitude for producing excellent typescripts from sometimes near illegible manuscripts. Defects and errors that remain are, of course, my own responsibility.

Sidney F. A. Kettle<br>Norwich, April, 1984

# CHAPTER 1
# *Theories in conflict*

## 1.1 INTRODUCTION

This book is concerned with the symmetry and structure of molecules. Of these, the latter — in the sense of either the geometric or electronic structure — has long been of concern to chemists. In this text we shall be interested in both these aspects of the structure of molecules and shall adopt the viewpoint that the geometric structure of a molecule tells us something about its electronic structure. The connection between the two will be provided by the molecular symmetry. Ultimately, however, this book is concerned with the chemical consequences of molecular symmetry, and these extend far beyond the problems of chemical bonding. Rather, we shall use the problem of chemical bonding as a particularly convenient — and important — way of introducing the concepts of symmetry and then extend the application of the concepts revealed in this way to other areas of chemistry. In an introductory text such as this there will be no attempt to cover all of the uses of symmetry — an objective which it would be difficult to achieve in any text. Instead, we shall detail some of the more important aspects and provide a sufficiently comprehensive cover of the basis of the subject to enable the reader to apply them in other areas. Further, we shall do this in an almost entirely non-mathematical manner; rather detailed but, hopefully, readable mathematical treatments are reserved for the appendices.

## 1.2 THE AMMONIA MOLECULE

The ammonia molecule provides a convenient starting point for our study and we shall use it to see the problem of chemical bonding in a rather unusual perspective, one that leads us to attempt to infer molecular bonding *from* molecular geometry (in contrast to the more common procedure of explaining molecular geometry *in terms of* chemical bonding). We shall first review several approaches to the bonding in the ammonia molecule. The reader may well not be familiar with all of them but should not spend too much time trying to master any new ones — we are only interested in generalities in the present context. However, references are given to enable the reader to explore any of the approaches in more detail, should he or she so wish.

### 1.2.1 The atomic orbital model

This model has an historic importance — it is the description to be found in many pre-*ca*. 1955 texts.[1] Before looking at it we shall first detail the facts. The ammonia molecule is pyramidal in shape; all three hydrogen atoms are equivalent, the HNH bond angle being 107° (Figure 1.1). The simplest, and oldest, explanation of the shape of this molecule follows from the recognition that the ground state electronic configuration of an isolated nitrogen atom is $(1s)^2\,(2s)^2\,(2p)^3$, each of the 2p electrons occupying a different p orbital. These 2p electrons may each be paired with the electron present in the 1s orbital of a hydrogen atom by placing one hydrogen atom at one end of each 2p orbital so that the nitrogen 2p and hydrogen 1s orbitals overlap. We thus obtain an ammonia molecule which has the correct, pyramidal, shape and which has all three hydrogen atoms equivalently bonded to the nitrogen (Figure 1.2). However, the angle between any pair of 2p orbitals is 90° so that a bond angle of 90° is predicted by this model. Agreement with the experimental value of 107° is obtained by postulating the existence of electrostatic repulsion forces between the hydrogen atoms which cause these atoms to repel each other so that the bond angles increase. If, as seems probable, each N—H bond is slightly polar with each hydrogen carrying a small positive charge, this repulsion is nuclear–nuclear in origin. The consequent modification of the original bonding scheme as a result of this distortion of the bond angle from 90° is not usually considered.

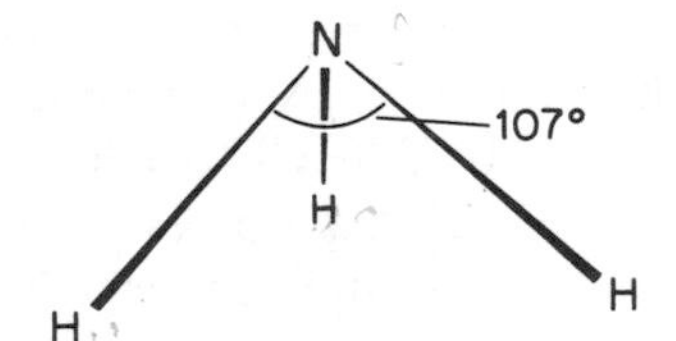


Figure 1.1 The ammonia molecule

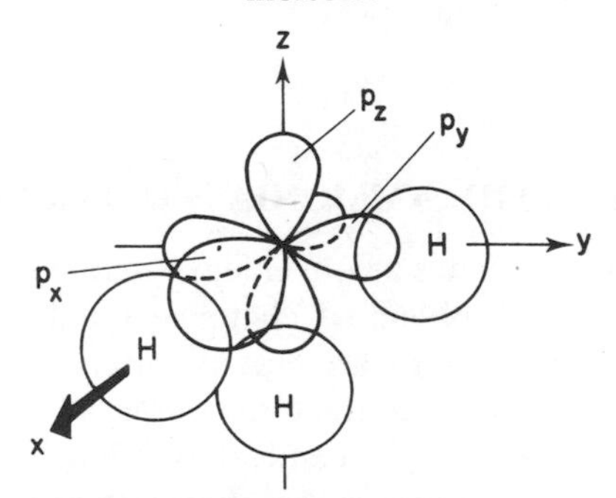


Figure 1.2 N—H bonding in $NH_3$ envisaged as resulting from the overlap of 2p orbitals of the nitrogen with 1s orbitals of the hydrogens

### 1.2.2 The hybrid orbital model

This is detailed in many post-*ca.* 1955 texts.[2] In this model an alternative description of the bonding in the ammonia molecule is obtained by hybridizing the valence shell orbitals of an isolated nitrogen atom, 2s, $2p_x$, $2p_y$ and $2p_z$, to give four, equivalent, $sp^3$ hybrid orbitals pointing towards the corners of a regular tetrahdron. Because there are five electrons in the valence shell of the nitrogen atom, three of these hybrid orbitals may be regarded as containing one electron whilst the fourth is occupied by two electrons. As in the previous model, 1s electrons from three hydrogen atoms pair with the unpaired electrons in the hybrid orbitals on the nitrogen to give a pyramidal ammonia molecule (Figure 1.3). Again, the three hydrogen atoms are equivalent but now the bond angle is predicted to be 109.5° (the angle between the axes of a pair of $sp^3$ hybrid orbitals). This value is in much closer agreement with experiment than that given by the previous model, but again some modification is necessary before the experimentally observed value is obtained.

This correction is usually made by invoking the effects of electron–electron repulsion. It is this electron–electron repulsion which forms the basis of a third model for ammonia and so the way that the 'hybrid orbital' model is modified to give agreement with experiment will become evident during the description of the next model.

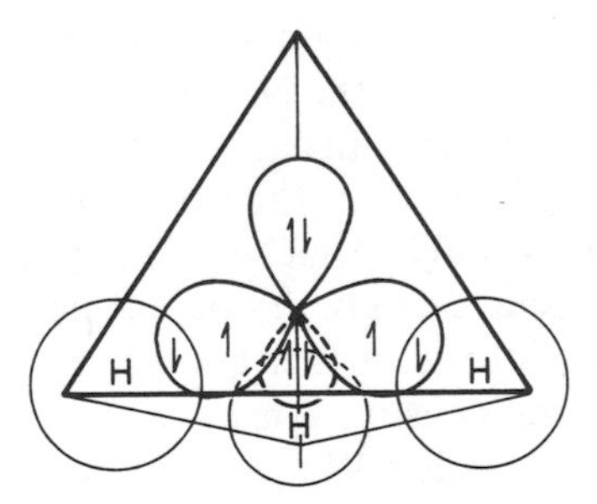


Figure 1.3 N—H bonding in $NH_3$ envisaged as resulting from the overlap of $sp^3$ hybrids of the nitrogen with 1s orbitals of the hydrogens

### 1.2.3 The electron repulsion model

This model is described in many current texts, a particularly complete treatment being given by Gillespie.[3] The first two models which we have considered seek to explain the structure of the ammonia molecule in terms of the bonding interactions between the constituent atoms. The atoms adopt that arrangement which makes bonding a maximum. In contrast, the next two models which we discuss explain the structure not in terms of bonding interactions (although these must exist to hold the atoms together) but rather by recognizing that electrons repel each other, and regard the structure as being determined by the condition that the interelectron repulsion energies are minimized. The first of these models which we shall consider is that originally due to Sidgwick and Powell and subsequently refined by Nyholm and Gillespie.

In the ammonia molecule there are four electron pairs involving the valence shell of the nitrogen atom. These are the three N—H bonding electron pairs and a non-bonding pair (in the first of the models we discussed the non-bonding electrons were those in the 2s orbital of the nitrogen whilst in the second they were placed in an $sp^3$ hybrid orbital). Because electrons repel each other we would expect these four pairs of electrons to be as far apart as possible consistent with still being bound to the nitrogen atom (three pairs are also, of course, bound to hydrogen atoms). It follows that the preferred orientation of these four electron pairs is that in which they point towards the corners of a regular tetrahedron. Remembering that three of the electron pairs are N—H bonding and that their orientation determines the positions of the hydrogen atoms, we are led to predict an HNH bond angle of 109.5°, just as we did for the second model. It is thought provoking to recognize that we can predict the same angle either by including bonding interactions or by ignoring them! If we consider the electron repulsion concept in more detail we recognize that there are two sorts of electron pairs, those involved in N—H bonding and those which are non-bonding and located on the nitrogen atom. The electron pairs which comprise the N—H bonds are each subject to strong electrostatic attractions from two nuclei, the nitrogen nucleus and that of one of the hydrogen atoms. In contrast, the non-bonding electrons are strongly attracted by one nucleus only, that of nitrogen. We would therefore, qualitatively, expect the centre of gravity of the electron density in the N—H bonds to be located at a distance further away from the nitrogen nucleus than that of the lone pair electron density. The recognition of this difference at once leads us to reconsider our model. We were led to a tetrahedral arrangement of electron pairs because we assumed that all electron pairs were equivalent. In the absence of such equivalance we cannot expect to find a regular tetrahedral arrangement. It seems reasonable that the repulsive forces occurring between electron pairs located in two N—H bonds will be less than the electrostatic repulsions between the non-bonding pair of electrons and a N—H bonding pair simply because the distance between the centres of gravity of electron density will be greater in the former case than in the latter. We would anticipate that this difference would lead the molecule to distort accordingly. It follows that we expect the HNH bond angle to be less than 109.5°. Although no quantitative prediction is possible with this simple model the qualitative prediction is in accord with experiment — the bond angle is 107°. Arguments similar to those that we have just used are also applied to the 'hybrid orbital' model (Section 1.2.2) to produce agreement between theory and experiment.

### 1.2.4 The electron-spin repulsion model

This is a little-used model, introduced by Linnett.[4] It differs from the preceding model principally in its recognition that electrons behave as individuals and repel each other as individuals rather than as pairs. Linnett suggested that it is more appropriate to consider eight electrons associated with the nitrogen atom, four with spin 'up' and four with spin 'down', than to think of there being four electron pairs (with no mention of spin). In the case of eight individual electrons we would expect the preferred orientation (in which the electrons are as far apart as possible) to be one in which the electrons are located at the corners of a cube. One of the general results of detailed quantum mechanics is that additional repulsion exists between electrons of like spin (compared with the repulsion between electrons of unlike spin) and so it would be expected that an electron of given spin would have electrons of the opposite spin as its nearest neighbours at the corners of the cube. It follows, therefore, that in the cubic orientation of electrons there would be four electrons with spin 'up' defining one tetrahedron and four with spin 'down' defining another. (If lines are drawn from one corner of a cube across the face diagonals to other corners and this procedure continued, just four corners are reached. These four corners define a regular tetrahedron. Another regular tetrahedron is defined by the four corners which remain — see Figure 1.4.) So far in this model we have regarded all of the electrons as being associated with the nitrogen atom and have really been thinking of $N^{3-}$, with eight valence shell electrons. When we introduce the hydrogen atoms we must, therefore, introduce them as bare protons. These protons attract the

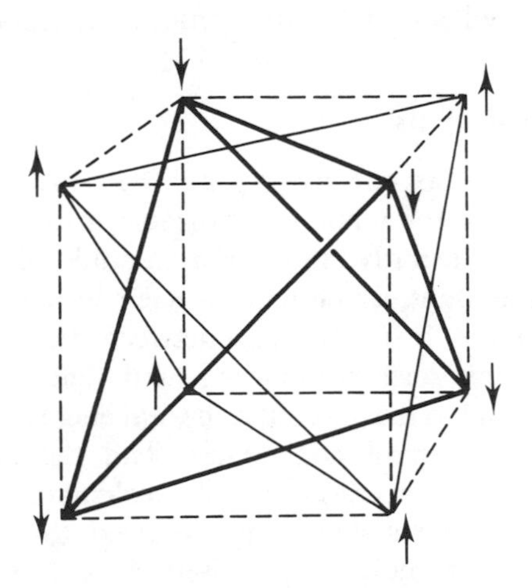

Figure 1.4 The two tetrahedra associated with a cube. Note the relationships between the spins of eight electrons placed at the corners of the cube and these tetrahedra in Linnett's model

eight electrons which we have associated with the nitrogen atom. The attraction between a proton and an electron does not depend upon whether the electron has its spin 'up' or 'down', although, of course, the extra repulsion between electrons of the same spin persists. The net result is that each proton attracts to its locality just one electron with spin 'up' and one with spin 'down'. This attraction brings the two distinct tetrahederal arrangements of electrons into coincidence to give a single tetrahdral arrangement. We are led to the conclusion that two electrons will be associated with each N—H bond and the remaining two will be non-bonding, which is the same as for the previous model. Evidently, Linnett's model also predicts a bond angle of $109.5^\circ$; it may be corrected in a manner similar to that descibed above for the electron pair model to again give qualitative agreement with experiment.

Although there is considerable overlap between the different models which we have considered, a survey of them does not lead us to any definite conclusion regarding the relationship between the structure of and the bonding in the ammonia molecule. Firstly, they are concerned with relatively fine points — bond angles; they say nothing about the more important point (in terms of energy) of bond lengths. Secondly, all start with the supposition that only valence shell electrons need be considered, but then diverge in their explanations. These explanations are not totally distinct, but what one model regards as the dominant factor another assumes to be relatively small. The first two models, effectively, say that the geometry is determined by the requirement that bonding interactions be maximized whilst the last two say that it is the consequence of the requirement that non-bonding repulsive forces be minimized. One point that they have in common, however, is the fact that none of them leads to a prediction that the ammonia molecule should be planar.

### 1.2.5 Accurate calculations

In 1970 Clementi and his coworkers published the results of some very accurate calculations on the ammonia molecule in the *Journal of Chemical Physics*. They were particularly interested in a study of a vibrational motion of the ammonia molecule in which it turns itself inside-out, like an umbrella in a high wind (Figure 1.5). Halfway between the two extremes of this umbrella motion the ammonia molecule is planar. The potential energy barrier for the inversion is equal to the difference in total energy between the ammonia molecule in its normal, pyramidal, shape and the planar configuration. Clementi therefore carried out rather detailed calculations for each geometry. The results were very surprising. They showed that the N—H bonding is greater in the planar molecule — there is a loss of N—H bonding energy of approximately $7.0 \times 10^2$ kJ mol$^{-1}$ (167 kcal mol$^{-1}$) in going from the planar to the pyramidal geometry. This loss is accompanied by a slight lengthening of the N—H bond. However, although bonding favours a planar ammonia molecule it is found that in a comparison of the most stable pyramidal and most stable planar geometries the electron–electron and

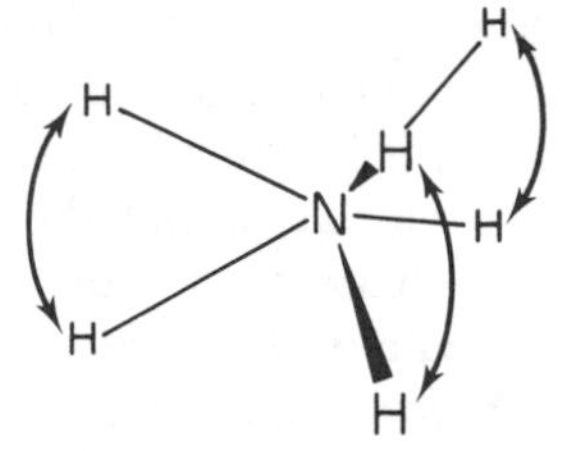


Figure 1.5 The 'umbrella' inversion motion of the ammonia molecule. As the hydrogens move up so the nitrogen moves down (and vice versa) so that the centre of gravity of the molecule remains in the same place

nuclear–nuclear repulsion energies favour the pyramidal over the planar by about $7.2 \times 10^2$ kJ mol$^{-1}$ (172 kcal mol$^{-1}$). That is, the bonding and repulsive energy changes between the two shapes almost exactly cancel each other; it is the slight dominance of the replusive forces by 20 kJ mol$^{-1}$ (5 kcal mol$^{-1}$) which leads to the equilibrium geometry of the ammonia molecule in its electronic ground state being pyramidal.

We are left with a most disturbing situation. There is no doubt that the strongest N—H bonding in the ammonia molecule is to be found when it is planar, yet two of the simple models which we considered earlier in this chapter explained its geometry by the assumption that this bonding is a maximum in the pyramidal molecule! Similarly, the models based on electron repulsion entirely ignored the fact that the changes in energy on which they were based were almost exactly offset by changes in bonding energy. This would not matter so much if there were some assurance that repulsion energy changes would outweigh the bonding in all molecules (we could then explain molecular geometries using a repulsion-based argument). Unfortunately, no such general assurance can be given. This can be seen if we extend our discussion of the ammonia molecule to include some related species. The molecules $NH_3$, $PH_3$, $NH_2F$, $PH_2F$, $NHF_2$, $PHF_2$, $NF_3$ and $PF_3$ all have similar, pyramidal, structures and would be treated similarly in all simple models. However, calculations by Schmiedekamp and coworkers, published in 1977 in the *Journal of Chemical Physics*, have shown that the first four owe their pyramidal geometry to the dominance of repulsive forces (bonding is stronger when they are planar) but the last four are pyramidal because the bonding is greatest in this configuration and dominates the repulsive forces (which now favour a planar arrangement)! Although this last sentence is marginally stronger than strictly permitted by the calculations, there is no doubt about the general conclusion. Although these eight compounds all have the same structure they do not all have it for the same reason, because of the close competition between repulsive and bonding forces. At present there are no rules to enable us to predict which will win the competition in a particular case.

Although simple explanations of molecular shape such as those we described in this chapter are very useful to the chemist — and are widely and fruitfully used — they can be considered only as guides because they are not infallible. It is for this reason, and because it happens to be particularly convenient for our purpose, that in this book we shall adopt the opposite strategy of using the experimentally determined shape of a molecule to infer details of the electronic structure of the molecule *in that shape*. We shall seldom attempt to explain why a molecule has a particular shape, although we shall perhaps understand in more detail why simple explanations may be unreliable.

**Problem 1.1** Show that each of the models described in Sections 1.2.1 and 1.2.4 predicts that the water molecule is non-linear (the bond angle is actually $104.5^\circ$).

**Problem 1.2** Hazard a guess at whether it is bonding or non-bonding forces which lead to $NCl_3$ having a pyramidal shape. An answer will be found in a paper by Faegri and Kosmus in the *Journal of the Chemical Society*[5] (be prepared for a surprise).

## REFERENCES


1. For example, E. de B. Barnett and C. L. Wilson, *Inorganic Chemistry*, p. 65, Longman Green, 1953.
2. For example, E. Cartmell and G. W. A. Fowles, *Valency and Molecular Structure*, p. 159, Butterworths, 1956.
3. R. J. Gillespie, *Molecular Geometry*, Van Nostrand-Reinhold, 1972.
4. J. W. Linnett, *The Electronic Structure of Molecules*, Methuen, 1964.
5. K. Faegri and W. Kosmus, *J. Chem. Soc.* (Far. II), **73**, 1602 (1977).

CHAPTER 2

# *The symmetry of the water molecule*

In this chapter we shall begin our investigation into the consequences of molecular symmetry. Following the discussion in the previous chapter it would be appropriate to develop our arguments with particular reference to the ammonia molecule. Unfortunately, this problem is, for the moment, too difficult and we shall defer it until Chapter 6. Instead we shall look at the water molecule. In so doing, we shall develop an approach which will subsequently be extended to more complicated species (one of which is ammonia).

## 2.1 SYMMETRY OPERATIONS AND SYMMETRY ELEMENTS

We take as our starting point a discussion of the meaning of the word 'symmetry'. When we say that a molecule has high symmetry we usually mean that within the molecule there are several atoms which have equivalent positions in space. Thus, the tetrahedral symmetry of the methane molecule is manifest in the fact that the four hydrogen atoms are equivalent (Figure 2.1). Suppose that you have in front of you a model of the methane molecule which is so well constructed that no minor blemishes serve to distinguish one hydrogen atom from another. If you were to momentarily close and then open your eyes you would have no means of telling whether someone had rotated the model so that, although each hydrogen atom had been moved, the final position of the model was indistinguishable from its starting position. Such questions provide a convenient approach to symmetry and is the approach which we shall follow in this book. The symmetry of a molecule is characterized by the fact that it

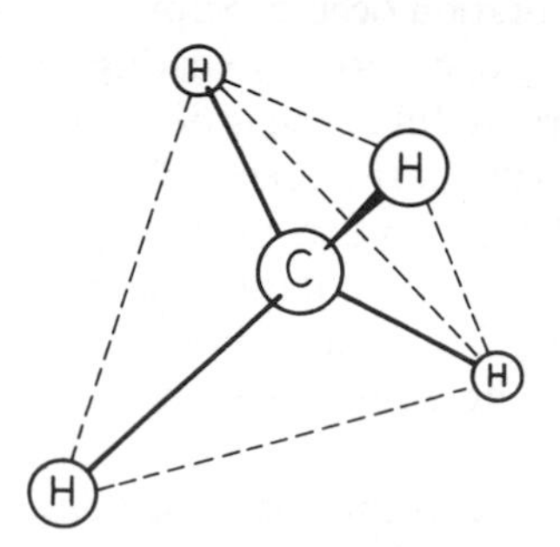


Figure 2.1 The methane molecule

is possible, hypothetically at least, to carry out *operations* which, whilst interchanging the positions of some (or all) of the atoms, give arrangements of atoms which are indistinguishable from the initial arrangement. We may note in passing that these symmetry operations need not always be physically possible in the way that a rotation is. Thus, the operation of time reversal is important in some aspects of theoretical chemistry, although it is not one which we shall consider in this book. Of the operations that we shall need to use in the following chapters only the rotation operations are physically possible. The others — such as the operation of reflection in a mirror plane or inversion in a centre of symmetry — are not. Those operations which we shall use which we cannot physically carry out are called 'improper rotations' in contrast to the 'proper rotations', which are physically possible. We shall meet a more precise definition of improper rotations later (page 87). Actually, the distinction between physically possible and impossible operations is something of a red herring because we are really concerned with a mathematics — the mathematics of group theory — and not with ball-and-stick models.

It is helpful to consider a particular example, and in this chapter we shall look in some detail at the symmetry (and, thus, the symmetry operations) of the water molecule. Our first task will be to obtain a list of those symmetry operations which turn the water molecule into a configuration indistinguishable from the initial one.

The most evident symmetry operation which turns the water molecule into itself is the act of rotation by 180° about an axis which bisects the HOH angle and lies in the molecular plane. Figure 2.2 shows the water molecule before, in the middle of and after completion of this operation. Apart from the arrows, which have been added for clarity, the first and third diagrams are indistinguishable. The effect of the operation is to interchange the two hydrogen atoms. We say that 'the two hydrogen atoms are symmetry related' or 'they are symmetrically equivalent'. A rotation operation is denoted by the letter C (which may be conveniently thought of as derived from the symbol ↻). Because it takes two successive rotations to return each atom to its original position, the rotation operation is called a twofold rotation operation and is denoted $C_2$. The same symbol, $C_2$, is used to denote the rotation operation and the axis about which the rotation occurs. Some authors distinguish between an axis and the corresponding operation by writing the latter in bold type — $\mathbf{C_2}$. The use of bold type is very useful if one is developing the mathematics of symmetry theory (group theory). In the present book, however, it will always be clear from the context whether an axis or operation is being discussed and we shall not use bold type. Twofold rotation operations are not the only ones which can exist; threefold ($C_3$), fourfold ($C_4$), fivefold ($C_5$) and sixfold ($C_6$) rotation operations are quite common in chemistry. We shall meet several examples later in this book.

In defining rotation axes (and the corresponding operations) it is necessary to require that the rotation which is repeated several times in order to return each atom to its original position is always carried out in the same sense

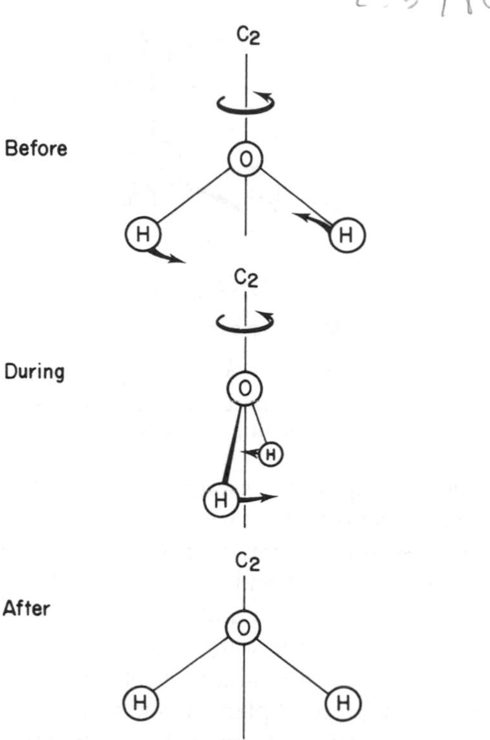


Figure 2.2 The conversion of the $H_2O$ molecule into an arrangement which is indistinguishable from the original by a rotation of 180° ($= 360/2 \equiv C_2$)

(clockwise or anticlockwise). For a $C_n$ axis, where $n$ rotations in the same sense are required to reproduce the starting arrangement, each operation involves a rotation of $360°/n$ about the $C_n$ axis. The larger the value of $n$ the higher the rotational symmetry of the axis.

The twofold rotation operation is not the only manifestation of symmetry in the water molecule. If the plane defining the water molecule were to be replaced by an infinitely thin mirror, as shown in Figure 2.3, then reflection in this mirror plane would have the effect of turning the water molecule into a configuration indistinguishable from the original one. This operation has the effect of turning the 'front' of the two hydrogen atoms and of the oxygen atom into the 'back', and vice versa. Mirror planes and the operation of reflection in them are both denoted by sigma, $\sigma$ (the operation sometimes being distinguished by bold type), and, just as for rotation axes, various subscripts are used. In the present case the subscript v is appended to give the symbol $\sigma_v$. This subscript arises when the axis of highest symmetry ($C_2$ in the present case) is arranged so as to be vertical, as in Figure 2.3; this causes the mirror

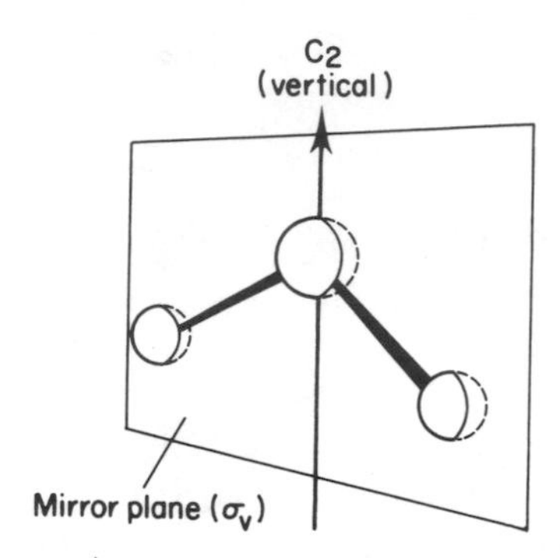


Figure 2.3 The mirror plane of symmetry in the molecular plane of the $H_2O$ molecule

plane also to be vertical. The subscript v on $\sigma_v$ is the initial letter of *v*ertical. Thus: a $\sigma_v$ mirror plane is vertical with respect to the axis of highest symmetry (this axis always lies in the $\sigma_v$ mirror plane). Other subscripts on $\sigma$ which will be met are h (for *h*orizontal) and d (for *d*ihedral). We shall discuss them in detail later in this book. The $C_2$ and $\sigma_v$ symmetry operations do not exhaust the symmetry possessed by the water molecule. Another feature of this symmetry is the existence of a second mirror plane. This mirror plane, which lies perpendicular to the molecular plane, is shown in Figure 2.4. Like the first, the second mirror plane contains the twofold axis (indeed, the line of intersection between the two mirror planes defines the twofold axis). It follows that, like the first mirror plane, the second is denoted $\sigma_v$. However, its effect on the molecule is quite different to that of the first — it has the effect of interchanging the two hydrogen atoms, for instance — and so it is necessary to distinguish between them. This is done by adding a prime to the symbol for the second mirror plane, thus: $\sigma_v'$. Had the water molecule possessed a third

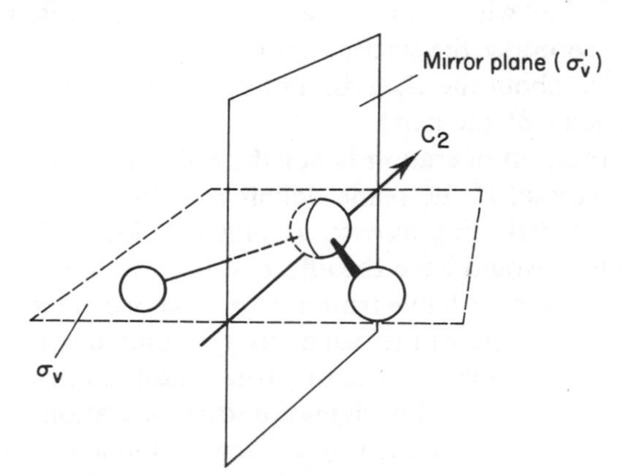


Figure 2.4 A second mirror plane of symmetry, perpendicular to the first, in the $H_2O$ molecule

type of vertical mirror plane (which it does not!) this would have been denoted $\sigma_v''$, and so on. Note that $\sigma_v$ is an improper symmetry operation — although you can see its effect you cannot actually carry it out.

In this section we have found it convenient to talk of rotation axes and mirror planes almost as if they were physical objects. Collectively, they are called *symmetry elements*. We shall meet examples of other symmetry elements — such as a centre of symmetry — later in this book. It cannot be emphasized too strongly that our real concern here is with symmetry operations. Symmetry elements are introduced simply to enable the corresponding operations to be more readily understood, notwithstanding the fact that symmetry elements may appear to have more physical reality than symmetry operations.

As can be shown by an abortive search, no other rotation axes or mirror planes exist in the water molecule, so it would seem that the three symmetry operations which we have recognized define the symmetry of the water molecule. This, however, is not strictly so. We have seen that the application of the $C_2$ operation twice over regenerates the original molecule with each atom restored to precisely its original position. It is easy to see that the same is true of the $\sigma_v$ and $\sigma_v'$ operations. That is, the end result of carrying out any of these symmetry operations twice is the same as that of leaving the molecule alone. The implication of this is that we should formally recognize the possibility that one way of turning a molecule into a configuration indistinguishable from the original is simply to leave the molecule alone! This so-called identity operation we shall denote by the letter E (some books use I). No matter how much or how little symmetry a molecule posesses the identity operation always exists for it.* It is the set of four symmetry operations, E, $C_2$, $\sigma_v$ and $\sigma_v'$, which completely defines the symmetry of the water molecule. One way of defining the symmetry of the water molecule would be to give this list. However, rather than give a complete listing of symmetry operations (which for some high symmetry molecules could be rather tedious) this information is compressed into a shorthand symbol, which for the set of operations of the water molecule is $C_{2v}$ (pronounced 'see-two-vee'). We talk of the water molecule as 'having $C_{2v}$ symmetry' or we talk of 'the symmetry operations of the $C_{2v}$ point group' (by which we mean E, $C_2$, $\sigma_v$ and $\sigma_v'$).

We have just used two new words, 'point' and 'group'. The word 'group' arises from the fact that if we apply two of the symmetry operations one after the other the result is equivalent to the application of just one of the operations of the group (which may be different from the two that we used). Thus, as we shall see later, for the case of the water molecule, following the $\sigma_v$ operation by $C_2$ gives the same result as the application of the $\sigma_v'$ operation. (Indeed, this combination method is sometimes a useful method of making sure that

*It is sometimes convenient to regard the identity operation as a $C_1$ rotation, i.e. a rotation of 360°. This interpretation highlights the fact that although there is an infinite number of choices of $C_1$ axis there is only one distinct $C_1$ (E) *operation*.

you have discovered all of the symmetry operations of a particular molecule.) There is a limit to the process, however. Eventually all of the symmetry operations of a particular molecule will have been obtained. The successive application of members of the set of operations will serve only to produce a result which is equivalent to another member of the set. Sets which are closed in this fashion are called *groups*. We are interested in *groups* of symmetry operations. For each group there has to be a method of combining the group elements — for symmetry operations it is applying them one after another. Other types of groups may have very different methods of combination.

The complete, formal, definition of a 'group' requires some mathematics and is reserved for Appendix 1, but it may help to give two more examples of groups. Consider the three numbers 1, 0, $-1$. Do these form a group under the operations of addition and subtraction? Whilst it is clear that all three numbers can be interrelated by these operations, it is equally clear that when we apply the operation $(+1)$ to the number 1 we generate the number 2. Similarly, $(-1)$ applied to the number $-1$ gives the number $-2$. Clearly, 1, 0, $-1$ do not comprise the entire group because 2 and $-2$ have to be included. In similar fashion we see that 3, $-3$ and, indeed, all integers between $\infty$ and $-\infty$ (plus and minus infinity) have to be included. This is an example of an infinite group. Groups such as this are of importance in the description of the translational symmetry found in crystal lattices.

As a second example consider the rectangular table shown in Figure 2.5. The top of the table has been divided into quarters and two of these are coloured black and two white. Were there no such colouration, the table would have the same symmetry as the water molecule, as shown in Figure 2.6. However, the presence of the coloured sections means that the $\sigma_v$ and $\sigma_v'$ operations are no longer symmetry operations unless they are each combined with quite a new type of symmetry operation, that of changing colour — black into white and white into black. If we call these (reflection and colour change) operations $(\sigma_v)$ and $(\sigma_v')$, then the operations E, $C_2$, $(\sigma_v)$ and $(\sigma_v')$ form a group. Later in this chapter you will be asked to demonstrate the truth of this statement.

As has been mentioned, a geometrical feature corresponding to a symmetry operation is called a symmetry *element*. Thus, corresponding to a rotation operation is a rotation axis and corresponding to a reflection operation is a

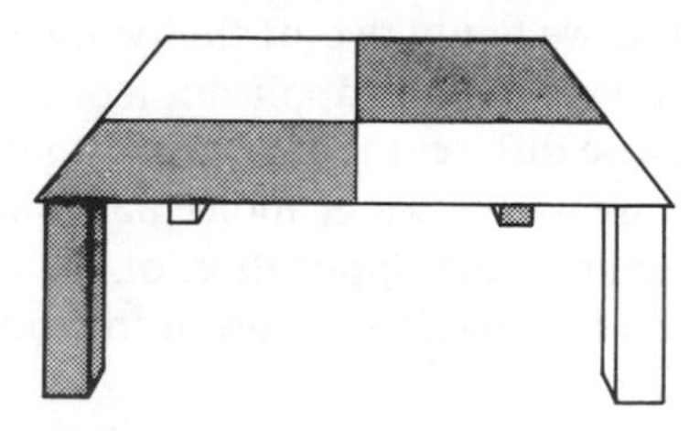

Figure 2.5 A table showing black and white colour change symmetry

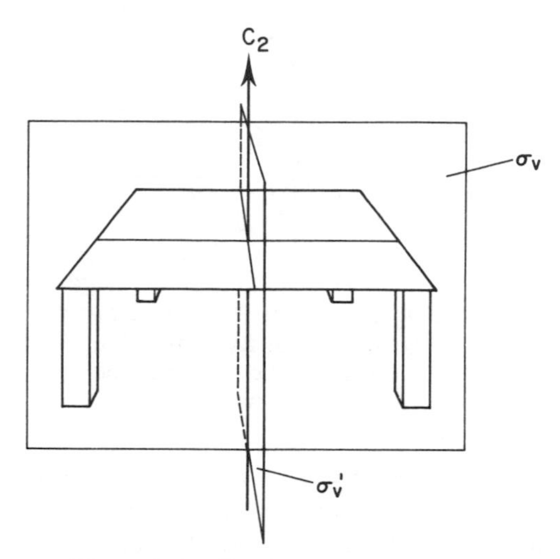


Figure 2.6 The $C_{2v}$ symmetry of the table of Figure 2.5 when uncoloured

mirror plane. Rotation axes and mirror planes (and also other similar things, such as a centre of symmetry) are examples of symmetry elements. For all molecules it is true that all the symmetry elements which they possess pass through a common point in the molecule (in the case of the $C_{2v}$ point group they pass through an infinite number of common points along the $C_2$ axis). This is the reason why all such groups (of operations) are called *point* groups. Put another way, there is always at least one point which is left invariant (unchanged) by all of the operations of a point group. The point may or may not be a point at which an atom is located.

## 2.2 MULTIPLIERS ASSOCIATED WITH SYMMETRY OPERATIONS

From the way we have defined them it is evident that the effect of each of the symmetry operations of the $C_{2v}$ point group when applied to the water molecule, considered as a whole, is to turn the molecule into itself. An alternative way of putting this is to say that the effect of each of the symmetry operations on the molecule is equivalent to multiplication by the number 1. That is, we can represent the effect of each of the operations by the following table:

| Symmetry operation | E | $C_2$ | $\sigma_v$ | $\sigma_v'$ |
|---|---|---|---|---|
| Effect of the operation on the water molecule (considered as a whole) | 1 | 1 | 1 | 1 |

The apparently pointless exercise of representing the effects of symmetry operations by the number 1 begins to acquire some significance when we ask

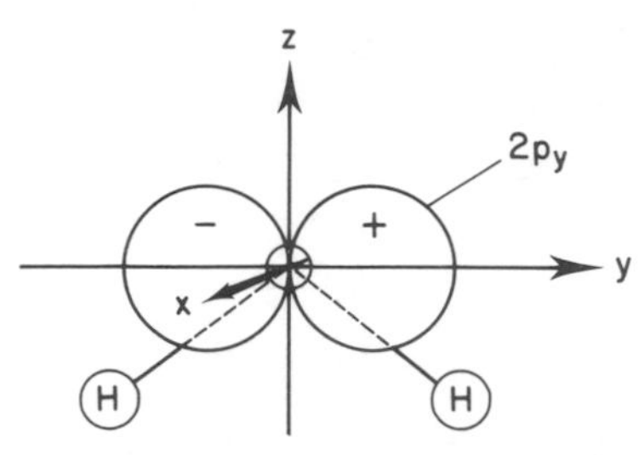


Figure 2.7 The $2p_y$ orbital of oxygen, in $H_2O$. By convention, the y axis is taken to lie in the plane of a planar molecule

ourselves whether all quantities associated with the water molecule are, like the water molecule itself, turned into themselves by the operations of the $C_{2v}$ point group. We shall see that they are not. Consider, for example, the oxygen $2p_y$ orbital shown in Figure 2.7. In Figure 2.8 we picture the effects of the symmetry operations of the $C_{2v}$ point group on this orbital. It is evident that, whilst the identity and $\sigma_v$ operations have the effects of regenerating the original orbital, the $C_2$ and $\sigma_v'$ operations have the effect of reversing the phases of the lobes of the orbital. It follows that whilst the result of the application of the E and $\sigma_v$ operations may be represented by the multiplication factor 1, the effects of the $C_2$ and $\sigma_v'$ operations have to be represented by the multiplicative factor $-1$. That is, the association between symmetry operations and multiplicative factors is:

| Symmetry operation | E | $C_2$ | $\sigma_v$ | $\sigma_v'$ |
|---|---|---|---|---|
| Effect on the oxygen $2p_y$ orbital | 1 | $-1$ | 1 | $-1$ |

Having looked at the effect of the symmetry operations of the $C_{2v}$ point group on the $2p_y$ orbital of the oxygen atom it is natural next to enquire into their effects on the $2p_x$ and $2p_z$ orbitals, and this we shall now do. The $2p_x$ orbital is shown in Figure 2.9, where we have tried to indicate by the perspective of the diagram the fact that the positive lobe is located above the plane of the page and the negative lobe beneath this plane. In order to avoid completely obscuring the negative lobe behind the positive we have shown the water molecule viewed from a slightly skew position. In Figure 2.10 we show the effects of the four symmetry operations of the $C_{2v}$ point group on the oxygen $2p_x$ orbital. It is evident that, whilst the application of the E and $\sigma_v'$ operations result in the phases of the lobes of the orbital being unchanged, the application of the $C_2$ and $\sigma_v$ operations leads to a reversal of these phases. In this case the numbers representing the effects of the symmetry operations are:

| Symmetry operation | E | $C_2$ | $\sigma_v$ | $\sigma_v'$ |
|---|---|---|---|---|
| Effect on the oxygen $2p_x$ orbital | 1 | $-1$ | $-1$ | 1 |

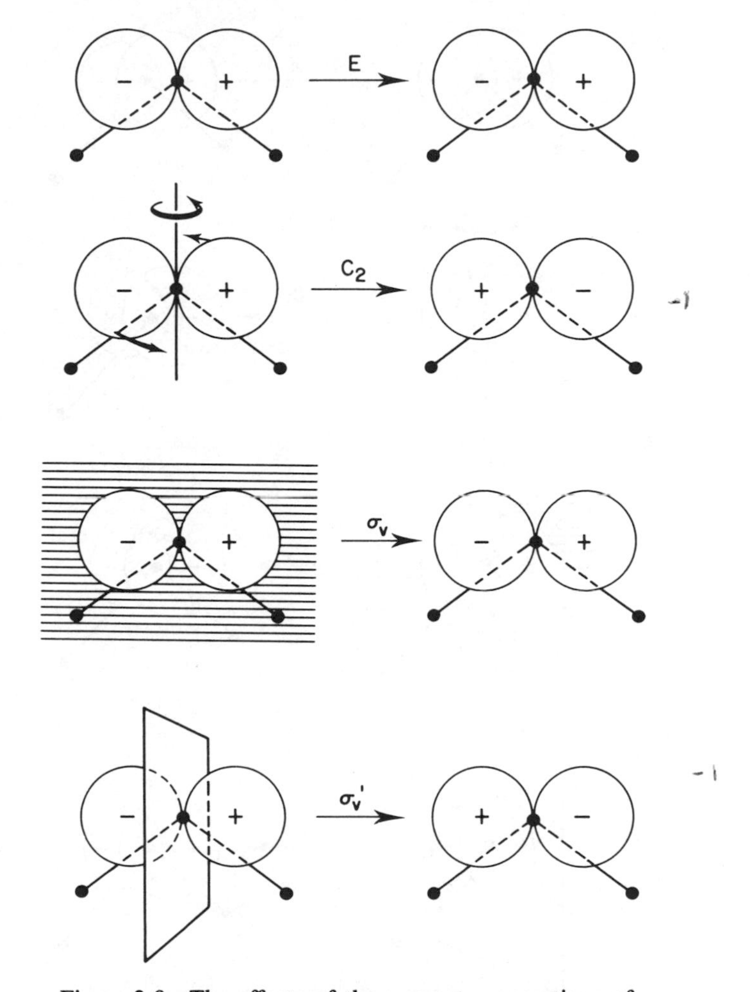


Figure 2.8 The effects of the symmetry operations of the $C_{2v}$ point group on the oxygen $2p_y$ orbital in the water molecule

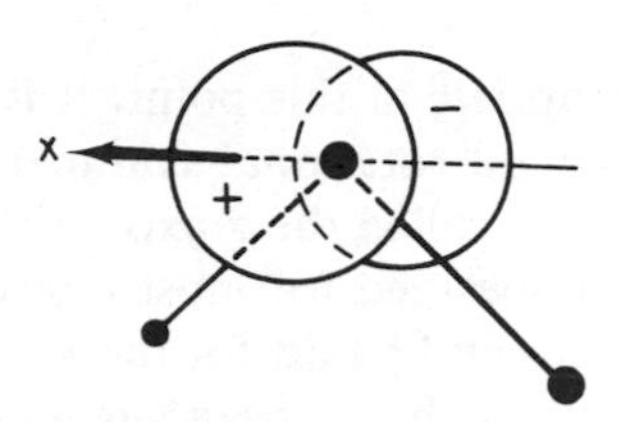


Figure 2.9 The $2p_x$ orbital of oxygen in $H_2O$. By convention, the x axis is taken to be perpendicular to the plane of a planar molecule

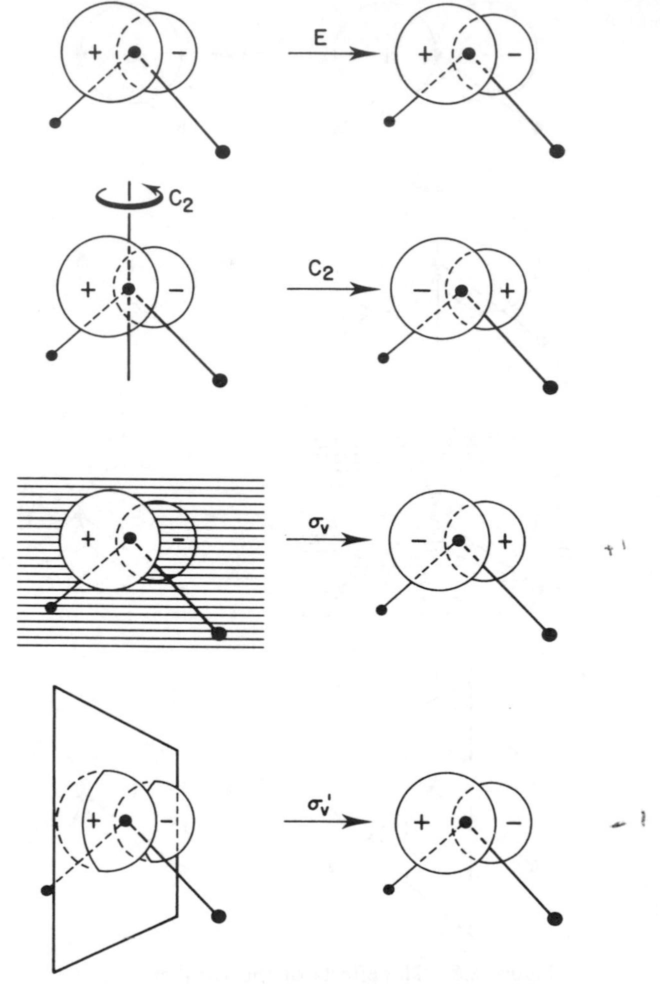


Figure 2.10 The effects of the symmetry operations of the $C_{2v}$ point group on the oxygen $2p_x$ orbital in the water molecule

A note of warning is appropriate at this point. It is a generally accepted convention that the axis of highest rotational symmetry in a molecule ($C_2$ in the case of the water molecule) is called the z axis. Although the direction of the z axis is therefore uniquely specified for most molecules by this convention it is seldom true that the same can be said for the x and y axes. In this book we are following a convention which has been suggested by Mulliken but is not always followed — that of requiring that a planar molecule lies in the yz plane. Therefore, the reader may find that, in the case of the water molecule, what we have called the x axis some authors will call the y axis (thus making the zx plane, rather than the yz, the molecular plane).

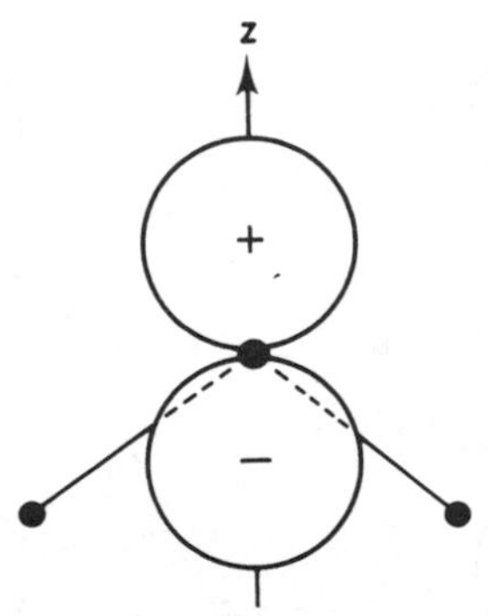


Figure 2.11 The $2p_z$ orbital of oxygen in $H_2O$. By convention, the z axis is taken to lie along the axis of highest rotational symmetry of a molecule (provided that this is a unique axis)

We now return to the problem of the symmetry properties of the $2p_z$ orbital of the oxygen atom. This orbital is shown in Figure 2.11 and its behaviour under the symmetry operations of the group in Figure 2.12. It is evident from Figure 2.12 that, although the symmetry operations may have the effect of turning one side of the orbital into the other, this change is always accompanied by the retention of the phase of the lobes of the orbital, so that the number representing the effect of each operation is 1:

| Symmetry operation | E | $C_2$ | $\sigma_v$ | $\sigma_v'$ |
|---|---|---|---|---|
| Effect on the oxygen $2p_z$ orbital | 1 | 1 | 1 | 1 |

This set of numbers is the same as that which we obtained earlier as a description of the symmetry properties of the whole molecule. We conclude that although it is possible for the behaviour of quantities associated with the water molecule to give rise to the same set of numbers as the molecule itself, other alternatives are possible (such as those found for the $2p_y$ and $2p_x$ oxygen orbitals).

We now come to the key point in the argument which we are developing. This is that the differing symmetry properties of, for example, the $2p_y$, $2p_x$ and $2p_z$ orbitals of the oxygen atom in the water molecule (i.e. the fact that their behaviour under the set of symmetry operations differs) may be *represented* by the sets of numbers which we have obtained. Quantities which have different symmetry properties give rise to different sets of numbers. Evidently, the next question which we have to consider is whether the sets of numbers which we have already generated comprise a complete list of the different types

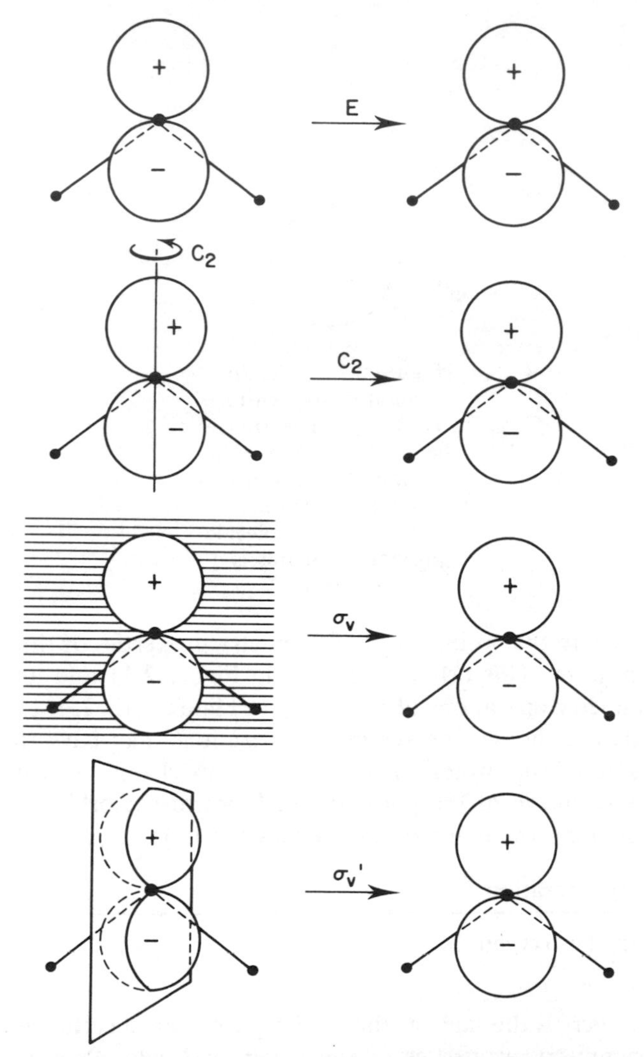


Figure 2.12 The effects of the symmetry operations of the $C_{2v}$ point group on the oxygen $2p_z$ orbital in the water molecule

of symmetry behaviour which may be shown by quantities (such as atomic orbitals) associated with the water molecule. The answer is 'no'; there is just one further type of symmetry behaviour (i.e. set of numbers) which we have to obtain. Let us consider the symmetry properties of the $3d_{xy}$ orbital of the oxygen atom. Although this orbital is not commonly included in elementary discussions of the electronic structure of oxygen-containing compounds (because it is not a valence shell orbital) it does nonetheless exist and would be included in most sophisticated calculations of the electronic structure of such molecules. It is shown in Figure 2.13. Note that where the product coordinate axes xy is

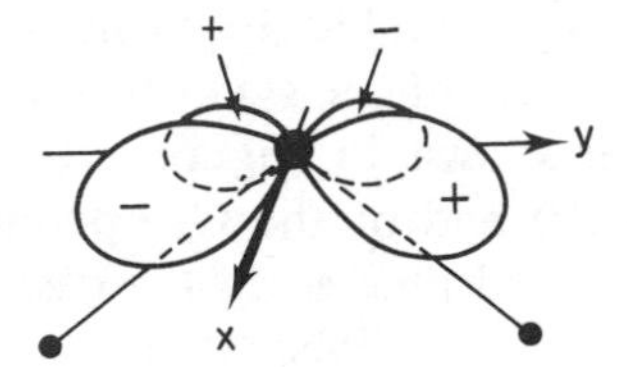


Figure 2.13 The $3d_{xy}$ orbital of oxygen in $H_2O$. Note that the phases of the lobes of the orbital are those of the product xy

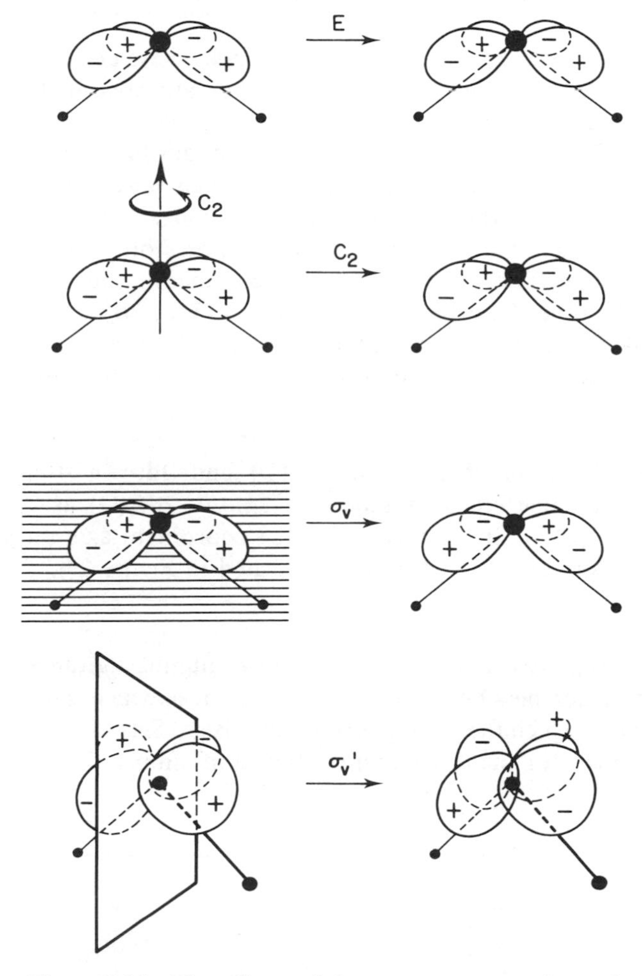


Figure 2.14 The effects of the symmetry operations of the $C_{2v}$ point group on the oxygen $3d_{xy}$ orbital in the water molecule

positive, the phase of the $3d_{xy}$ orbital is also positive (this is implicit in the use of the xy subscript). The effects of the symmetry operations of the $C_{2v}$ point group on this orbital are shown in Figure 2.14. This figure shows that the effect of the identity (E) and of the $C_2$ operations is to regenerate the original orbital with unchanged phases. In the case of the $\sigma_v$ and $\sigma_v'$ operations, however, the phase of each lobe of the orbital is reversed. The appropriate multiplicative factors representing the effects of the operations in this case are therefore:

| Symmetry operation | E | $C_2$ | $\sigma_v$ | $\sigma_v'$ |
|---|---|---|---|---|
| Effect on the oxygen $3d_{xy}$ orbital | 1 | 1 | −1 | −1 |

We have now generated four sets of numbers. These are collected together in Table 2.1, although the order in which they are presented is not that in which we obtained them.

Against each row of numbers is shown the orbital which we used to generate it, the (O)'s indicating that the orbitals considered were those of the oxygen atom. We now assert that it is impossible to find an atomic orbital of the oxygen atom which will generate a set of numbers other than one of those given in this table. This can be checked by considering the transformation of the 2s and the other 3d orbitals of the oxygen ($3d_{z^2}$, $3d_{x^2-y^2}$, $3d_{zx}$ and $3d_{yz}$). Besides providing a partial proof of our assertion this exercise will provide the reader with experience which will prove invaluable as our discussion develops.

**Problem 2.1** Show that 2s(O), $3d_{z^2}$(O) and $3d_{x^2-y^2}$(O) individually generate the same set of numbers as $2p_z$(O), $3d_{zx}$(O) as $2p_x$(O) and $3d_{yz}$(O) as $2p_y$(O). In each case use the coordinate axis set shown in Figure 2.7 and operation labels as indicated in Figure 2.4.

The assertion that we cannot add to the table of numbers given above suggests that this set of numbers has properties beyond those which might be expected from the way in which they were derived. This is so. Sets of such numbers will provide the basis for the discussion contained in almost all of the remainder

Table 2.1

| E | $C_2$ | $\sigma_v$ | $\sigma_v'$ | |
|---|---|---|---|---|
| 1 | 1 | 1 | 1 | $2p_z$(O) |
| 1 | 1 | −1 | −1 | $3d_{xy}$(O) |
| 1 | −1 | 1 | −1 | $2p_y$(O) |
| 1 | −1 | −1 | 1 | $2p_x$(O) |

of this book. As an illustration of the unexpected properties of these numbers, we make what might appear to be a digression to discuss the effects of applying two of the symmetry operations of the water molecule in succession.

## 2.3 GROUP MULTIPLICATION TABLES

Earlier in this chapter we asserted that the effect of the successive application of symmetry operations of a group was always equivalent to the effect of some single operation of the group. We now investigate this in detail for the $C_{2v}$ point group by considering, in turn, each operation and the effect of following it with one of the four symmetry operations of the group. It will be helpful to focus our attention on a particular molecule. The water molecule is inconvenient for this purpose (because of the apparent equivalence of the effects of applying different symmetry operations — a phenomenon encountered several times already in this chapter) and instead we shall consider the molecule ethylene oxide, $(CH_2)_2O$, as an example. This molecule is shown in Figure 2.15. In this figure we have labelled the hydrogen atoms with the suffixes a, b, c or d; in order to study the effects of the symmetry operations on this molecule all that we have to do is to see how these labels are rearranged. The effects of the operations of the $C_{2v}$ point group on these labels are shown in Figure 2.16. It will be noted that each symmetry operation gives rise to a different final arrangement of labels.

Because the identity operation does not change the distribution of the labels at all, it is evident that any operation preceded or followed by the identity operation gives rise to the same final arrangement as that operation on its own. We immediately conclude that:

E followed by E $\equiv$ E
E followed by $C_2$ $\equiv C_2$
E followed by $\sigma_v$ $\equiv \sigma_v$
E followed by $\sigma_v'$ $\equiv \sigma_v'$
$C_2$ followed by E $\equiv C_2$
$\sigma_v$ followed by E $\equiv \sigma_v$
$\sigma_v'$ followed by E $\equiv \sigma_v'$

We now have to consider the result of the successive application of pairs of operations from the set $C_2$, $\sigma_v$ and $\sigma_v'$. We have mentioned earlier in this chapter that any one of these operations followed by itself gives rise to the initial arrangement and so the sequence is equivalent to the identity operation. That is:

$C_2$ followed by $C_2$ $\equiv$ E
$\sigma_v$ followed by $\sigma_v$ $\equiv$ E
$\sigma_v'$ followed by $\sigma_v'$ $\equiv$ E

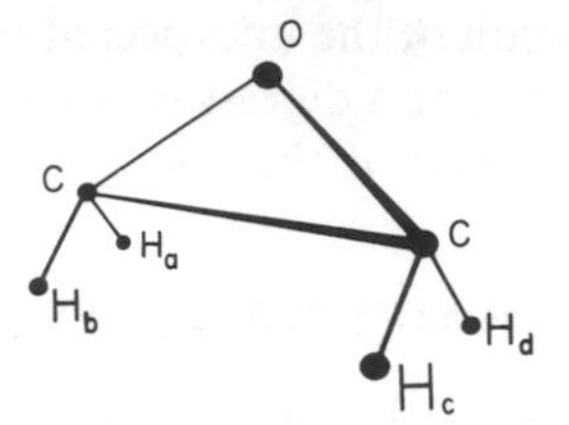


Figure 2.15 The ethylene oxide molecule, $C_2H_4O$

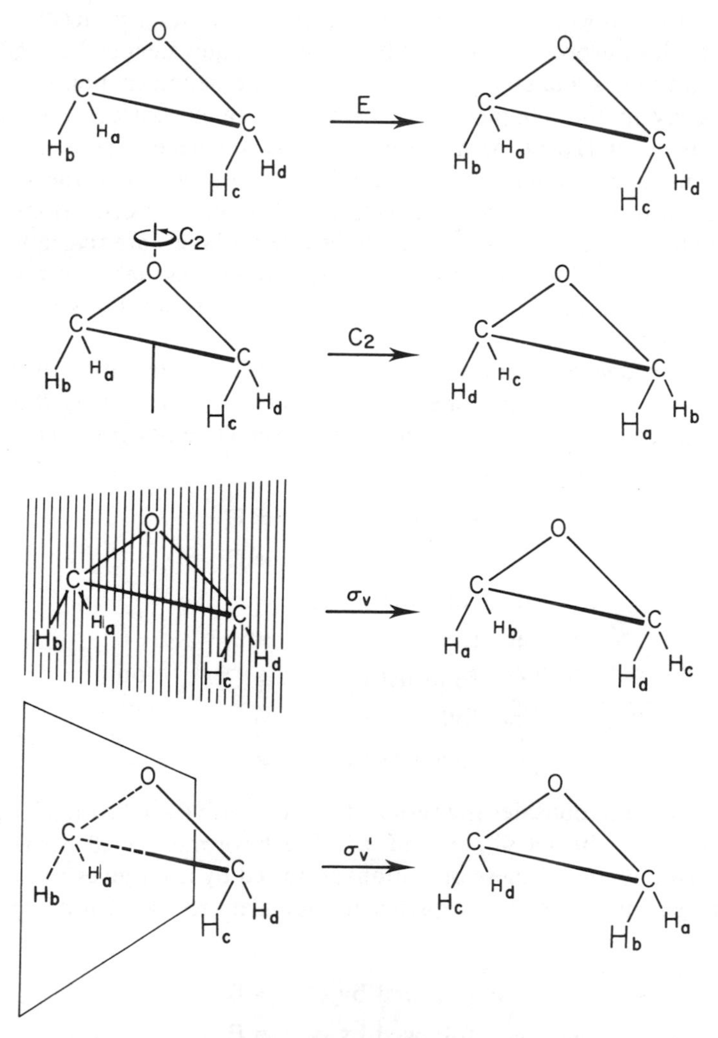


Figure 2.16 The effects of the symmetry operations of the $C_{2v}$ point group on the four hydrogen atoms of ethylene oxide

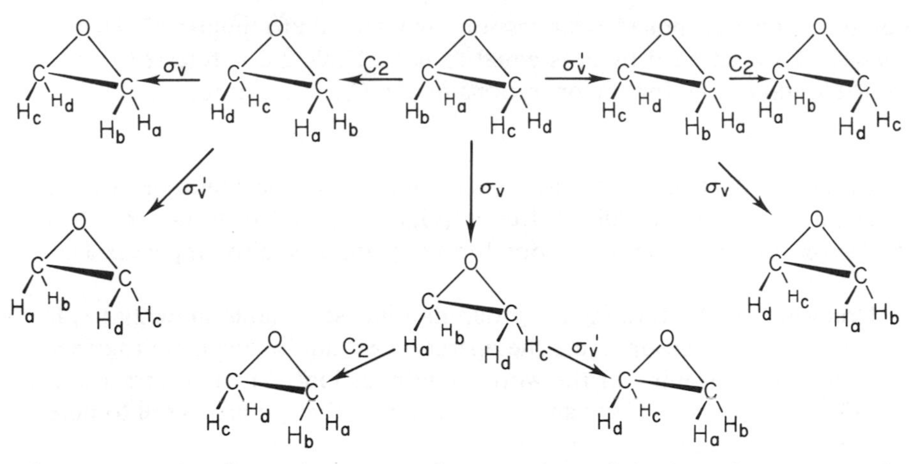


Figure 2.17 The effects of two successive operations of the $C_{2v}$ point group on the four hydrogen atoms of ethylene oxide

In Figure 2.17 we illustrate the remaining combinations of operations and the reader may, by comparison with Figure 2.16, determine which single operation is equivalent to each combination. We conclude that:

$$C_2 \text{ followed by } \sigma_v \equiv \sigma_v'$$
$$C_2 \text{ followed by } \sigma_v' \equiv \sigma_v$$
$$\sigma_v \text{ followed by } C_2 \equiv \sigma_v'$$
$$\sigma_v \text{ followed by } \sigma_v' \equiv C_2$$
$$\sigma_v' \text{ followed by } C_2 \equiv \sigma_v$$
$$\sigma_v' \text{ followed by } \sigma_v \equiv C_2$$

These results are collected together in Table 2.2.

It is usual in the mathematical theory of groups to refer to the law of combination of group elements as 'multiplication' (although only rarely does the operation have anything to do with ordinary arithmetical or algebraic multiplication). Thus, in the present case, where two symmetry operations

Table 2.2

| | | First operation | | | |
|---|---|---|---|---|---|
| | $C_{2v}$ | E | $C_2$ | $\sigma_v$ | $\sigma_v'$ |
| Second operation | E | E | $C_2$ | $\sigma_v$ | $\sigma_v'$ |
| | $C_2$ | $C_2$ | E | $\sigma_v'$ | $\sigma_v$ |
| | $\sigma_v$ | $\sigma_v$ | $\sigma_v'$ | E | $C_2$ |
| | $\sigma_v'$ | $\sigma_v'$ | $\sigma_v$ | $C_2$ | E |

combine by being applied in succession, they are said to 'multiply'. Therefore we say: '$C_2$ multiplied by $\sigma_v$ is equal to $\sigma_v'$.' Table 2.2 is referred to as the *multiplication table* for the operations of the $C_{2v}$ point group.

**Problem 2.2** Construct a multiplication table for the operations E, $C_2$, ($\sigma_v$) and ($\sigma_v'$) of the coloured table (Figure 2.5 and page 14) and thus demonstrate that the operations form a group (see also Appendix 1).

**Problem 2.3** By forming the group multiplication table show that 1, i, $-$i, $-1$ form a group under the operation of multiplying them together in the usual meaning of the word 'multiplication'. The numbers 1 and $-1$ are ordinary numbers and $i = \sqrt{-1}$ (it is perhaps more useful to note that $i^2 = -1$ and $-i^2 = 1$).

**Problem 2.4** Before using ethylene oxide as the example for Section 2.3 the author considered urea, $CO(NH_2)_2$, as an alternative. He decided against it because although urea could have a structure with four hydrogen atoms arranged similarly to Figure 2.16, in the crystal the molecule is planar, with two pairs of equivalent hydrogen atoms. Show that in both of these arrangements, planar and non-planar, urea has $C_{2v}$ symmetry.

## 2.4 CHARACTER TABLES

We embarked on the digression in the previous section in order to illustrate some of the properties of the sets of numbers contained in Table 2.1. We do this by combining Tables 2.1 and 2.2 in the following way. Let us choose any row of Table 2.1, say the second. Abstracting this row from the table we have the association between symmetry operations and numbers shown below:

| E | $C_2$ | $\sigma_v$ | $\sigma_v'$ |
|---|---|---|---|
| 1 | 1 | $-1$ | $-1$ |

Turning to Table 2.2, everywhere that we see the operation E listed in this table we replace it by the number with which it is associated in our chosen row of Table 2.1. That is, in the present case, we replace it by the number 1. Similarly, wherever $C_2$ appears in Table 2.2 it is replaced by 1, whilst both $\sigma_v$ and $\sigma_v'$ are replaced by $-1$. When these replacements have been made we arrive at Table 2.3.

The interesting thing about this table is that, if it is looked upon simply as a table in which numbers multiply each other, arithmetically, then the products are all correct. It is left to the reader to demonstrate that this statement is true no matter which row of numbers is selected from Table 2.1. Further,

Table 2.3

| | | | | |
|---|---|---|---|---|
| | 1 | 1 | −1 | −1 |
| 1 | 1 | 1 | −1 | −1 |
| 1 | 1 | 1 | −1 | −1 |
| −1 | −1 | −1 | 1 | 1 |
| −1 | −1 | −1 | 1 | 1 |

we now assert that it is impossible to find any other set of numbers to represent the operations of the $C_{2v}$ point group (apart from the trivial one in which each operation is represented by the number 0), for which it is possible, after appropriately substituting into Table 2.2, to obtain a table corresponding to Table 2.3 which is arithmetically correct.

Because of the close relationship between the multiplication of the operations of the $C_{2v}$ point group (given in Table 2.2) and the multiplication of the numbers in the rows of Table 2.1, each set of numbers may be regarded as representing (i.e. behaving in an analogous way to) the set of symmetry operations.* We shall speak of each row of Table 2.1 as being a *representation* of the symmetry operations. Further, we shall call them *irreducible representations* (the significance of the word 'irreducible' will not become evident until the next chapter, when we introduce the concept of reducible representations). In our subsequent discussion we shall often need to refer to the individual rows in Table 2.1 and it is convenient to circumvent the necessity of writing each one out in full by giving each a label. The labels commonly used are those shown in Table 2.4.

Thus, the set of numbers given at the beginning of this section would be referred to as 'the $A_2$ irreducible representation of the $C_{2v}$ point group'. Because the association between the symmetry operations and irreducible representations given in Table 2.4 is unique to the $C_{2v}$ point group, we indicate this by including the group label in the top left-hand corner of the table.

There is some system about the choice of the labels $A_1$, $A_2$, $B_1$ and $B_2$ in Table 2.4. Firstly, the A's are distinguished from the B's by the fact that they

Table 2.4

| $C_{2v}$ | E | $C_2$ | $\sigma_v$ | $\sigma_v'$ | |
|---|---|---|---|---|---|
| $A_1$ | 1 | 1 | 1 | 1 | $2p_z(O)$ |
| $A_2$ | 1 | 1 | −1 | −1 | $3d_{xy}(O)$ |
| $B_1$ | 1 | −1 | 1 | −1 | $2p_y(O)$ |
| $B_2$ | 1 | −1 | −1 | 1 | $2p_x(O)$ |

*Note that the word 'multiplication' in this sentence does not have the same meaning when applied to symmetry operations as it does when it refers to numbers.

have numbers of +1 for the $C_2$ operation whereas the B's have −1. $A_1$, by convention, is the so-called 'totally symmetric' irreducible representation and has +1 for all of its numbers. It is called totally symmetric because *all* the operations of the group turn something of $A_1$ symmetry into itself. This distinction can be extended to the B's by noting that irreducible representations with the suffix 1 are symmetric (character +1) under the $\sigma_v$ operation whereas those with suffix 2 are antisymmetric (character −1). However, in the case of the B's this distinction is marred by the fact that the distinction between $\sigma_v$ and $\sigma_v'$ is somewhat arbitrary — interchange the use of these labels and the labels $B_1$ and $B_2$ would have to change too. In practice this means that it is advisable to check the notation used by each author — one worker's notation may not be the same as the next. As long as one is consistent in the notation used for a particular problem there is no ambiguity about the final answer obtained. Similar considerations apply to many of the groups commonly used in chemistry. Although the totally symmetric representation may not be labelled $A_1$ it is always the first A listed (it could be $A_g$ or A′ for instance).

Just as the set of operations (E, $C_2$, $\sigma_v$, $\sigma_v'$) may be represented by any of the irreducible representations $A_1$, $A_2$, $B_1$, $B_2$, so, too, individual symmetry operations, such as $C_2$, are characterized in each irreducible representation by a particular number (which, in general, varies from one irreducible representation to the next). These individual numbers are termed *characters* and tables such as Table 2.4 are called *character tables.* As we have already indicated, character tables are of prime importance for the topics discussed in this book. We shall meet an example of this when, in the next chapter, we discuss the bonding in the water molecule.

On the right-hand side of Table 2.4 we have indicated the oxygen orbital which we used to generate a particular irreducible representation. Functions which have the property of generating an irreducible representation are commonly listed alongside character tables in this way. Such functions are called 'basis functions'. We have seen that the transformations of the oxygen $2p_y$ orbital under the operations of the $C_{2v}$ point group lead to the $B_1$ set of characters — the oxygen $2p_y$ orbital is a basis function for the generation of the $B_1$ characters. Thus, 'the oxygen $2p_y$ orbital is a basis for the $B_1$ irreducible representation of the $C_{2v}$ point group'. Alternatively, and more simply, 'the oxygen $2p_y$ orbital has $B_1$ symmetry in the $C_{2v}$ point group'.

Finally, the $C_{2v}$ point group is an *Abelian* point group. Abelian groups have multiplication tables which are symmetric about their leading diagonal (top left to bottom right); inspection of Table 2.2 shows that this is true for the $C_{2v}$ group. That is, the result of multiplying two operations is independent of the order in which they are multiplied — of which operation comes first and which comes second. The reason that we have deferred consideration of the ammonia molecule is that the result of multiplying some of the elements of its group does depend on the order in which we take them. An alternative (but equivalent) definition of an Abelian point group is to regard such point groups as those for which the character tables contain only numbers like 1 and $-1$. In Chapter 11 we shall see that they can also contain complex numbers, such as i and $-i$, which are such that some power of them equals 1 (thus $i^4 = 1$, for instance). The character tables of Abelian groups never contain numbers such as 3, $-3$, 2, $-2$ and 0.

**Problem 2.5** Compare the multiplication table you obtained in answer to Problem 2.2 with Table 2.2. Thus, or otherwise, construct the character table for the group of Problem 2.2.

**Problem 2.6** Repeat Problem 2.5 using the group of Problem 2.3. In this case you will probably have to use the 'or otherwise' method and may only be able to generate an incomplete character table. The complete character table has the same set of characters as one that we shall meet in Chapter 11 (Table 11.1).

**Problem 2.7** Show that the dipole moment of the water molecule forms a basis for the $A_1$ irreducible representation of the $C_{2v}$ group. (*Hint*: the dipole moment must, by symmetry, be directed along the z axis.)

**Problem 2.8** Show that the translation of the entire water molecule along the y axis forms a basis for the $B_1$ irreducible representation of the $C_{2v}$ point group. (*Hint:* represent the translation as an arrow and consider the transformations of this arrow and the direction in which it points. The y direction is given in Figure 2.7.)

**Problem 2.9** Repeat Problem 2.8 but now consider a translation along the x axis and show that it transforms as $B_2$.

## 2.5 SUMMARY*

In a molecule the axis of highest symmetry is conventionally chosen to be the z axis; recommendations for the choice of x and y exist (page 18). We are concerned with point group symmetry operations (page 13), which are named according to a conventional nomenclature (page 10). These operations form a group (page 14 and Appendix 1). In the present and the next two chapters, for simplicity we restrict our discussion to Abelian point groups (page 28). For the cases that we meet there, individual quantities — such as atomic orbitals on a central atom (page 22) — that are transformed into themselves

*Page numbers refer to the page in the chapter on which a full discussion commences. Sometimes in the summaries words are used in a way that should be evident from the context but which will be discussed in detail in later chapters, e.g. 'isomorphism'.

under the operations of the point group may have these transformations described by characters (page 28). (An example of an Abelian group which shows a more complicated behaviour will be met in Chapter 11.) A complete collection of these characters is called a character table (page 28). The sets of characters are called irreducible representations (page 27) and the individual quantities used to generate them are said to be bases for the irreducible representations (page 28). Characters multiply together in a way that is isomorphous (page 26) to the way that the operations of the point group multiply (page 26, but see Appendix 2). Irreducible representations are given labels in a systematic, but not always unambiguous, way (page 28).

# CHAPTER 3

# *The electronic structure of the water molecule*

In the last chapter we showed that it is possible to obtain sets of numbers (characters) — which we called irreducible representations — by a study of the transformational properties of the atomic orbitals of the oxygen atom in the water molecule. Atomic orbitals are not the only things which may serve as bases for the generation of irreducible representations. In Problems 2.7, 2.8 and 2.9 you were asked to consider the transformational properties of other quantities associated with the water molecule, such as its dipole moment. In the following chapters we shall meet a variety of bases; for instance, when studying the vibrations of a molecule we shall use small displacements of the individual atoms along Cartesian axes as bases. Sometimes the sets of numbers — the representations — generated by the transformation properties of a basis set appear in the character table. This happens when an irreducible representation is generated. More commonly, however, the representation generated does not appear in the character table. In such cases the representation is a reducible one. One of the representations encountered in the present chapter is a reducible representation; by studying it we shall develop a method of breaking up reducible representations into simpler, irreducible, representations. Firstly, however, we return to the character table of the $C_{2v}$ point group.

## 3.1 ORTHONORMAL PROPERTIES OF IRREDUCIBLE REPRESENTATIONS

As indicated in Chapter 2, the sets of characters comprising the $C_{2v}$ character table have properties beyond those which one might reasonably expect from the way that we derived them. Before using this character table we will give one further example of this. Consider any irreducible representation of the $C_{2v}$ point group (Table 3.1) and multiply its individual characters by the corresponding characters of any other irreducible representation. We then sum the products of characters. Thus, if we consider as an example the $A_2$ and $B_1$ irreducible representations, the sum of the products of characters is $(1 \times 1) + (1 \times -1) + (-1 \times 1) + (-1 \times -1) = 0$. In this case, and for *all* others

Table 3.1

| $C_{2v}$ | E | $C_2$ | $\sigma_v$ | $\sigma_v'$ | |
|---|---|---|---|---|---|
| $A_1$ | 1 | 1 | 1 | 1 | 2s(O), $2p_z$(O), $\psi(A_1)$ |
| $A_2$ | 1 | 1 | −1 | −1 | |
| $B_1$ | 1 | −1 | 1 | −1 | $2p_y$(O), $\psi(B_1)$ |
| $B_2$ | 1 | −1 | −1 | 1 | $2p_x$(O) |

in which the characters of two *different* irreducible representations of the $C_{2v}$ point group are multiplied together, the sum is zero. If, however, we choose the same irreducible representation each time (i.e. square the individual characters) then the sum of products is equal to four — a number which happens to be the number of operations in the $C_{2v}$ point group. This is no accident. Because the number of operations in a group is an important quantity, it is given a name — it is called 'the order of the group'. Thus, we say, 'The $C_{2v}$ point group is of order four'.

If, instead of choosing a row of the character table, we had worked with columns, a similar result would have been obtained — the sum of the products of the characters in two columns is equal to zero when the characters come from two different columns. If the *same* column is chosen, i.e. the characters squared, then the sum of squares is equal to the order of the group. These results are known as the character table orthonormality relationships; we shall discuss a more general form of them in Chapter 5 where we shall show that they may be used to derive character tables as an alternative to the procedure we used in Chapter 2. It is in large measure the existence of these relationships which enables symmetry considerations to simplify so many problems in the physical sciences. We shall use them frequently in this book.

The word 'orthonormal' is a composite of the words 'orthogonal' and 'normal', and embodies both. 'Orthogonal' here means 'independent'. When two things are orthogonal it means that one behaves — and can be discussed — without automatically requiring a change to the other. Thus, all the wavefunctions associated with the orbitals of an atom are orthogonal to each other and, in the present case we can talk of different irreducible representations quite independently of each other. 'Normal' or 'normalized' means 'weighted equally', and equal weighting usually means being given unit weight. This concept is most easily seen for two one-electron wavefunctions of an atom. The wavefunctions are (mathematically) normalized if, when we (mathematically) ask the question 'How many electrons may each wavefunction describe?', we obtain (mathematically) the answer 'one'. If we obtained the answer 'one' for the first wavefunction but some different answer, say '1.83', for the second, we would say that the second was not normalized and we would have to modify it with a multiplicative scale factor so that we did, indeed, get the answer 'one'. This wavefunction, also, would then be said to be normalized. Later in this chapter we shall be, effectively, normalizing irreducible representations when we divide the number 4 (obtained by simple arithmetic) by the order of the $C_{2v}$ point group (the total number of operations in the group), which is also 4, to give the number 1. As implied above, the orthonormality relationships are best expressed mathematically and this is done in Appendix 2.

## 3.2 THE TRANSFORMATION PROPERTIES OF ATOMIC ORBITALS IN THE WATER MOLECULE

In the present chapter we shall show that the $C_{2v}$ character table may be used to greatly simplify a discussion of the bonding in the water molecule. We know that it is possible to think of this bonding as arising from the interaction of orbitals located on the oxygen atom with those on the two hydrogen atoms. We shall largely confine our discussion to the valence shell atomic orbitals of these atoms. That is, we shall consider the oxygen 2s, $2p_z$, $2p_x$ and $2p_y$ orbitals and the two hydrogen 1s orbitals. The transformational properties of the oxygen orbitals have already been discussed (Section 2.2 and Problem 2.1) and the results are summarized on the right-hand side of Table 3.1; we shall discuss the hydrogen 1s orbitals shortly. In this book we shall find it convenient to use phrases like 'orbitals of $A_1$ symmetry', by which, in the present example, we would mean the 2s and $2p_z$ orbitals of the oxygen together with any orbitals of this symmetry which we may subsequently discover (we shall find such an orbital arising from the hydrogen 1s orbitals). In a similar way we shall refer to the $2p_y$ orbital of the oxygen as 'an orbital of $B_1$ symmetry', by which we mean that the characters of the $B_1$ irreducible representation describes its transformations under the operations of the $C_{2v}$ point group.

So far we have taken care to consider only the transformational properties of individual orbitals. However, there is no reason why we should not have considered, for instance, all three 2p orbitals of the oxygen atom together. Indeed, had we chosen to place the z axis in an arbitrary direction in the molecule rather than along the twofold axis (and similarly placed no symmetry constraints on the x and y axes) we would have had to treat the orbitals as a set because, for example, a $2p_z$ orbital pointing in an arbitrary direction would not be turned neatly into itself by all of the symmetry operations of the group. Evidently, a careful choice of direction for coordinate axes can simplify symmetry discussions! Had we persisted in choosing arbitrary (but, of course, mutually perpendicular) directions for our axes the final result would have been the same — we would have ended up with 2p orbitals transforming as

$A_1$, $B_1$ and $B_2$. However, the work involved would have been more difficult and until Appendix 2 has been read the reader will not be equipped to prove the assertion that we have just made.

When we turn to the two hydrogen 1s orbitals we are immediately confronted with a similar problem. Should we consider them as individuals or as a pair in order to determine their symmetry properties? The answer to this question is simple (and covers the case of oxygen 2p orbitals oriented along arbitrary Cartesian axes). Whenever one (or more) operations of the point group has the effect of interchanging or mixing orbitals (or, as sometimes happens, a bit of both) then all of the orbitals which are scrambled must be considered together. Although, in the present context, we are talking about atomic orbitals, precisely the same is true for other quantities. We shall meet many other examples in this book. For instance, the small atomic displacements used in the study of molecular vibrations are bases which are quite often scrambled by symmetry operations.

We return now to the specific problem of the transformation of the two hydrogen 1s orbitals in the water molecule. In Figure 3.1 we show the behaviour of the two hydrogen 1s orbitals (which we shall denote $h_1$ and $h_2$) under the symmetry operations of the $C_{2v}$ point group. For the $C_2$ and $\sigma_v'$

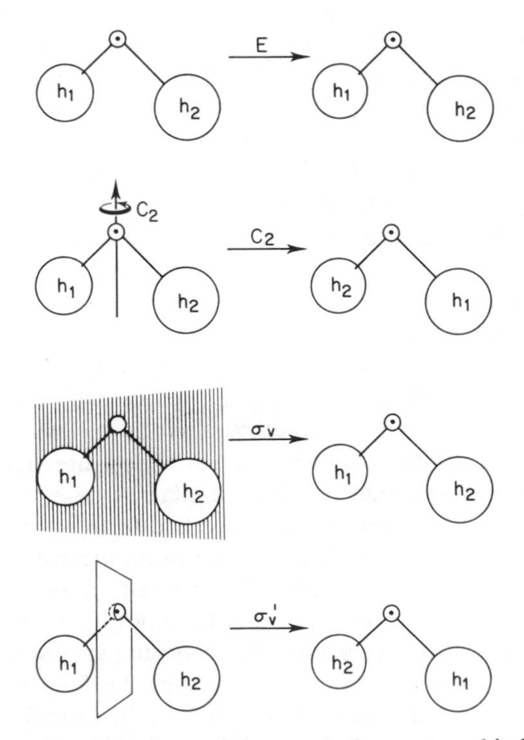


Figure 3.1 The behaviour of the two hydrogen 1s orbitals of $H_2O$ under the symmetry operations of the $C_{2v}$ point group

operations the two hydrogen 1s orbitals interchange, but under E and $\sigma_v$ each remains itself. When something remains unchanged under an operation we obtain a character of one (the numbers which were introduced as multiplicative factors in Chapter 2 will now be referred to as characters) and so, when considering two things together, each of which remains unchanged and each of which, therefore, makes a contribution of one to the character, we obtain an aggregate character of two. We can generalize this:

> When the transformation of several things is being considered together the character which they together generate under a symmetry operation is the sum of the characters which they generate as individuals.

For both the E and $\sigma_v$ symmetry operations we therefore obtain characters of two. However, in the case of the $C_2$ and $\sigma_v'$ operations we encounter a situation which we have not previously met, because the orbitals $h_1$ and $h_2$ interchange under these operations. We have to represent the fact that $h_1$, for instance, *disappears* from its original position under these operations by some multiplicative factor. The only evident way of doing this is by using a multiplicative factor (character) of zero. The same is true for $h_2$. We can generalize this result:

> Symmetry operations which lead to all of the members of a set interchanging with each other give rise to a resultant character of zero. If a symmetry operation results in some members interchanging whilst others remain in the same position, it is only the latter which make non-zero contributions to the character.

This discussion contains no explicit recognition of the fact that $h_1$ and $h_2$ *interchange* under the $C_2$ and $\sigma_v'$ operations — only that they disappear from their original positions; this aspect of the problem is explored in Appendix 2. The more detailed treatment given there provides proof of the two important rules given in boxes above and, in particular, shows that the transformation of $h_1$ and $h_2$ under $C_2$ and $\sigma_v'$ leads each to contribute zero to the aggregate character.

The set of characters which we have just obtained is shown below:

| E | $C_2$ | $\sigma_v$ | $\sigma_v'$ |
|---|---|---|---|
| 2 | 0 | 2 | 0 |

This set has properties which are rather different to those of corresponding sets which we obtained earlier (and which we called irreducible representations). For instance, as is readily shown, when these characters are substituted for the corresponding symmetry operations in the group multiplication table (Table 2.2), the multiplication table obtained is not arithmetically correct.

**Problem 3.1** Substitute characters for the corresponding operations in Table 2.2 using the correspondence

| $E$ | $C_2$ | $\sigma_v$ | $\sigma_v'$ |
|---|---|---|---|
| 2 | 0 | 2 | 0 |

and check that the table obtained is not arithmetically correct.

**Problem 3.2** Show that the 1s orbitals of the four hydrogen atoms of the ethylene oxide molecule, discussed in Section 2.3 and, in particular, Figures 2.15, 2.16 and 2.17, form a basis for the following representation of the $C_{2v}$ point group:

| $E$ | $C_2$ | $\sigma_v$ | $\sigma_v'$ |
|---|---|---|---|
| 4 | 0 | 0 | 0 |

## 3.3 A REDUCIBLE REPRESENTATION

Although the set of characters which we generated at the end of the previous section is not identical to any of the irreducible representations of the $C_{2v}$ point group, it is equal to a sum of them. If we add together the corresponding characters of the $A_1$ and $B_1$ irreducible representations, as shown below, we obtain the same set of characters as those obtained using $h_1$ and $h_2$ as a basis:

| | $E$ | $C_2$ | $\sigma_v$ | $\sigma_v'$ |
|---|---|---|---|---|
| $A_1$ | 1 | 1 | 1 | 1 |
| $B_1$ | 1 | $-1$ | 1 | $-1$ |
| $A_1 + B_1$ | 2 | 0 | 2 | 0 |

That is, the representation which we generated can be decomposed into a sum of irreducible representations. A representation which can be reduced to a sum of other representations is, reasonably enough, called a *reducible* representation. The use of the name 'irreducible representation' for the representations appearing in the character table should now be clear. These representations cannot be reduced further; they are irreducible. There are similarities between reducible representations and irreducible representations, although there are also important differences (one of which we have already seen — for the $C_{2v}$ group a reducible representation does not multiply arithmetically to give a multiplication table in which there is a consistent correspondence between the numbers it contains and the operations in the corresponding group multiplication table). A most important similarity between reducible and irreducible representations is found when we investigate the applicability of the orthonormality relationships (Section 3.1) to the connection between reducible and irreducible representations. Thus, if we multiply the individual characters of our reducible representation

| $E$ | $C_2$ | $\sigma_v$ | $\sigma_v'$ |
|---|---|---|---|
| 2 | 0 | 2 | 0 |

by the corresponding characters of one of the irreducible representations of the $C_{2v}$ character table and sum the products, we find that for the $A_2$ and $B_2$ irreducible representations we obtain the answer zero. For the $A_2$ irreducible representation, for example, we obtain

$$(2 \times 1) + (0 \times 1) + (2 \times -1) + (0 \times -1) = 0$$

We know that our reducible representation is a sum of the $A_1$ and $B_1$ irreducible representations and so, when we form a similar sum of products using the $A_1$ irreducible representation rather than $A_2$, we are really forming the product of $A_1$ with $(A_1 + B_1)$. That is, we are forming products between $A_1$ and $A_1$ and between $A_1$ and $B_1$ simultaneously. We already know that the first of these gives a sum which is equal to the order of the group, whilst the second sums to zero. That is, we only obtain non-zero answers when we select from the character table those irreducible representations which are contained in the reducible one. This fact leads to a general method for reducing reducible representations into their irreducible components.

Let us use this method to show, systematically, that our reducible representation contains a $B_1$ component. The steps involved are:

(a) Write down the reducible representation

| E | $C_2$ | $\sigma_v$ | $\sigma_v'$ |
|---|---|---|---|
| 2 | 0 | 2 | 0 |

(b) Write down the $B_1$ irreducible representation:

| | | | |
|---|---|---|---|
| 1 | −1 | 1 | −1 |

(c) Multiply the characters in the same column:

| | | | |
|---|---|---|---|
| 2 | 0 | 2 | 0 |

(d) Add these products together and then divide the sum by the number four (the order of the group):

$$4/4 = 1$$

We conclude that our reducible representation contains the $B_1$ irreducible representation (had it contained $B_1$ twice the final answer would have been 2, and so on). The reader can similarly demonstrate that the reducible representation contains an $A_1$ component.

**Problem 3.3** Use the method described above to reduce the following reducible representations of the $C_{2v}$ point group:

| | E | $C_2$ | $\sigma_v$ | $\sigma_v'$ |
|---|---|---|---|---|
| (a) | 2 | 0 | −2 | 0 |
| (b) | 3 | −1 | −1 | −1 |
| (c) | 4 | 0 | 0 | 0 |

**Problem 3.4** Reduce the following reducible representations of the $C_{2v}$ point group and for each check your answer by adding together the characters of the irreducible representations to regenerate those given below (there is an aspect of the irreducible representations in this problem which distinguishes them from those in Problem 3.3 and which makes this check worthwhile):

| | $E$ | $C_2$ | $\sigma_v$ | $\sigma_v'$ |
|---|---|---|---|---|
| (a) | 3 | −1 | −3 | 1 |
| (b) | 4 | 0 | 0 | −4 |
| (c) | 6 | −4 | −2 | 0 |

## 3.4 SYMMETRY-ADAPTED COMBINATIONS

What is the significance of the fact that the reducible representation generated by the transformation of the two hydrogen 1s orbitals $h_1$ and $h_2$ may be reduced into a sum of $A_1$ and $B_1$ irreducible representations? As we have seen, an irreducible representation such as $A_1$ describes the transformation properties of a single orbital as, too, does the $B_1$ irreducible representation. We therefore anticipate that it is possible to derive from the orbitals $h_1$ and $h_2$ one orbital whose transformations are described by the $A_1$ irreducible representation and a second orbital which transforms as $B_1$. Evidently, we now have to investigate the form of these orbitals to find out what they look like. There is a systematic method of deriving such orbitals but we shall not introduce this until Chapter 4. For the present example a rather simpler argument will suffice.

The reader will recall that in a discussion of the electronic structure of the hydrogen molecule, $H_2$, two hydrogen 1s orbitals combine to give bonding and antibonding combinations. If we, hypothetically, remove the oxygen atom from a water molecule we are left with a hydrogen molecule, albeit with a rather stretched H—H bond. It would be reasonable to expect that the combinations of hydrogen 1s orbitals in this stretched $H_2$ molecule would be related to the correct combinations of hydrogen 1s orbitals in the water molecule. With neglect of overlap between the two atomic orbitals the bonding and antibonding combinations of hydrogen 1s orbitals in the $H_2$ molecule have the form

$$\psi\,(\text{bonding}) = \frac{1}{\sqrt{2}}(h_1 + h_2)$$

$$\psi\,(\text{antibonding}) = \frac{1}{\sqrt{2}}(h_1 - h_2)$$

where we have used the same labels for the hydrogen atomic orbitals as in the water molecule. Let us consider the transformation of these bonding and antibonding combinations under the operations of the $C_{2v}$ point group. The transformations of the bonding combination is shown in Figure 3.2 where, to

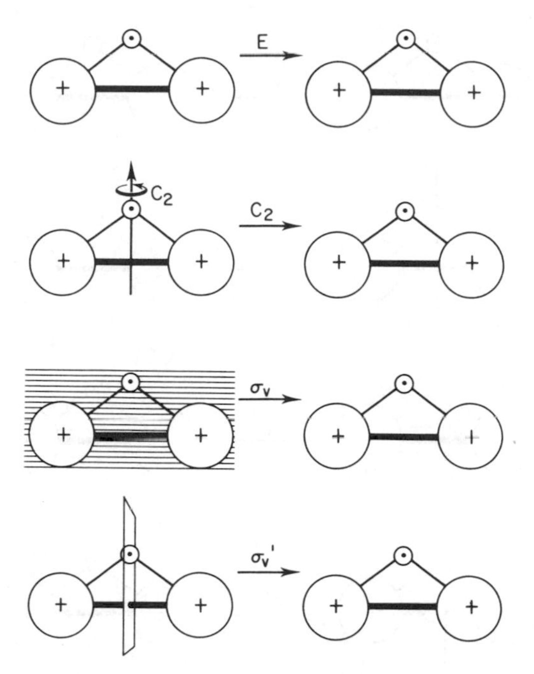


Figure 3.2 The transformations of the H—H bonding orbital of $H_2$ under the symmetry operations of the $C_{2v}$ point group

emphasize the fact that we are considering the transformation of a single orbital, the component hydrogen 1s orbitals have been joined together. It is evident from this figure that under all of the operations of the $C_{2v}$ point group this orbital is transformed into itself, giving charcters of $+1$ for each operation. That is, the combination $(1/\sqrt{2})(h_1 \times h_2)$ is of $A_1$ symmetry in the $C_{2v}$ point group.

In Figure 3.3 we show the transformational properties of the antibonding combination of hydrogen 1s orbitals in the hydrogen molecule. In this figure, to emphasize the fact that we are considering the transformation of a single orbital, the two parts of the orbital are again joined together. The characters generated by the action of the operations of the $C_{2v}$ point group on this orbital are

| E | $C_2$ | $\sigma_v$ | $\sigma_v{}'$ |
|---|---|---|---|
| 1 | $-1$ | 1 | $-1$ |

so this orbital is of $B_1$ symmetry.

We have, then, the fact that the bonding and antibonding 1s molecular orbitals of the hydrogen molecule have, in the $C_{2v}$ point group, the symmetries $A_1$ and $B_1$ respectively. As we saw at the beginning of this section, we expect there to be $A_1$ and $B_1$ combinations of hydrogen 1s orbitals in the water molecule but were not able to say what they looked like. Clearly, the molecular

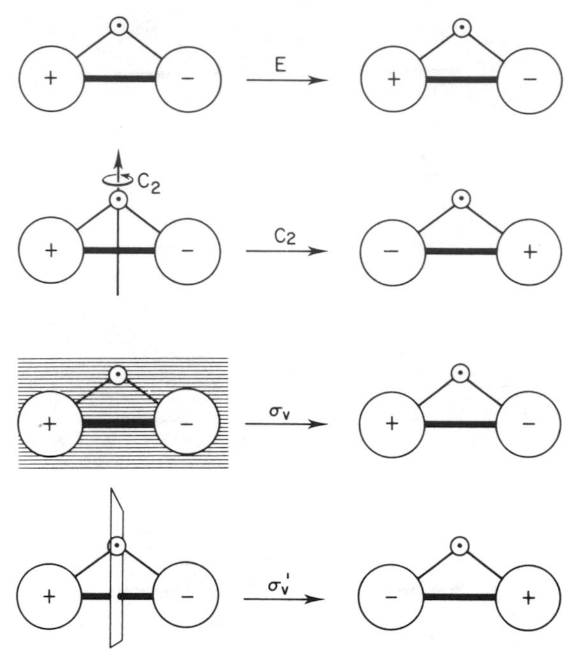


Figure 3.3 The transformations of the H—H antibonding orbital of $H_2$ under the symmetry operations of the $C_{2v}$ point group

orbitals of $H_2$ provide the answer to our search. We conclude that, in the water molecule, rather than considering the two hydrogen 1s orbitals separately, we may, instead, consider two combinations, one of $A_1$ symmetry and one of $B_1$ symmetry. The $A_1$ combination is

$$\psi(A_1) = \frac{1}{\sqrt{2}}(h_1 + h_2)$$

and the $B_1$ combination is

$$\psi(B_1) = \frac{1}{\sqrt{2}}(h_1 - h_2)$$

The argument we have used to obtain $\psi(A_1)$ and $\psi(B_1)$ was based on plausibility rather than mathematical rigour. The mathematical method will be developed in Chapter 4.

## 3.5 THE BONDING INTERACTIONS IN $H_2O$ AND THEIR ANGULAR DEPENDENCE

We have now found the two linear combinations of hydrogen 1s orbitals in the water molecule which transform as the $A_1$ and $B_1$ irreducible representations. It is to be emphasized that although the mathematical form we have given these orbitals is one which neglects overlap between $h_1$ and $h_2$, the presence or absence of overlap in no way affects their symmetry species.

We now come to a vital point in our argument. It involves as the key step an assertion which, for the moment, the reader is asked to take to some extent on trust. A proof will be given in Chapter 10 although a partial justification is included here. The assertion is that:

> Interactions between orbitals transforming as different irreducible representations are always zero.*

That is, in a discussion of the bonding in a molecule we can break up our argument into quite separate discussions, one for each irreducible representation. In the case of the water molecule, for example, the only orbital of $B_2$ symmetry is the $2p_x$ orbital of the oxygen. There is no hydrogen 1s combination of this symmetry and so the oxygen $2p_x$ orbital does not interact with any other orbital in the molecule. That is, it is a non-bonding orbital located on the oxygen atom.

Figure 3.4 provides some justification for the assertion that we can confine our consideration of bonding interaction to those between orbitals of the same symmetry. It shows the overlap between the 2s orbital of the oxygen — of $A_1$ symmetry — and the $B_1$ combination of hydrogen 1s orbitals, $\psi(B_1)$. It is evident that although these orbitals overlap with one another the overlap *integral* is zero since the regions of positive overlap are exactly cancelled by the regions of negative overlap. The zero overlap integral between $2p_x$ and the $A_1$ or $B_1$ combinations of hydrogen 1s orbitals $\psi(A_1)$ or $\psi(B_1)$ can be similarly demonstrated.

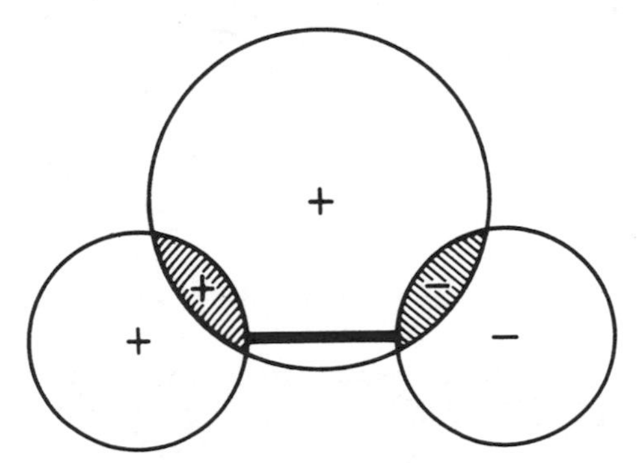


Figure 3.4 The zero overlap *integral* between an orbital of $A_1$ symmetry (oxygen 2s) and an orbital of $B_1$ symmetry (a linear combination of hydrogen 1s orbitals)

*We have in mind here the one-electron terms in the Hamiltonian. Analogous statements may be made to cover the two-electron terms; the form of such statements will become evident in Chapter 10.

The basis functions associated with the $C_{2v}$ character table (Table 3.1) provide a list of all the orbitals which interact with one another. The orbitals of $A_1$ symmetry which we must discuss are the 2s and $2p_z$ orbitals on the oxygen, each of which interacts with the hydrogen 1s combination $\psi(A_1)$. There are no orbitals of $A_2$ symmetry and only two of $B_1$ symmetry — the $2p_y$ orbital of oxygen and a hydrogen 1s combination $\psi(B_1)$. When two (or more) orbitals of the same symmetry species interact, the final molecular orbitals are of the same symmetry species as the initial ones. We shall first consider the $B_1$ interactions qualitatively but in some detail.

The interaction between the $2p_y$ orbital of the oxygen and the $B_1$ combination of hydrogen 1s orbitals will lead to bonding and antibonding molecular orbitals. A schematic representation of the overlap between these orbitals and the form of the resultant bonding and antibonding molecular orbitals is shown in Figure 3.5. The bonding molecular orbital is an out-of-phase combination of $2p_y$ and $\psi(B_1)$ whilst the antibonding molecular orbital is an in-phase combination (if this seems strange, compare the relative phases of $2p_y$ and $\psi(B_1)$ in Figure 3.5).

The $B_1$ bonding orbital is occupied by two electrons in the water molecule and so contributes to the molecular stability. There is an important point which must be made concerning this bonding molecular orbital of $B_1$ symmetry. Let us consider, qualitatively, the dependence of the molecular stabilization derived from this orbital upon the HOH bond angle, $\theta$. For the (hypothetical!) case of very small $\theta$, shown in Figure 3.6, the lobes of $\psi(B_1)$

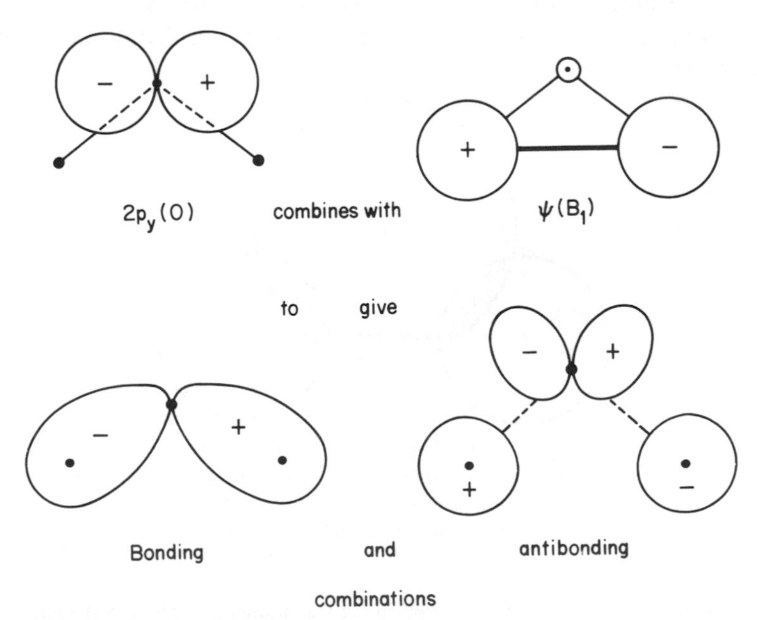


Figure 3.5 Interactions between orbitals of $B_1$ symmetry leading to bonding and antibonding combinations

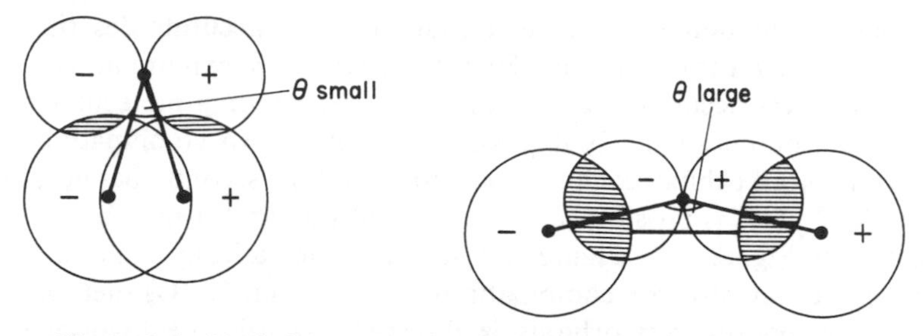


Figure 3.6 Variation of overlap integral of the $B_1$ orbitals with variation of bond angle in $H_2O$

overlap with $2p_y$ in a way that leads to a relatively small value of the overlap integral; the overlap integral decreases as the bond angle decreases. When $\theta$ is very small, then, the interaction between the two orbitals of $B_1$ symmetry, which varies roughly as the overlap integral, will be small and the $B_1$ bonding molecular orbital will make little contribution to the molecular stability. It is qualitatively evident from Figure 3.6 (and is confirmed by more detailed calculations) that as $\theta$ increases (keeping the OH bond length constant) the interaction between $2p_y(O)$ and $\psi(B_1)$ smoothly increases with $\theta$ and reaches a maximum for a bond angle of 180°. We conclude that, were this interaction the only thing determining the geometry of the water molecule, $H_2O$ would be a linear molecule!

We now turn our attention to the more difficult case of the interaction between the three orbitals of $A_1$ symmetry. In the case of the $B_1$ interaction which we have just considered, the final molecular orbitals were mixtures of the two orbitals with which we started. Similarly, we would expect our final $A_1$ molecular orbitals to be mixtures of $2s(O)$, $2p_z(O)$ and $\psi(A_1)$. An important point to note here is that although we have introduced $2s(O)$ and $2p_z(O)$ as separate functions they will be mixed (i.e. contribute to the same molecule orbitals) by virtue of their mutual interaction with $\psi(A_1)$. Detailed calculations show that this mixing is not very large and so we have in the water molecule an oxygen 2s orbital mixed with a small amount of oxygen $2p_z$ together with a second orbital which is largely $2p_z$ but mixed with a little bit of 2s, both interacting with the same hydrogen 1s orbital combination. That is, there are two molecular orbitals of $A_1$ symmetry which are each filled with electrons and

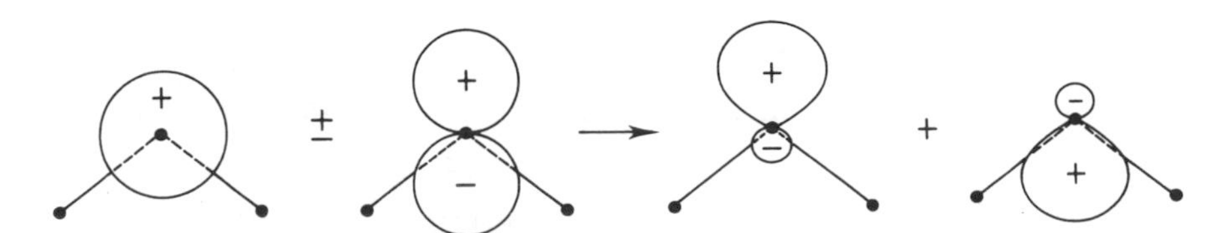


Figure 3.7 In-phase and out-of-phase combinations of oxygen 2s and $2p_z$ orbitals give two (sp) hybrid orbitals

contribute to the bonding. This is a qualitative but accurate description and we shall briefly return to it later. Firstly, however, we explore an alternative, simpler, but less accurate, description of the $A_1$ bonding molecular orbitals.

Consider the question, 'Is it possible to obtain two combinations of the 2s(O) and $2p_z$(O) orbitals such that one does, and the second does not, interact with $\psi(A_1)$?' This is possible, and the general way that it may be achieved is indicated in Figure 3.7. Figure 3.7 shows, schematically, that if we take in-phase and out-of-phase combinations of 2s(O) and $2p_z$(O) then one of the resulting mixed (hybrid) orbitals is directed towards the hydrogen atoms whereas the second combination is largely located in a region remote from these atoms. This second combination would be essentially non-bonding and we may, as a first approximation, ignore its interactions with $\psi(A_1)$. It is convenient to choose 2s(O) and $2p_z$(O) combinations which have a form which simplifies the pictorial representation of the problem and, therefore, to assume that they are sp hybrids of the form:

$$\frac{1}{\sqrt{2}}\,[2s(O) + 2p_z(O)] \qquad \text{(non-bonding)}$$

$$\frac{1}{\sqrt{2}}\,[2s(O) - 2p_z(O)] \qquad \text{(involved in bonding)}$$

and it is these which are shown, qualitatively, in Figure 3.7.

We have now simplified our problem so that it is analogous to that discussed earlier for the case of the $B_1$ interactions — we have only the interaction between two orbitals, $\psi(A_1)$ and the second given above, to consider. These two orbitals will combine to give in-phase and out-of-phase combinations which are, respectively, bonding and antibonding molecular orbitals. These orbitals are shown schematically in Figure 3.8.

To complete our picture we consider the relationship between the magnitude of the stabilization resulting from the interactions between the various orbitals of $A_1$ symmetry and the value of the HOH bond angle. For this discussion we consider interactions involving the 2s(O) and $2p_z$(O) orbitals separately. It is evident, from Figure 3.9, that the magnitude of the interaction between 2s(O) and the hydrogen 1s combination $\psi(A_1)$ does not depend upon the HOH bond angle. Because the oxygen orbital is spherically symmetrical the overlap integral is independent of the bond angle.

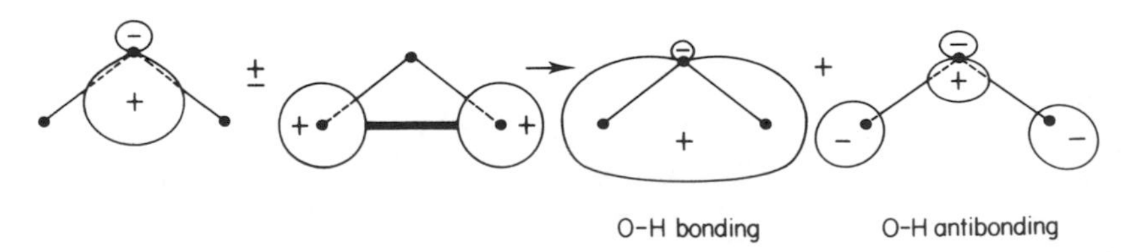


Figure 3.8 Interaction betweeen orbitals of $A_1$ symmetry leading to bonding and antibonding combinations

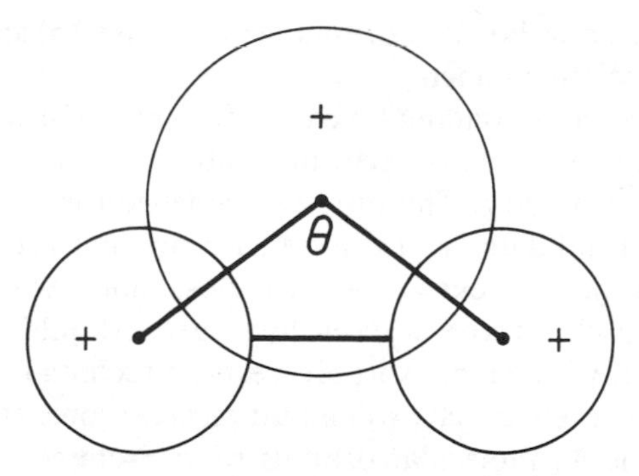


Figure 3.9 An $(A_1)$ overlap integral which does not depend on bond angle (cf. Figure 3.6)

The $2p_z(O)-\psi(A_1)$ interaction is shown schematically in Figure 3.10. It is evident from Figure 3.10 that when $\theta = 180°$ there is a zero overlap integral and so no interaction between $2p_z$ and $\psi(A_1)$. This result has its origin in molecular symmetry. The symmetry of a linear water molecule is no longer $C_{2v}$ but that of a different point group (called $D_{\infty h}$). In this latter group the $2p_z(O)$ orbital and $\psi(A_1)$ are of different symmetries; it follows that they do not interact. Evidently, the interaction between $2p_z(O)$ and $\psi(A_1)$ increases smoothly as the HOH bond angle decreases from a value of 180° and reaches a maximum when $\theta = 0$. Because the interaction of $\psi(A_1)$ with 2s(O) is independent of bond angle it is the interaction of $\psi(A_1)$ with $2p_z(O)$ which determines the angular variation of the interaction of $\psi(A_1)$ with mixtures of 2s(O) and $2p_z(O)$ orbitals; this interaction will tend to be a maximum at $\theta = 0$ (although, of course, because it involves a superposition of the two hydrogen atoms, this limit will never be reached!).

In summary, we conclude that of the bonding interactions in the water molecule, those of $A_1$ symmetry favour a bond angle $\theta \rightarrow 0°$ and that of $B_1$ symmetry leads to a stabilization which maximizes as $\theta \rightarrow 180°$. There is only

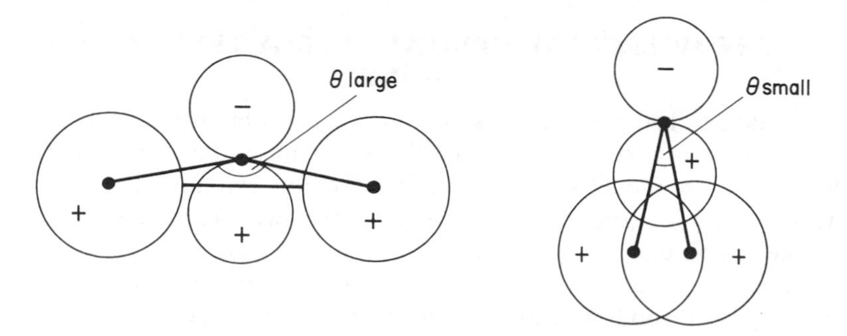


Figure 3.10 An $(A_1)$ overlap integral which varies with bond angle (cf. Figure 3.9)

one unambiguously non-bonding orbital. This, a pure $2p_x$ atomic orbital of the oxygen atom, is of $B_2$ symmetry.

A second entirely non-bonding lone pair does not exist in the isolated water molecule, although, as we have seen, in a rather less accurate model, we can create one of $A_1$ symmetry. The physical evidence for two lone pairs comes largely from structural data. Thus, in ice each oxygen is roughly tetrahedrally surrounded by four hydrogens — two close and two distant. It seems reasonable that each of the distant hydrogens should be associated, by hydrogen bonding, with a lone pair. However, attaching two more protons to the water molecule, even loosely, will modify its electronic structure. So, for instance, each of the $A_1$ molecular orbitals of an isolated water molecule will also be involved in the bonding of these additional protons (just as is the case in methane). Further, the reader should be able to show that the 1s orbitals of the distant hydrogens give rise to a combination of $B_2$ symmetry, so that even the — genuinely — non-bonding $p_x$ orbital of $H_2O$ may become weakly involved in bonding in ice.

**Problem 3.5** The individual oxygen atoms in ice are surrounded by a distorted tetrahedron of hydrogen atoms. That is, they resemble the carbon atom of Figure 2.1 but two of the oxygen–hydrogen bonds are longer than the other two. The closely bonded pair are those discussed in Section 3.3. Show that the transformations of the 1s orbitals of the more distant hydrogen atoms give rise to a reducible representation with $A_1 + B_2$ components.

**Problem 3.6** In a discussion of the bonding in $H_2S$ (bond angle 93°) the valence shell orbitals on the sulphur are 3s, 3p and 3d. Inclusion of the 3d orbitals would increase the number of possible interactions with $\psi(A_1)$ and $\psi(B_1)$ which would have to be considered. List all of the sulphur valence shell orbitals which could interact with each of them (use Table 2.4 and the results of Problem 2.1).

## 3.6 THE MOLECULAR ORBITAL ENERGY LEVEL DIAGRAM FOR $H_2O$

We now have to bring our discussion together and obtain a schematic molecular orbital energy level diagram for the water molecule. Rather than work with the accurate model we shall consider the approximate. This is because, as we shall see, it is the model that is consistent with relatively simple ideas about the bonding in the molecule — it is the one most likely to be produced by simply following chemical intutition. Notwithstanding its approximate nature, it gives a good prediction of the *relative* ordering of molecular orbital energies and so gives hope that similar approximate models will also be of value for other molecules. We proceed by presenting a schematic energy level

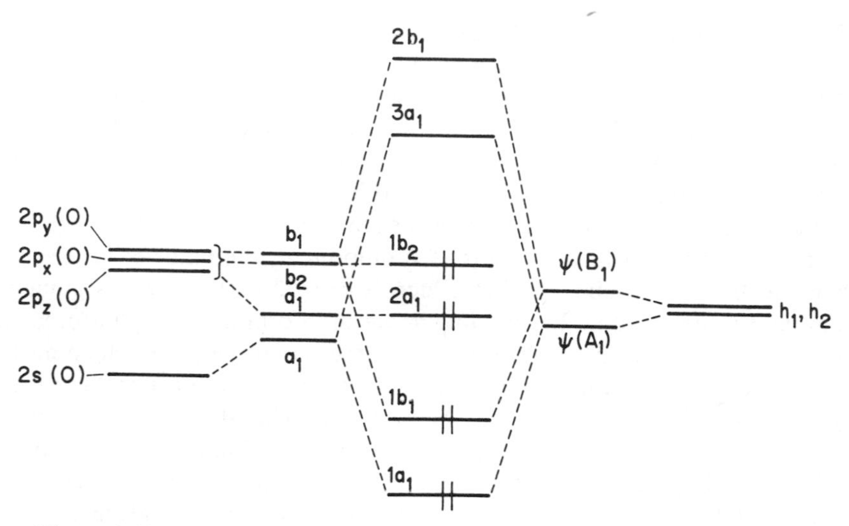


Figure 3.11 A schematic molecular orbital energy level diagram for $H_2O$

scheme in Figure 3.11 and will shortly detail the arguments used in its derivation. Before doing this we note that, in contrast to the discussion in the text, we have used lower case symbols in Figure 3.11. In fact, lower case symbols like this are used to describe wavefunctions. Each electron in a molecule has a unique wavefunction so if two electrons have the same spatial wavefunction they must differ in their spin wavefunctions. It is the spatial wavefunctions which are referred to by a symbol such as $b_1$ so there can be two electrons with this wavefunction (loosely, one with spin 'up' and one with spin 'down'). Thus, although a one-electron wavefunction, reasonably enough, characterizes a single electron, whereas an orbital can be occupied by two electrons, the distinction in usage is a rather fine one and we shall talk of 'orbitals' — either 'the orbital of $A_1$ symmetry' or 'the $a_1$ orbital'. We shall generally use upper case symbols in this text to avoid switching back and forth between the two usages and largely confine lower case symbols to diagrams. One important context in which lower case symbols are invariably used in the literature is when orbital occupancies are specified. Thus, Figure 3.11 shows an orbital occupancy, starting with the lowest energy first and labelling orbitals of the same symmetry sequentially, which is written as $1a_1{}^2$, $1b_1{}^2$, $2a_1{}^2$, $1b_2{}^2$. Here the superscripts indicate the number of electrons in each orbital.

We now return to the details of Figure 3.11 and explain the arguments leading to the orbital energy sequence shown there. Firstly, we have used a nodal plane criterion, which experience shows to be reliable.[1] We recall that in the simple model of the water molecule there are just two bonding molecular orbitals, one of $A_1$ and one of $B_1$ symmetry. Figures 3.5 and 3.8 reveal that although for both $A_1$ and $B_1$ bonding molecular orbitals the oxygen–hydrogen interactions are entirely bonding, in the $B_1$ case the hydrogen–hydrogen interactions are antibonding — $\psi(B_1)$ corresponds to the antibonding combination of 1s atomic orbitals in the hydrogen molecule. The corresponding component in the $A_1$ bonding molecular orbital is bonding (Figure 3.8) and so we anticipate that this orbital will be the more stable. That is, the orbital with the fewest nodal planes is usually expected to be the most stable. Secondly, to obtain the probable order of the two non-bonding molecular orbitals (of $A_1$ and $B_2$ symmetries) we note that the 2s orbital of the isolated oxygen atom is of a lower energy than the 2p's. It seems reasonable, therefore, that the non-bonding A, orbital, which contains a 2s component, should be of lower energy than the $B_2$ non-bonding orbital which is pure $2p_x$ (the same argument is relevant to the $A_1$ and $B_1$ bonding molecular orbitals and reinforces our previous conclusions about their order). There is another point which must be made in connection with Figure 3.11. In this figure the interaction between the hydrogen 1s orbitals, $h_1$ and $h_2$, is shown as removing the degeneracy of these two orbitals. This splitting corresponds to the separation between the bonding and antibonding molecular orbitals of $H_2$ (much reduced in the present case because of the large separation between the two hydrogen atoms). On the other hand, the mixing of the 2s and 2p orbitals of the oxygen (induced by their mutual overlap with the hydrogen 1s combination $\psi(A_1)$) is shown as bringing them closer together in energy. This is because if the combination has the idealized sp hybrid form which we gave them earlier they would be precisely equivalent (although differently orientated in space). It follows that if we were to try to work out energies associated with these two hybrids then we would expect to obtain the same result for each. We conclude, therefore, that when orbitals on the same atom are mixed to give general — not idealized — hybrid orbitals these hybrids should be regarded as having energies intermediate between those of their components.

In the water molecule there are eight valence electrons available to be allocated to the four lowest molecular orbitals shown in Figure 3.11 (the electron configuration of the oxygen atom is $1s^2\ 2s^2\ 2p^4$ and contributes six valence electrons; each hydrogen is $1s^1$). It follows that the lowest four orbitals, two non-bonding and two bonding molecular orbitals, are occupied.

## 3.7 COMPARISON WITH EXPERIMENT

Is there any experimental test of the model which we have just developed? The most pertinent test would be the observation of individual orbital energy levels. Such data are provided by photoelectron spectroscopy measurements in which electrons are ejected from individual molecules by high energy monochromatic radiation in a high vacuum, the difference between the (measured) kinetic energy of an ejected electron and the energy of the incident photons being the energy required to remove the electron from the molecule. A variety of electron energies results corresponding to a variety of molecular ionization energies. These ionization energies correspond very closely to the usual definition of orbital energy. An orbital energy is defined as the energy required to remove an electron from a molecule, subject to the restriction that the orbitals of the other electrons in the molecule are unchanged. Evidently, this is a theoretical, rather than practical, definition — some readjustment of the orbitals of the residual electrons would be expected. Fortunately, however, the effect of these readjustments is usually rather small. We can therefore fairly confidently use the ionization energies given by photoelectron spectroscopy to test our model.

The photoelectron spectrum of water shows four peaks (of energies 12.62, 13.78, 17.02 and 32.2 eV). Qualitatively, then, the photoelectron measurements support the energy scheme given in Figure 3.11. There are four different ionization energies arising from valence shell electrons. A detailed analysis of the photoelectron spectrum can also give some idea of the symmetry species of the molecular orbitals. In the case of the water molecule the 12.62 eV peak is probably associated with ionization from a $B_2$ orbital, the 13.78 and 32.2 eV peaks with ionization from $A_1$ orbitals and the 17.02 eV peak with ionization from a $B_1$ orbital, in agreement with the qualitative prediction we have made. Agreement with the more accurate model is even better. The ionization from the most stable orbital is from a largely 2s(O) orbital; ionizations from the other orbitals are from orbitals which have considerable 2p(O) contributions and so are relatively close together in energy.

With the present generation of computers it is possible to do accurate calculations on molecules with fewer than, perhaps, fifty electrons. The water molecule with a total of only ten electrons should, therefore, be amenable to quite precise theoretical investigation. All of the accurate calculations which have been performed on this molecule lead to roughly the same orbital energies, and demonstrate the presence of molecular orbitals of $A_1$ symmetry at *ca.* 14 and 30 eV, a $B_1$ at *ca.* 17 eV and one of $B_2$ symmetry at *ca.* 12.5 eV. The agreement between these data and the photoelectron results is very good.

It is particularly encouraging to find that the qualitative symmetry-based arguments which we have used to discuss the electronic structure of the water molecule should give results which are in excellent qualitative agreement both with those obtained by experiment and those obtained by detailed calculations. Hopefully, we may apply the same techniques to other molecules and obtain similar sensible results. It is obvious that as molecular complexity increases the difficulty in arriving at an unambiguous energy level scheme will also increase. However, the symmetry of the water molecule is not particularly high and we may hope that for larger but higher symmetry molecules the increase in molecular complexity will be compensated for by the increase in molecular symmetry.

## 3.8 WALSH DIAGRAM FOR TRIATOMIC DIHYDRIDES

We are now in a position to reconsider a problem which we first encountered in Chapter 1 — that of the significance which can be placed upon the observation that the bond angle in the electronic ground state of the water molecule is 104.5°. As we have seen the bonding interactions responsible for the stability of the water molecule are maximized at quite different bond angles. The $B_1$ interaction involving $2p_y(O)$ maximizes at a bond angle of 180° whereas the $A_1$ interactions involving $2p_z(O)$ maximize at a small bond angle. The observed bond angle represents a compromise, showing that interactions involving both $2p_y(O)$ and $2p_z(O)$ are of importance. However, the total bonding is unlikely to show a strong dependence on bond angle because although a change in $\theta$ will reduce the stabilization resulting from interactions involving orbitals of one symmetry species it will increase the stabilization accruing from the other. Put another way, if water had been found to be linear we could conclude that the major contribution to the molecular bonding resulted from interactions between those orbitals which we have identified as being of $B_1$ symmetry. Conversely, if the bond angle were very small, say 60°, then it would be fair to conclude that most of the stabilization resulted from the $A_1$ interaction. A diagram showing schematically this behaviour is given in Figure 3.12, where we have again adopted the convention of denoting orbitals by lower case symbols. In this diagram we have also recognized the fact that at the $\theta = 180°$ limit the orbital which we have, qualitatively, called the $A_1$ non-bonding orbital loses its 2s component and becomes a pure 2p non-bonding orbital. This latter orbital is therefore degenerate with (has the same energy as) the non-bonding orbital which we have labelled $B_2$. Conversely, the lower, bonding, $A_1$ orbital is pure oxygen 2s at 0° and so is of lower energy at this limit. The non-bonding $B_2$ orbital remains unchanged in energy as the bond angle changes (actual calculations show that it increases slightly in energy at the 180°

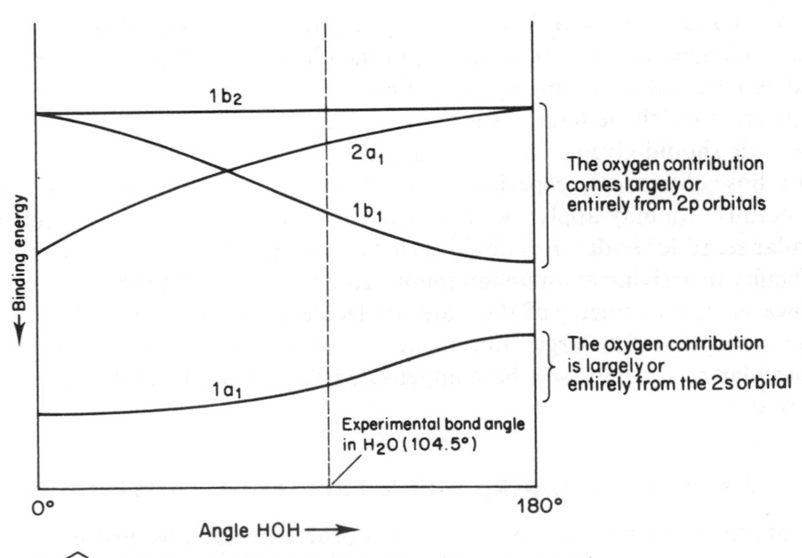


Figure 3.12 A Walsh diagram for $H_2O$

Table 3.2 The effects of occupancy on bond angle. (Data from Wasserman[3] and Takahata[4], although note that in the latter $B_1$ and $B_2$ are interchanged compared with here)

| Molecule | Orbital occupancy | | Bond angle (degrees) |
|---|---|---|---|
| | $A_1$ | $B_2$ | |
| $BH_2$ (excited) | 0 | 1 | 180 |
| $BH_2$ | 1 | 0 | 131 |
| $CH_2$ (triplet excited) | 1 | 1 | 136 |
| $CH_2$ (singlet excited) | 1 | 1 | 140 |
| $NH_2^+$ (excited) | 1 | 1 | 144 |
| $BH_2^-$ | 2 | 0 | 100 |
| $CH_2$ | 2 | 0 | 102 |
| $CH_2^-$ | 2 | 1 | 99 |
| $NH_2$ | 2 | 1 | 103 |
| $OH_2^+$ | 2 | 1 | 107 |
| $OH_2$ | 2 | 2 | 105 |
| $NH_2^-$ | 2 | 2 | 104 |

limit, but as this is caused by the effects of electron repulsion we have not included it in Figure 3.12). Figure 3.12 is specifically drawn for $H_2O$, but its general form is applicable to all $MH_2$ molecules for M atoms which have similar valence shell orbitals to oxygen. Diagrams of this type were first introduced by Walsh[2] and are therefore commonly known as Walsh diagrams.

We cannot directly test the effect of the $B_2$ interaction but we can relate the observed geometries of first-row $MH_2$ molecules (in electronic ground and excited states) to the occupancy of the highest $B_2$ and $A_1$ orbitals in Figures 3.11 and 3.12. Occupancy of the former, which is non-bonding, would be expected to have little effect on the bond angle but the lower the occupancy of the latter the larger we expect the bond angle to be (and vice versa). Table 3.2 details those data currently available and shows that when there are two electrons in the highest $A_1$ orbital an angle of *ca.* 103° results; one electron in this orbital leads to a bond angle of *ca.* 140° whilst when this orbital is empty a linear molecule results. The bond angle shows little sensitivity to $B_2$ orbital occupancy, as expected.

**Problem 3.7** The following species have been the subject of theoretical investigations but their bond angles have yet to be determined experimentally. Predict approximate values for these angles:

$$FH_2^{3+},\ BeH_2^{3-},\ CH_2^{2-},\ BH_2^{3-},\ NH_2^+$$

## 3.9 SIMPLE MODELS FOR BONDING IN $H_2O$

We conclude the present chapter by investigating the relationship between the picture of the electronic structure of the water molecule which we have

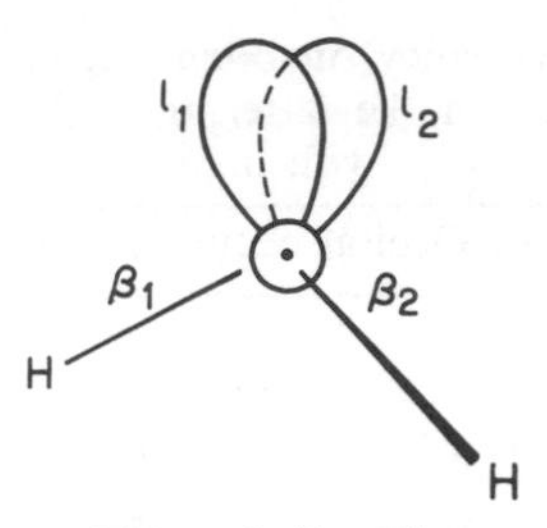


Figure 3.13 The tetrahedral arrangement of bonding electron pairs ($\beta$) and lone pairs (l) in $H_2O$

developed in this chapter and that given by models such as those discussed for ammonia in Chapter 1. The first model which we consider is that in which the oxygen atom in the water molecule is regarded as being tetrahedrally surrounded by electron pairs, two bonding and two non-bonding. The concordance between this model and the general pattern of energy levels shown in Figure 3.11 is easy to show. In Figure 3.13 we depict, schematically, the bonding electron pairs which we call $\beta_1$ and $\beta_2$, and the lone pair electrons $l_1$ and $l_2$. It is easy to show that the transformations of the two bonding orbitals $\beta_1$ and $\beta_2$, considered as a pair, generate the reducible representation:

| E | $C_2$ | $\sigma_v$ | $\sigma_v'$ |
|---|---|---|---|
| 2 | 0 | 2 | 0 |

which has $A_1$ and $B_1$ components. These, of course, are precisely the same symmetries as possessed by the bonding molecular orbitals shown in Figure 3.11. It is also easy to show that the transformations of the lone pair orbitals, $l_1$ and $l_2$, generate the reducible representation

| E | $C_2$ | $\sigma_v$ | $\sigma_v'$ |
|---|---|---|---|
| 2 | 0 | 0 | 2 |

which is also easily shown to be the sum of the $A_1$ and $B_2$ irreducible representations. We conclude that the lone pair orbitals may be combined to give orbitals transforming as $A_1$ and as $B_2$.

**Problem 3.8** Generate the two reducible representations discussed above. Use Figure 3.13.

Let us consider the lone pair orbitals in more detail. Since they are localized on the oxygen atom we conclude that they must be derived from the valence shell atomic orbitals of the oxygen atom. However, the only $B_2$ valence shell oxygen atomic orbital is $2p_x(O)$, so the orbital of this symmetry which is a combination of lone pair orbitals must be identical to that obtained from our symmetry-based discussion. Similarly, the $A_1$ combination must be some mixture of 2s(O) and $2p_z$(O), again in qualitative agreement with our earlier result.

Study of the O—H bonding orbitals leads to a similar agreement with the model we have developed in this chapter. Because $\beta_1$ and $\beta_2$ are bases for $A_1$ and $B_1$ irreducible representations we conclude that 2s(O), $2p_z$(O) and $\psi(A_1)$ may contribute to the $A_1$ combination. Similarly, $2p_y$(O) and $\psi(B_1)$ contribute to the $B_1$ combination — conclusions identical to those reached earlier. Further, the statement that $\beta_1$ and $\beta_2$ are *bonding* means that the $A_1$ and $B_2$ combinations must be in-phase, bonding, combinations of oxygen and hydrogen orbitals. We conclude that the simple two-bonding, two-non-bonding orbital picture of the water molecule may readily be reinterpreted in the language which we have used in this chapter. Further, by using the energy level criteria based on the number of nodal planes and the relative energies of atomic orbitals, discussed in the present chapter, we can again arrive at the qualitative energy level diagram shown in Figure 3.11.

The second simple model of the water molecule which we shall consider is one that is sometimes quickly discarded. This is the model in which only the 2p orbitals of the oxygen atom are considered as being involved in O—H bonding, the 2s orbital being, implicitly, regarded as non-bonding. In this model, the oxygen 2p orbitals which lie in the plane of the water molecule overlap with the hydrogen 1s orbitals. The relevant oxygen 2p orbitals, which we shall take to be $2p_z$(O) and $2p_y$(O), are orientated so as to point directly at the hydrogen atoms, as shown in Figure 3.14, and so a bond angle of 90° is predicted. The oxygen $2p_x$ orbital, which is perpendicular to the plane of the molecule, is non-bonding, precisely the role which we found it to play. The two non-bonding orbitals are, therefore, 2s(O) and $2p_x$(O), which are of $A_1$ and $B_2$ symmetries, just as we found (their relative energies — one very stable $A_1$ orbital and one high energy $B_2$ orbital — also agree with the more accurate model). It is easy to show that the two bonding orbitals which result from the

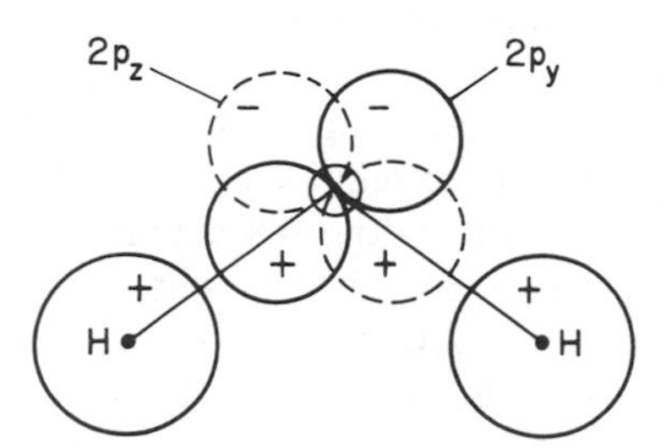


Figure 3.14 Bonding in $H_2O$ with 2p as the only oxygen orbitals involved. Note that the axis labels y and z do not follow the convention adopted elsewhere in this book

overlap of the $2p_z(O)$ and $2p_y(O)$ orbitals (those in the plane of the molecule) with the hydrogen 1s orbitals give rise to a reducible representation with $A_1$ and $B_1$ components. Arguments analogous to those developed above for $\beta_1$ and $\beta_2$ in the previous model demonstrate that the qualitative forms of the corresponding $A_1$ and $B_1$ molecular orbitals are those which we deduced earlier in this chapter.

It is pertinent to comment that this particular simple model gets closer to the results of accurate quantum mechanical calculations than does any other. It predicts a low-lying non-bonding 2s orbital, one pure 2p non-bonding oxygen orbital and two bonding molecular orbitals involving oxygen 2p orbitals.

**Problem 3.9** Show that the $2p_z(O)$ and $2p_y(O)$ orbitals shown in Figure 3.14 transform together as $A_1 + B_1$.

## 3.10 A RAPPROCHMENT BETWEEN SIMPLE AND SYMMETRY MODELS

It is necessary at this point to review the development of the arguments in this book. In Chapter 1 we concluded that simple models of molecular bonding are not expected to be infallible predictors of molecular geometry. However, in the present chapter we have shown that these simple models may, at least for the case of the water molecule, be reinterpreted in such a way as to show that the bonding descriptions which they present are equivalent, qualitatively, to a wholly symmetry-based description. The latter has shown that the relationship between molecular geometry and the contribution to the bonding from the various bonding molecular orbitals is not necessarily a simple one. The stabilization resulting from one interaction may be independent of geometry; others may be very sensitive to geometry. Different interactions may make a maximum contribution to molecular stability at quite different patterns of bond angles.

It is here that the circle closes. Simple pictures of molecular bonding are perhaps more reliable predictors of relative energies of molecular orbitals than they are of molecular geometries (although usually used for the latter rather than the former!). When a simple picture fails to give a correct molecular orbital energy level pattern it is usually because there are some interactions in the molecule involving orbitals other than those considered in the simple model. In such cases the simple models are, nonetheless, usually good starting points for a detailed discussion. Finally, we note that, despite their apparent differences, when there is a variety of simple approaches to the bonding in a molecule, they commonly lead to the same qualitative energy level diagram. Again, the only exceptions occur when different models include different interactions, but here the differences are themselves illuminating.

What is the particular attraction of a symmetry-based approach which leads us to refer all other models to it? We have already indicated a computational advantage — interactions are only non-zero between wavefunctions of the same symmetry species. There is another important reason. Whenever excited electronic states or ionized species are considered it becomes essential to use a symmetry-based approach. This is because it is only such an approach which allows a simple connection between the discussion of the ground and excited or ionized states of a molecule. One illustration will make the point. Suppose we excite an electron in the water molecule from a bonding orbital to some high-lying, non-bonding, orbital and suppose that the excited electron comes from a single O—H bond (as it would apparently have to, according to all of the simple models of the bonding in the water molecule in which those electrons associated with one bond are not associated with the other). In the excited state the two O—H bonds would differ — one has only one bonding electron whilst the other has two. This is in contradiction to the observation that in all stable excited states of the water molecule the O—H bonds are equivalent (we exclude unstable states from which dissociation into H + OH occurs). For a symmetry-based description, in which the bonding electron comes from a molecular orbital spread equally over both hydrogen atoms, the observed equivalence of the two hydrogen atoms in excited or ionized states follows naturally. We are thus led to prefer a symmetry-based description for both ground and excited states.

## 3.11 SUMMARY

The irreducible representations which appear in character tables are orthonormal (page 31) — each component is independent of the others and carries equal weight. This property enables reducible representations (page 36) to be reduced systematically to their irreducible components (page 37). In the context of molecular bonding this enables the interactions between orbitals of each symmetry type to be discussed separately (page 40). Such discussions, together with simple nodal-plane criteria (page 47), enable qualitative molecular orbital energy level diagrams to be constructed (page 47) and the angular variation of each bonding interaction assessed (page 43 and 45). This latter information may be conveniently represented as a Walsh diagram (page 50). Asymmetry analysis of simple picture of molecular bonding reveals that they have similarities with each other and with the symmetry-based approach (page 51).

## REFERENCES

1. E. B. Wilson, *J. Chem. Phys.*, **63**, 4870 (1975).
2. A. D. Walsh, *J. Chem. Soc.*, **1963**, 2250 (1963).
3. E. Wasserman, *Chem. Phys. Letters*, **24**, 18 (1974).
4. Y. Takahata, *Chem. Phys. Letters*, **59**, 472 (1978).

# CHAPTER 4

# *The $D_{2h}$ character table and the electronic structures of ethylene and diborane*

The present chapter has several objectives, in addition to those indicated by its title. The first is to introduce a new symmetry group and its character table. The group has been chosen because it is related to the $C_{2v}$ group with which the reader is now familiar. It is not the simplest group that we could have used to discuss the bonding in ethylene and diborane — the simplest would be the group $D_2$, a group which we shall meet shortly — but this discussion itself is only part of the objective of the present chapter. Use of the more complicated $D_{2h}$ (pronounced 'dee-two-aich') group will enable us to begin to explore relationships between groups and the corresponding character tables. A second objective is to present and to use the rather important technique of projection operators. Despite their somewhat off-putting name, these provide a very simple method of obtaining functions transforming as a particular irreducible representation.

## 4.1 THE SYMMETRY OF THE ETHYLENE MOLECULE

The effect of bringing two $CH_2$ units, each of $C_{2v}$ symmetry, together to form an ethylene molecule has the effect of generating additional symmetry elements. All of the symmetry elements of the ethylene molecule are shown in Figure 4.1. As shown in this figure, each $CH_2$ fragment is turned into itself by the operations that we met when discussing the $C_{2v}$ point group — that is, there are two perpendicular mirror planes, the intersection between them defining a twofold axis, common to the two $CH_2$ units. As Figure 4.1 shows, the union of the two $CH_2$ units to form ethylene has the effect of generating two new $C_2$ rotation axes, the three twofold axes being mutually perpendicular. This immediately suggests that we should also use them as our Cartesian coordinate axes, a suggestion which we shall adopt. Note that each of the three $C_2$ axes is unique. This is rather important because later in this book we shall meet sets of twofold axes which are not unique in the same way. In ethylene, each twofold axis is unique because there is no operation in the group which interchanges any pair of them. This assertion can be checked after reading the next

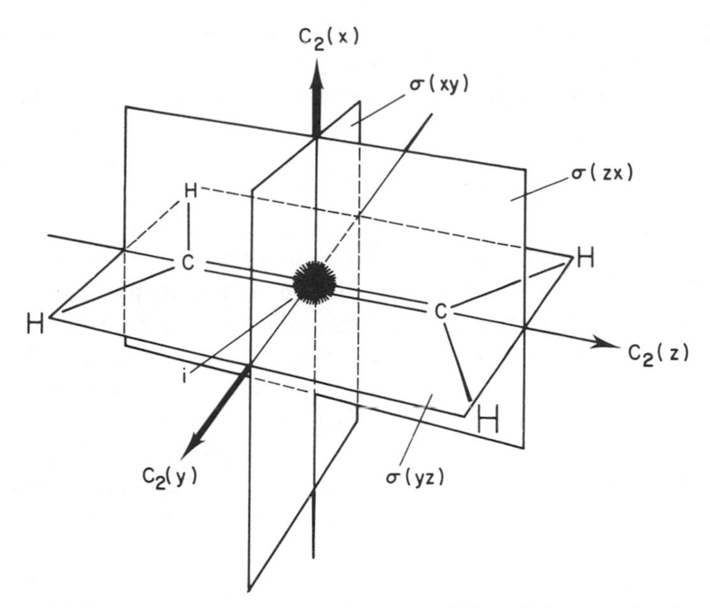


Figure 4.1 The symmetry elements of the ethylene molecule

few paragraphs and we have obtained a complete list of the ethylene symmetry operations. In high-symmetry molecules it is quite common for there to be a set of rotation axes (or mirror plane reflections or other operations) which are either interchanged or mixed* by other operations of the group. In such cases there is a corresponding complication in the character table and it is this complication which we seek to avoid here — we shall meet it in the next chapter — by working with another example of an Abelian point group (the $C_{2v}$ group of Chapters 2 and 3 is Abelian).

When, as in this case, there are several apparently equally good choices for the z axis, it is usual to choose that axis which contains the largest number of atoms and so we shall take as z that twofold axis which passes through the two carbon atoms. It follows that the z axis of each $CH_2$ fragment (of $C_{2v}$ symmetry) is coincident with the molecular z axis. Just as in the $C_{2v}$ case, the labels of the other coordinate axes are determined by the convention that the planar molecule lies in the yz plane.

As is evident from Figure 4.1, the mirror planes of each $CH_2$ fragment persist as symmetry elements in the complete molecule. A third mirror plane exists in the ethylene molecule. This passes through the mid-point of the carbon–carbon bond and is perpendicular to that bond. Figure 4.1 shows that each of the mirror planes is perpendicular to one of the coordinate axes; equally, it lies in the plane defined by the other two coordinate axes. Rather than use a $\sigma_v$ notation for the mirror planes, each mirror plane is conventionally

*We include the word 'mixed' here to make the statement rigorous. Its meaning will only become clear after reading Chapters 5 and 6.

labelled by the molecular coordinate axes which it contains: $\sigma(xy)$, $\sigma(yz)$ and $\sigma(zx)$. We shall have more to say about these labels shortly.

All of the symmetry elements we have listed so far are similar to those which we encountered in the $C_{2v}$ group. Additionally, however, the ethylene molecule contains a centre of symmetry, a point such that inversion of the whole molecule through it gives a molecule which is indistinguishable from the starting one. This centre of symmetry is indicated by the star-like point at the centre of Figure 4.1. More strictly, a centre of symmetry is such that inversion of any point of the molecule in it gives an equivalent point. Pictorially, if we draw a straight line from any point (the starting point) in the molecule to the centre of symmetry and then extend the line an equal length beyond the centre of symmetry, the terminal point of the line is symmetry-equivalent to the starting point. This element and the corresponding operation are conventionally denoted by the lower case symbol i. A centre of symmetry, if there is one, is always at the centre of gravity of a molecule. A molecule may possess several rotation axes and several mirror planes but it can never possess more than one centre of symmetry.

The symmetry elements of the ethylene molecule provide a better example of the use of the word 'point' when talking about a point group than does the water molecule. As is evident from Figure 4.1 all of the symmetry elements have one point in common, located at the centre of gravity of the molecule (in this context the identity element is best thought of as corresponding to a $C_1$ rotation axis).

In summary, then, and talking in terms of symmetry operations rather than symmetry elements, the symmetry operations which turn the ethylene molecule into itself are

$$E \quad C_2(x) \quad C_2(y) \quad C_2(z) \quad i \quad \sigma(yz) \quad \sigma(zx) \quad \sigma(xy)$$

This group of symmetry operations is commonly given the shorthand label $D_{2h}$. We shall defer a detailed discussion of such shorthand labels until Section 7.5 because until then we will not have met all the symmetry operations on which the classification is based; until then we shall have to discuss each label individually. It is clear that the label $D_{2h}$ requires some explanation before we proceed. Point groups which contain a principal $C_n$ axis and, perpendicular to this principal axis, $n$ twofold axes, are called dihedral point groups and hence carry the label $D_n$ (D for dihedral). If we were to slightly rotate one $CH_2$ group of the ethylene molecule about the z axis (so that the molecule becomes non-planar) then we would destroy all of the mirror planes and the centre of symmetry and be left with a molecule of $D_2$ symmetry.

**Problem 4.1** Show that when the two $CH_2$ groups of ethylene are rotated (twisted) by equal and opposite amounts about the z axis of Figure 4.1 the molecule is of $D_2$ symmetry (with operations E, $C_2(x)$, $C_2(y)$, $C_2(z)$). Note that for the simpler rotation described in the text (rotate one $CH_2$ group by $\theta°$) it is necessary to rotate the x and y axes, and the $C_2$ rotations associated with them, by $\theta°/2$ about the z axis in the same sense.

If, perpendicular to the principal rotation axis — that of the highest $n$ value in $C_n$ — in a molecule, there is a mirror plane which is perpendicular to this axis, i.e. *h*orizontal with respect to it (the principal axis being vertical), then this is denoted $\sigma_h$ (recall that in Chapter 2 we called a mirror plane which is *v*ertical with respect to the principal rotation axes, $\sigma_v$). If a $D_n$ group also contains a $\sigma_h$ mirror plane then the point group is labelled $D_{nh}$. The present point group falls into this category once we have overcome a difficulty — that we have refrained from calling any mirror plane $\sigma_h$ (or $\sigma_v$, for that matter). The reason for this is that we have three $C_2$ axes, any one of which might equally well be chosen as the principal axis. The mirror plane which should be labelled $\sigma_h$ would depend upon which particular $C_2$ axis we nominated as our principal axis. In this particular case, where all three mirror planes are equally good candidates for being labelled $\sigma_h$, the egalitarian solution is to give none of them this label but, rather, designate them as we have done above. However, egality cannot alter the claim of the group to be recognized as one of the $D_{nh}$ type; accordingly it is labelled $D_{2h}$.

## 4.2 THE CHARACTER AND MULTIPLICATION TABLES OF THE $D_{2h}$ GROUP

In order to proceed further we must obtain the character table of the $D_{2h}$ point group. We could follow the procedure which we adopted for the $C_{2v}$ case to generate the $D_{2h}$ character table but because the procedure is entirely analogous we shall not devote space to it but invite the reader to derive it himself in the following problem.

**Problem 4.2** Derive the character table of the $D_{2h}$ point group (Table 4.1). In Chapter 2 we generated the irreducible representations of the $C_{2v}$ character table by considering the transformations of the orbitals of a unique atom (the oxygen in $H_2O$). In order to use this technique in the present problem it is necessary to first have a unique atom. This can be done by placing a hypothetical atom at the centre of gravity of the ethylene molecule. Using just the familiar s, p and d orbitals it is not possible to generate the $A_u$ irreducible representation of the $D_{2h}$ point group. This irreducible representation can be generated using one of the f orbitals, the $f_{xyz}$ orbital, a diagram of which is shown in Figure 4.2. To provide a check, we have given at the right-hand side of Table 4.1 the atomic orbital(s) of the hypothetical atom which generates each irreducible representation.

Table 4.1

| $D_{2h}$ | E | $C_2(z)$ | $C_2(y)$ | $C_2(x)$ | i | $\sigma(xy)$ | $\sigma(zx)$ | $\sigma(yz)$ | |
|---|---|---|---|---|---|---|---|---|---|
| $A_g$ | 1 | 1 | 1 | 1 | 1 | 1 | 1 | 1 | s, $d_{z^2}$, $d_{x^2-y^2}$ |
| $B_{1g}$ | 1 | 1 | −1 | −1 | 1 | 1 | −1 | −1 | $d_{xy}$ |
| $B_{2g}$ | 1 | −1 | 1 | −1 | 1 | −1 | 1 | −1 | $d_{zx}$ |
| $B_{3g}$ | 1 | −1 | −1 | 1 | 1 | −1 | −1 | 1 | $d_{yz}$ |
| $A_u$ | 1 | 1 | 1 | 1 | −1 | −1 | −1 | −1 | $f_{xyz}$ |
| $B_{1u}$ | 1 | 1 | −1 | −1 | −1 | −1 | 1 | 1 | $p_z$ |
| $B_{2u}$ | 1 | −1 | 1 | −1 | −1 | 1 | −1 | 1 | $p_y$ |
| $B_{3u}$ | 1 | −1 | −1 | 1 | −1 | 1 | 1 | −1 | $p_x$ |

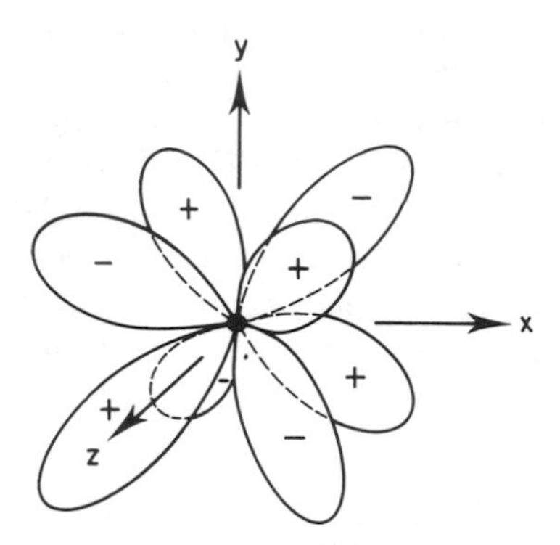


Figure 4.2 The $f_{xyz}$ orbital of a hypothetical atom placed at the centre of gravity of the ethylene molecule

For every group there exists a group multiplication table; that for the $D_{2h}$ group is given in Table 4.2. Its derivation is analogous to the derivation of the $C_{2v}$ group multiplication table. Just as in the $C_{2v}$ case, it will be found that when the appropriate substitution of characters for the corresponding sym-

Table 4.2

| | | First operation | | | | | | | |
|---|---|---|---|---|---|---|---|---|---|
| | $D_{2h}$ | E | $C_2(z)$ | $C_2(y)$ | $C_2(x)$ | i | $\sigma(xy)$ | $\sigma(zx)$ | $\sigma(yz)$ |
| Second operation | E | E | $C_2(z)$ | $C_2(y)$ | $C_2(x)$ | i | $\sigma(xy)$ | $\sigma(zx)$ | $\sigma(yz)$ |
| | $C_2(z)$ | $C_2(z)$ | E | $C_2(x)$ | $C_2(y)$ | $\sigma(xy)$ | i | $\sigma(yz)$ | $\sigma(zx)$ |
| | $C_2(y)$ | $C_2(y)$ | $C_2(x)$ | E | $C_2(z)$ | $\sigma(zx)$ | $\sigma(yz)$ | i | $\sigma(xy)$ |
| | $C_2(x)$ | $C_2(x)$ | $C_2(y)$ | $C_2(z)$ | E | $\sigma(yz)$ | $\sigma(zx)$ | $\sigma(xy)$ | i |
| | i | i | $\sigma(xy)$ | $\sigma(zx)$ | $\sigma(yz)$ | E | $C_2(z)$ | $C_2(y)$ | $C_2(x)$ |
| | $\sigma(xy)$ | $\sigma(xy)$ | i | $\sigma(yz)$ | $\sigma(zx)$ | $C_2(z)$ | E | $C_2(x)$ | $C_2(y)$ |
| | $\sigma(zx)$ | $\sigma(zx)$ | $\sigma(yz)$ | i | $\sigma(xy)$ | $C_2(y)$ | $C_2(x)$ | E | $C_2(z)$ |
| | $\sigma(yz)$ | $\sigma(yz)$ | $\sigma(zx)$ | $\sigma(xy)$ | i | $C_2(x)$ | $C_2(y)$ | $C_2(z)$ | E |

metry operation is made in Table 4.2, any row of characters appearing in the $D_{2h}$ character table turns Table 4.2 into a multiplication table which is arithmetically correct.

**Problem 4.3** By combining (multiplying) pairs of operations of the $D_{2h}$ character table show that Table 4.2 is correct. Some help in this problem is provided by Section 2.3. Indeed, if the hypothetical molecule considered there — $O(CH_2)_2$ — is flattened symmetrically so that it becomes planar, then we have a molecule of the same symmetry as ethylene which also contains the unique atom of Problem 4.2. A further tip on how to do this problem is provided by Figure 4.3. Take a general point in space (indicated by the solid star), perform the first operation (in Figure 4.3, $\sigma(zx)$) to give the cross-hatched star and follow it with the second operation (in Figure 4.3, i) to give the open star. Then ask, 'What single operation turns the solid star into the open one?' (in Figure 4.3, $C_2(y)$). In our case, then, $\sigma(zx)$ followed by i is equivalent to $C_2(y)$.

**Problem 4.4** Take any four of the irreducible representations of Table 4.1 and by substituting the appropriate character for each operation in Table 4.2 show that in each case an arithmetically correct multiplication table is obtained. If needed, Section 2.4 will provide guidance on this problem.

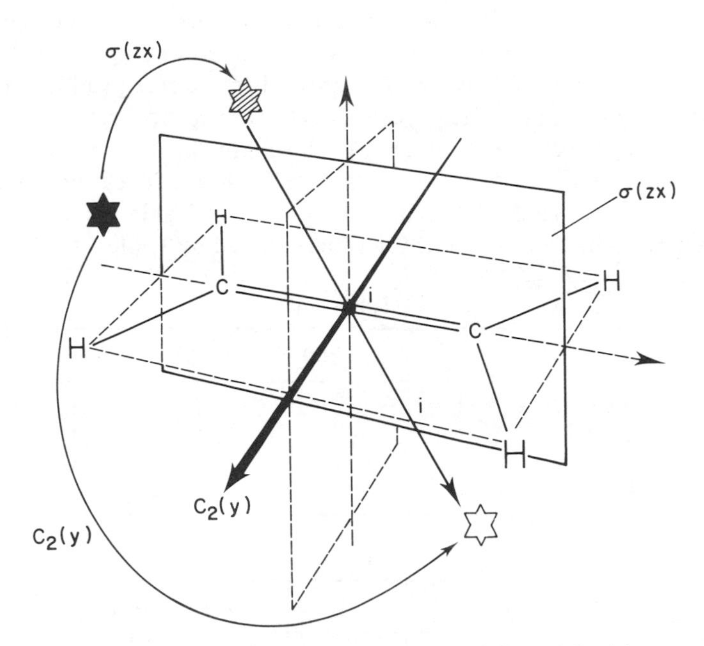


Figure 4.3 An illustration that $\sigma(zx)$ followed by i is equivalent to $C_2(y)$

## 4.3 DIRECT PRODUCTS OF GROUPS

There are several interesting features of Table 4.2 — e.g. it is symmetric about either diagonal. Another is the way that it may be broken into four smaller blocks, pairs of which are identical. Similarly, the $D_{2h}$ character table (Table 4.1) may also be broken into four blocks, but now three of the blocks are identical and in the fourth the same set of characters appear, but with all signs reversed. There is a simple reason for these patterns. As is evident from Table 4.2, the operation $\sigma(xy)$ is equivalent to $C_2(z)$ followed by the inversion i. Similarly, $\sigma(yz)$ equals $C_2(x)$ followed by i and $\sigma(zx)$ equals $C_2(y)$ followed by i. It follows that the operations of the $D_{2h}$ group may be rewritten as follows:

| E $C_2(z)$ $C_2(y)$ $C_2(x)$ | i $\sigma(xy)$ $\sigma(zx)$ $\sigma(yz)$ |
|---|---|
| are equivalent to | are equivalent to |
| E $C_2(z)$ $C_2(y)$ $C_2(x)$ | E $C_2(z)$ $C_2(y)$ $C_2(x)$ |
| followed by E | followed by i |

That is, the operations of the $D_{2h}$ group may be obtained by forming all possible products of members of the set E, $C_2(z)$, $C_2(y)$, $C_2(x)$, which, as we have already seen, form the $D_2$ group, with members of the set E, i — two operations which together form a group usually called the $C_i$ group (pronounced 'cee-eye'); we shall return to this group shortly. Technically, one says that 'the $D_{2h}$ group is the *direct product* of the $D_2$ and $C_i$ groups', where 'direct product' means 'form all possible products of the members of one set with members of the other'. When the operations of a group may be expressed as direct products in this way so too may the corresponding character tables. That is, the character table of the $D_{2h}$ group is the direct product of those of the $D_2$ and $C_i$ groups (and the phrase 'direct product' now refers to characters but its meaning is otherwise unaltered). The character table of the $D_2$ group is given in Table 4.3 and that of the $C_i$ group in Table 4.4. They should, together, be compared with Table 4.1. The four blocks in Table 4.1 are simply the characters given in Table 4.3 with signs determined by Table 4.4. Thus in three

Table 4.3

| $D_2$ | E | $C_2(z)$ | $C_2(y)$ | $C_2(x)$ |
|---|---|---|---|---|
| A | 1 | 1 | 1 | 1 |
| $B_1$ | 1 | 1 | −1 | −1 |
| $B_2$ | 1 | −1 | 1 | −1 |
| $B_3$ | 1 | −1 | −1 | 1 |

Table 4.4

| $C_i$ | E | i |
|---|---|---|
| $A_g$ | 1 | 1 |
| $A_u$ | 1 | −1 |

of the blocks Table 4.3 reappears with unchanged sign and in the fourth all of the characters are multiplied by $-1$.

**Problem 4.5** Multiply the character table, Table 4.3, by the character table, Table 4.4, and thus generate Table 4.1. Note that 'multiply' here means different things for operations and characters. For the latter it means simple arithmetic multiplication but for the former it means 'carry out the operations one after the other'. The way that the relevant operations multiply is indicated in this section; in order to generate Table 4.1 it is necessary to maintain the correct correspondence between products of characters and products of operations.

An interesting thing about Table 4.3 is that the sets of characters that appear in it are the same as those of the $C_{2v}$ point group (Table 2.4), although the operations and irreducible representation labels are not the same in the two groups. Groups which have character tables containing identical corresponding sets of characters are said to be *isomorphous* groups. Isomorphous groups need have no operation in common — except, of course, the identity operation, which appears in all groups. However, isomorphism between character tables means that there is a close connection between the groups. Thus, something true in one group has a counterpart in an isomorphous group. An illustration of this is given in Section 9.4.

In the $C_i$ character table (Table 4.4) the only distinction between the irreducible representations is the behaviour of the quantities they describe under the operation of inversion in the centre of symmetry, i. Both irreducible representations are denoted by A but something which is transformed into itself (= is symmetric) under the inversion operation is distinguished from one which is turned into minus itself (= is antisymmetric) by the subscripts 'g' (from *gerade*, German for 'even') and 'u' (from *ungerade*, German for 'odd') respectively; it is always true that a g suffix indicates an irreducible representation which describes something which is centrosymmetric whilst the suffix u describes something which is antisymmetric with respect to inversion in a centre of symmetry. The reader will find it helpful to compare the use of g and u suffixes in the irreducible representation labels of Table 4.1 with the characters under the i operation.

**Problem 4.6** For each irreducible representation in Table 4.1 which carries a g suffix (e.g. $A_g$) list the character under the i operation. Repeat this exercise for each irreducible representation carrying a u suffix (e.g. $A_u$). Compare your result with Table 4.4.

There is one final point to note. We have included dotted lines in Tables 4.1, 4.2 and 4.4 to clarify the discussion in the text. Normally they are omitted and, indeed, columns in these tables are sometimes permuted so that the pattern which is apparent from the way that we have written them becomes less evident.

## 4.4 THE SYMMETRIES OF THE CARBON ATOMIC ORBITALS IN ETHYLENE

The character table of the $D_{2h}$ group given in Table 4.1 will now be used in a qualitative discussion of the electronic structure of the ethylene molecule. At first sight one might expect that this discussion would be more complicated than that for the water molecule because we now have six atoms to consider. On the other hand, instead of working with a group with only four irreducible representations we now have eight, so we may hope that the increase in symmetry will offset the greater molecular complexity. The first step is, as always, an investigation of the transformation properties of the various sets of atomic orbitals. We then form linear combinations of these orbitals which transform as irreducible representations of the $D_{2h}$ group. Finally, the interaction between orbitals of the same symmetry species will be included and a qualitative molecular orbital energy level diagram obtained. The valence shell atomic orbitals that we must consider are the 2s and $2p_x$, $2p_y$ and $2p_z$ orbitals of the two carbon atoms and the four 1s orbitals of the terminal hydrogen atoms. Not one of these orbitals is unique — there is always at least one other, symmetry-related, atom in the molecule with a similar orbital. This means that, in a sense, the present discussion must start at the point to which our discussion of the water molecule took us. Just as for the hydrogen 1s orbitals in the water molecule, the transformations of corresponding orbitals of symmetry related atoms must be considered together. As a simple example we consider the 2s orbitals of the two carbon atoms (Figure 4.4). Each of these orbitals remains itself under the $C_2(z)$ rotation, the $\sigma(zx)$ and $\sigma(yz)$ reflection operations and, of course, under the identity operation. For all of the other symmetry operations of the group the two orbitals are interchanged. Now if an orbital is unchanged by a symmetry operation it makes a contribution of unity to the resultant character, whilst if it goes into another member of the same set it contributes zero, so the characters describing the transformation of the carbon 2s orbitals are:

| E | $C_2(z)$ | $C_2(y)$ | $C_2(x)$ | i | $\sigma(xy)$ | $\sigma(zx)$ | $\sigma(yz)$ |
|---|---|---|---|---|---|---|---|
| 2 | 2 | 0 | 0 | 0 | 0 | 2 | 2 |

Either by trial and error, or by systematic use of the group orthonormal relationships which we met in Chapter 3 and which we shall describe in more detail in Chapter 5, we conclude that this reducible representation has $A_g + B_{1u}$ components.

**Problem 4.7** Use the orthonormality theorem in the way described in detail towards the end of Section 3.3 to show that the two 2s orbitals of the carbon atoms in ethylene transform as $A_g + B_{1u}$. Your solutions to Problems 3.3 and 3.4 should give any additional guidance you may need.

If we forget the four hydrogen atoms and consider only the 2s orbitals of the two carbon atoms in the ethylene molecule, it is evident that they resemble the two hydrogen 1s orbitals in the hydrogen molecule (or in the water molecule). It is reasonable, therefore, to anticipate that the linear combinations of these orbitals which transform as $A_g$ and $B_{1u}$ will be similar to those which we obtained when discussing the water molecule. That is, if we call the carbon 2s orbitals 2s(a) and 2s(b), as shown in Figure 4.4, then the correct linear combinations are of the form:

$$\frac{1}{\sqrt{2}}\,[2s(a) + 2s(b)]$$

and

$$\frac{1}{\sqrt{2}}\,[2s(a) - 2s(b)]$$

Later in this chapter we shall obtain a systematic way of deriving such linear combinations and these functions can then be checked. Actually, whenever there are just two symmetry-related orbitals to be considered, the correct combinations are sum and difference combinations — like those above — irrespective of the details of the symmetry. Of the two combinations given above it is easy to demonstrate that the first has $A_g$ symmetry and the second $B_{1u}$.

**Problem 4.8** By drawing diagrams of them and considering their transformations under the operations of the $D_{2h}$ point group show that the two linear combinations given above transform as $A_g$ and $B_{1u}$ respectively. If you get the answers the wrong way round read the next paragraph and compare the phases you have chosen with those in Figure 4.4

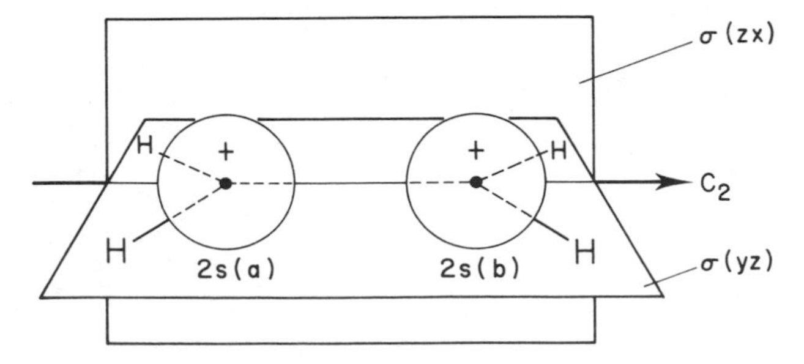


Figure 4.4 Those symmetry operations under which (together with the identity operation) the two carbon 2s orbitals of ethylene are not interchanged

There is a rather subtle aspect of this. Suppose that instead of choosing in Figure 4.4 to give the two carbon 2s orbitals the same phase we have given them opposite phases. The first combination above would, in this case, be an out-of-phase combination of the two orbitals, notwithstanding the + sign. The solution to this paradox is that in this case the first combination would have $B_{1u}$ symmetry and not $A_g$ whilst the second would be the $A_g$ combination. The systematic method of obtaining such functions takes account of our arbitrary choices of orbital phases and corrects for them. We must note, however, that we cannot work with combination functions like the two given above unless we know the phases chosen for the component atomic orbitals. One might think that the simple way would be to chose all orbitals to be of the same phase. Unfortunately such a simplification is not always possible. Thus, there are two alternative ways of drawing the $2p_z$ orbitals on the two carbon atoms; these are shown in Figure 4.5. In Figure 4.5(a) the $2p_z$ orbitals are chosen so that the phasing of the $2p_z$ orbitals coincides with that of the molecular coordinate axis system — the positive lobes point towards positive z and the negative lobes towards negative z. In Figure 4.5(b) the phase on one centre is reversed. This latter choice of phases has the advantage that under, say, the $C_2(x)$ rotation operation the $2p_z$ orbitals are simply interchanged whereas in the choice of Figure 4.5(a) they are not only interchanged but the phases of their lobes are also reversed. Here there is no simplification offered by convention; different people may, with equal validity, choose the phases differently. It follows that we must take care to check the basic choice of phases used by each person writing on the subject. If we choose the phases indicated in Figure 4.5(a) then the sum and difference combinations

$$\frac{1}{\sqrt{2}}\,[2p_z(a) + 2p_z(b)]$$

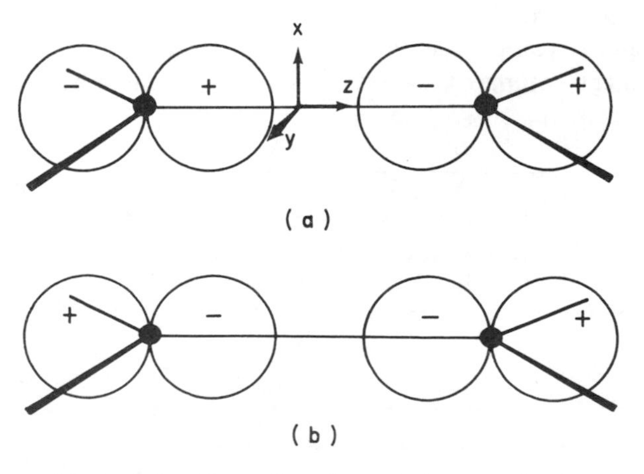


Figure 4.5 Alternative phase choices for the $2p_z$ orbitals of the carbon atoms in ethylene

and

$$\frac{1}{\sqrt{2}}\,[2p_z(a) - 2p_z(b)]$$

are, respectively, the $B_{1u}$ and $A_g$ (C—C $\sigma$-antibonding and -bonding respectively) combinations of carbon $2p_z$ orbitals.

**Problem 4.9** Because the transformation of the carbon $2p_z$ orbitals gives rise to the same irreducible representations as do the transformations of the carbon 2s orbitals it follows that they transform in the same way (the jargon statement is 'they transform isomorphously'). This means that the $2p_z$ orbitals in both Figure 4.5(a) and (b) give the reducible representation given in the text just before Problem 4.7. Show that this is indeed so, irrespective of which set of phases is chosen for the $2p_z$ orbitals (Figure 4.5a and b).

The $B_{1u}$ and $A_g$ combinations of carbon $2p_z$ orbitals are shown schematically in Figure 4.6. Also in this figure we show the $A_g$ and $B_{1u}$ combinations of carbon 2s orbitals. In these diagrams we have indicated that the atomic orbitals on the two carbon atoms overlap each other, although we have ignored this overlap in the expressions which we gave above. We are inconsistent in this way because it gives mathematical simplicity together with diagrammatic clarity.

Although detailed calculations did not fully justify it, in our discussion of the water molecule we found it convenient to mix together the oxygen 2s and $2p_z$ orbitals; since they had the same symmetry this mixing is allowed and the resultant picture that emerged was closely related to simple ideas on the bonding in the water molecule and so helped us to look at them in more detail. For the same reason we shall mix carbon 2s and 2p orbitals in our present description, forming, effectively, carbon sp hybrids. We chose simple sp hybrids, as for water, on grounds of simplicity. If we had formed these hybrids as a first step a simpler discussion would have resulted. Unfortunately this was not possible because we had not at that stage established that the carbon $2p_z$ and 2s orbitals transform in the same way. Instead of going back to the start of our argument and working with sp hybrids we shall simply combine the $A_g$ combination of carbon 2s orbitals with the Ag combination of $2p_z$ (and similarly for the $B_{1u}$ — the end result is the same. The result of mixing together — essentially, taking sum and difference combinations of — the two $A_g$ orbitals of Figure 4.6 and (separately) the two $B_{1u}$ orbitals is shown schematically in Figure 4.7. Two of these four orbitals will carry through, unmodified, into our final description of the ethylene molecule. These are an $A_g$ combination which is to be identified with the C—C $\sigma$-bonding orbital (Figure 4.7a) and a $B_{1u}$ combination which is the corresponding C—C $\sigma$-antibonding orbital (Figure 4.7b). We shall find that the other $A_g$ and $B_{1u}$ combinations (Figure 4.7c and d respectively), which are largely directed away from the C—C bond, are involved in interactions with the terminal hydrogen atoms.

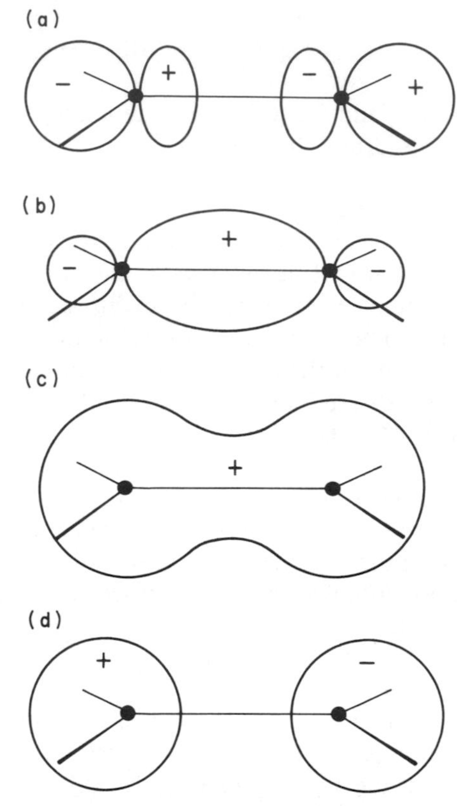


Figure 4.6 $B_{1u}$ and $A_g$ bonding and (antibonding) combination of $2p_z$ and 2s orbitals in ethylene:

(a) $B_{1u}$ (antibonding) combination of $2p_z$ orbitals,
(b) $A_g$ (bonding) combination of $2p_z$ orbitals,
(c) $A_g$ (bonding) combination of 2s orbitals and
(d) $B_{1u}$ (antibonding) combination of 2s orbitals

The other 2p orbitals of the carbon atoms are readily dealt with. For the pairs of $2p_x$ and $2p_y$ orbitals a similar phase ambiguity exists as for the $2p_z$ orbitals, although it is usually found to be less troublesome. We have chosen the phases shown in Figures 4.8 and 4.9. These orbitals transform as follows:

$$2p_x : B_{2g} + B_{3u}$$

$$2p_y : B_{3g} + B_{2u}$$

**Problem 4.10** Show that the transformations of the carbon $2p_x$ orbitals of Figure 4.8 and the $2p_y$ orbitals of Figure 4.9 give rise to the following

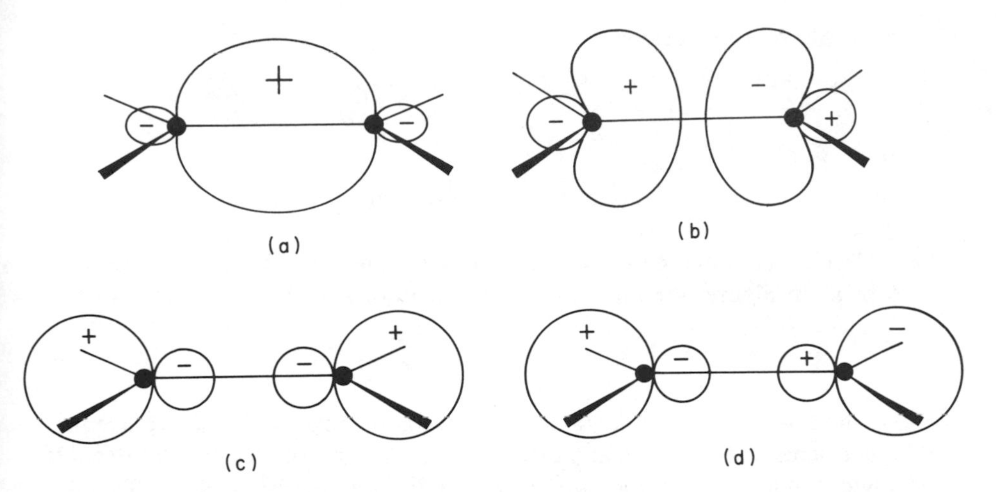


Figure 4.7 (a) The $A_g$ C—C σ-bonding orbital in ethylene (this is, essentially, Figure 4.6(b) + Figure 4.6(c)). (b) The $B_{1u}$ C—C σ-antibonding orbital in ethylene (essentially, Figure 4.6(a) + Figure 4.6(d)). (c) The $A_g$ carbon-based orbital involved in C—H bonding in ethylene (essentially, Figure 4.6(c) − Figure 4.6(b)). (d) The $B_{1u}$ carbon-based orbital involved in C—H bonding in ethylene (essentially, Figure 4.6(d) − Figure 4.6(a))

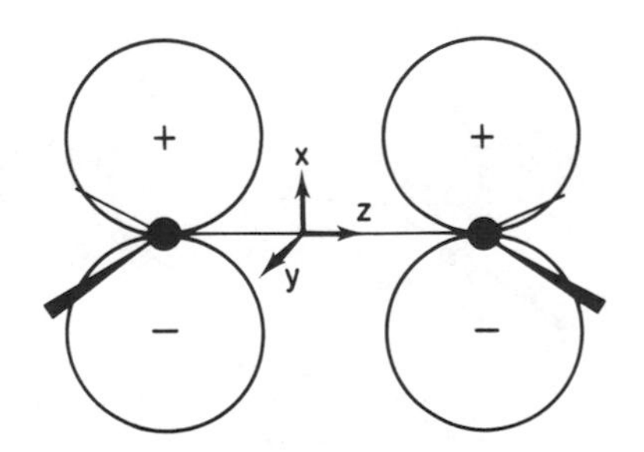


Figure 4.8 Carbon $2p_x$ orbitals in ethylene

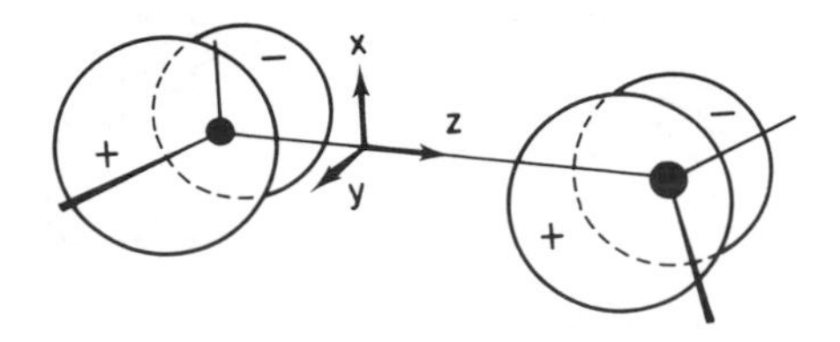


Figure 4.9 Carbon $2p_y$ orbitals in ethylene

reducible representations:

| | E | $C_2(z)$ | $C_2(y)$ | $C_2(x)$ | i | $\sigma(xy)$ | $\sigma(zx)$ | $\sigma(yz)$ |
|---|---|---|---|---|---|---|---|---|
| $2p_x$: | 2 | −2 | 0 | 0 | 0 | 0 | 2 | −2 |
| $2p_y$: | 2 | −2 | 0 | 0 | 0 | 0 | −2 | 2 |

and then, using the orthonormality theorem method of Section 3.3, that these reduce to the irreducible components given in the text.

Check that interchanging the phases of the lobes of one of the p orbitals in Figures 4.8 and 4.9 does not lead to any change in the above results.

Symmetry-correct linear combinations transforming as the above irreducible representations are sum and differences of the carbon $2p_x$ and $2p_y$ orbitals of Figures 4.8 and 4.9 and are shown in Figures 4.10 and 4.11. The $B_{3u}$ combination of carbon $2p_x$ orbitals shown in Figure 4.10 is immediately identified as the carbon–carbon $\pi$-bonding orbital and the $B_{2g}$ combination as the carbon–carbon $\pi$-antibonding orbital. Both of these will be carried through to the final energy level diagram.

This is a suitable point at which to define the labels $\sigma$ and $\pi$. It is convenient to think of just two bonded atoms (which may be part of a larger molecule) and of a line which connects their nuclei. If an orbital — be it bond-

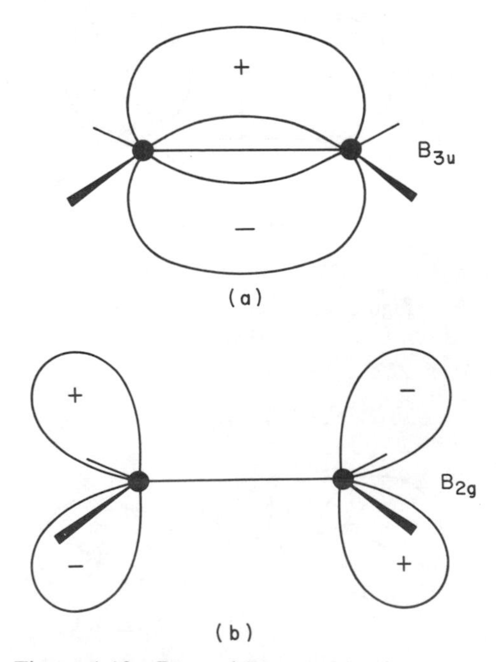


Figure 4.10 $B_{3u}$ and $B_{2g}$ combinations of carbon $2p_x$ orbitals in ethylene

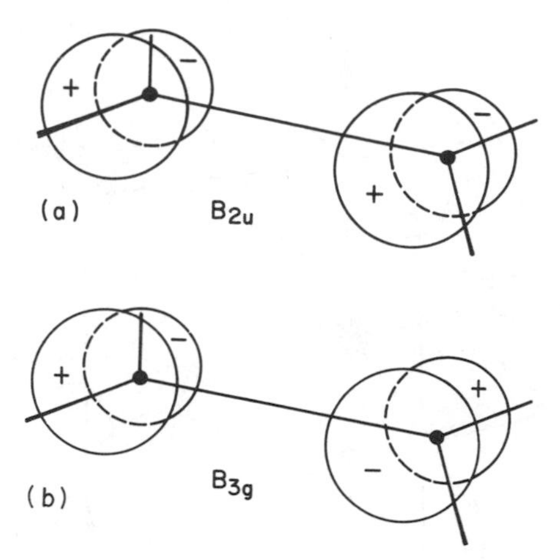


Figure 4.11 $B_{2u}$ and $B_{3g}$ combinations of carbon $2p_y$ orbitals in ethylene

ing or antibonding, localized or delocalized — has no nodal planes lying in the internuclear line then it involves a $\sigma$ interaction between the two nuclei. If there is a single nodal plane then the interaction is of the $\pi$ type; if there are two nodal planes then it is $\delta$. If a molecule is planar then the $\sigma/\pi$ distinction extends over the entire molecule and one can correctly distinguish $\sigma$ molecular orbitals from $\pi$ molecular orbitals (the former are symmetric and the latter antisymmetric with respect to reflection in the molecular plane).

There is an element of inconsistency in Figures 4.10 and 4.11. The only difference between the 2p orbitals shown in Figures 4.8 and 4.9 is that the former are rotated through 90° relative to the latter. One would therefore expect to find that Figure 4.11 is identical to Figure 4.10 except for this same rotation. In anticipation that the primary interactions involving the $B_{2u}$ orbitals of Figure 4.11 is with the terminal hydrogen atoms, whereas there is no such interaction involving the orbitals of Figure 4.10, we have chosen to ignore the carbon–carbon overlap in Figure 4.11.

## 4.5 THE SYMMETRIES OF THE HYDROGEN 1s ORBITALS IN ETHYLENE

We now turn our attention to the four hydrogen atoms of the ethylene molecule and consider the 1s orbital on each (which we shall take to have the same phase). These orbitals are all equivalent to one another — they may be interconverted by the symmetry operations of the group — and so we must consider all four together. They are shown in Figure 4.12, together with the symmetry elements of the $D_{2h}$ group. Of the entire set of symmetry operations only the identity operation and the $\sigma(zx)$ operation leave any of the hydrogen 1s orbitals in their original positions and each of these operations leaves all four orbitals unmoved; all other operations interchange all of them.

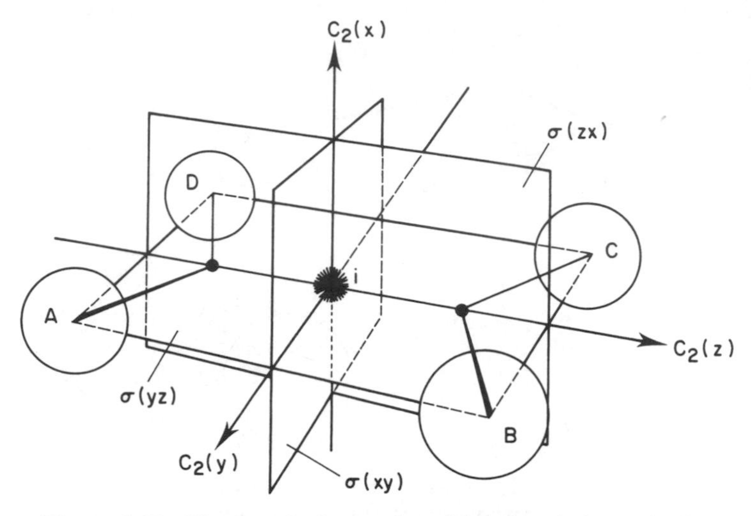


Figure 4.12 The four hydrogen 1s orbitals in ethylene together with the symmetry elements of the $D_{2h}$ group

The transformations of the four hydrogen 1s orbitals therefore generate the reducible representation:

| E | $C_2(z)$ | $C_2(y)$ | $C_2(x)$ | i | $\sigma(xy)$ | $\sigma(zx)$ | $\sigma(yz)$ |
|---|---|---|---|---|---|---|---|
| 4 | 0 | 0 | 0 | 0 | 0 | 0 | 4 |

We reduce this representation by the method described in Section 3.3. First we select an irreducible representation of the $D_{2h}$ group and multiply each character of our reducible representation by the corresponding character of our selected irreducible representation. We add these products together and then divide by the order of the group (eight in the present case). The integer which results* is the number of times the selected irreducible representation appears in the reducible representation. Thus, if we select the $A_g$ irreducible representation we obtain:

| | E | $C_2(z)$ | $C_2(y)$ | $C_2(x)$ | i | $\sigma(xy)$ | $\sigma(zx)$ | $\sigma(yz)$ |
|---|---|---|---|---|---|---|---|---|
| | 4 | 0 | 0 | 0 | 0 | 0 | 0 | 4 |
| $A_g$ | 1 | 1 | 1 | 1 | 1 | 1 | 1 | 1 |
| Products | 4 | 0 | 0 | 0 | 0 | 0 | 0 | 4 |

so that the sum of products is 8. Division by the order of the group yields the result that the $A_g$ irreducible representation appears once. Proceeding in this way we conclude that the irreducible representation has components

$$A_g + B_{3g} + B_{1u} + B_{2u}$$

*If a nonsense answer is obtained (e.g. a fraction) then either an arithmetical mistake has been made or the reducible representation has been wrongly generated. This is one way in which such mistakes are commonly discovered.

**Problem 4.11** Show that the above reducible representation also contains $B_{3g}$, $B_{1u}$ and $B_{2u}$ components.

## 4.6 THE PROJECTION OPERATOR METHOD

Although not essential to a qualitative discussion of the bonding in the ethylene molecule, we now seek combinations of the four hydrogen 1s orbitals, which, separately, transform as $A_g$, as $B_{3g}$, as $B_{1u}$ and as $B_{2u}$. As a bonus we shall obtain some idea of the C—H bonding molecular orbitals.

We first have to consider the transformations of the individual hydrogen 1s orbitals in much greater detail. Previously, we have only been concerned with whether or not a hydrogen 1s orbital was turned into itself under a particular symmetry operation. If it did not do this the destiny of the hydrogen atom did not concern us. This is no longer the case. We shall now look in detail at one of the four hydrogen 1s orbitals and determine the precise effect of each symmetry operation on this chosen orbital. We label the hydrogen 1s orbitals as shown in Figure 4.12 and then consider the transformation of the orbital which is labelled A. Under the identity operation A remains itself, under the $C_2(z)$ rotation it becomes the orbital we have labelled D, under the $C_2(y)$ rotation it becomes B, and so on. A complete list of its transformations is given in Table 4.5; it is important that the reader checks that this table is correct.

**Problem 4.12** Use Figure 4.12 to check that the table of transformations (Table 4.5) is correct.

We are now in a position to generate symmetry-correct linear combinations of the hydrogen orbitals. We know that the set A, B, C and D gives rise to a $B_{1u}$ combination and we shall now generate this combination.

Consider orbital A and the effect of the $C_2(y)$ operation. Table 4.1 shows that under this operation a function transforming as $B_{1u}$ changes sign. It follows, therefore, that orbitals A and B appear in the desired $B_{1u}$ linear combination in the form (A − B) since this expression changes sign under the $C_2(y)$ operation. By similarly considering the $C_2(x)$ and $C_2(z)$ operations — under which A interchanges with C and D respectively — it is evident that C and D must appear as −C and +D (because a $B_{1u}$ function changes sign under $C_2(x)$ but retains its sign under $C_2(z)$). We conclude that the $B_{1u}$ combination is of

Table 4.5

| | E | $C_2(z)$ | $C_2(y)$ | $C_2(x)$ | i | $\sigma(xy)$ | $\sigma(zx)$ | $\sigma(yz)$ |
|---|---|---|---|---|---|---|---|---|
| Under the operation A becomes | A | D | B | C | C | B | D | A |

the (normalized) form:

$$\tfrac{1}{2}(A - B - C + D)$$

It is a simple matter to check that this combination does indeed transform correctly (as $B_{1u}$) under all of the operations of the group. The general method is at once evident. In order to generate a required linear combination we simply take the entries in Table 4.5 and multiply each entry by the corresponding character. The sum of the answers so obtained is the desired linear combination. As an illustration of this method let us generate the $B_{3g}$ linear combination of hydrogen 1s orbitals, by this, the *projection operator*, method:

| | E | $C_2(z)$ | $C_2(y)$ | $C_2(x)$ | i | $\sigma(xy)$ | $\sigma(zx)$ | $\sigma(yz)$ |
|---|---|---|---|---|---|---|---|---|
| Under the operation | | | | | | | | |
| A becomes | A | D | B | C | C | B | D | A |
| $B_{3g}$ | 1 | −1 | −1 | 1 | 1 | −1 | −1 | 1 |
| Multiply | A | −D | −B | C | C | −B | −D | A |

$$\text{Sum} = 2A - 2B + 2C - 2D$$

The linear combination which we have generated by this procedure is $2A - 2B + 2C - 2D$. This function is not normalized since the sum of squares of coefficients appearing is 16, not 1; to normalize we have to divide by $\sqrt{16} = 4$ and so obtain the normalized $B_{3g}$ combination:

$$\tfrac{1}{2}(A - B + C - D)$$

The $A_g$ and $B_{2u}$ combinations are obtained in a precisely similar way. All four linear combinations are given in Table 4.6 and are shown in Figure 4.13. Such combinations are often referred to as 'symmetry adapted combinations'.

Table 4.6

| Symmetry species | Linear combination of 1s orbitals of hydrogen atoms in ethylene |
|---|---|
| $A_g$ | $\tfrac{1}{2}(A + B + C + D)$ |
| $B_{3g}$ | $\tfrac{1}{2}(A - B + C - D)$ |
| $B_{1u}$ | $\tfrac{1}{2}(A - B - C + D)$ |
| $B_{2u}$ | $\tfrac{1}{2}(A + B - C - D)$ |

**Problem 4.13** Use the projection operator method to obtain the (normalized) $A_g$ and $B_{2u}$ combinations of hydrogen 1s orbitals.

It is to be emphasized that each of the four diagrams in Figure 4.13 shows one orbital and *not* four. An instructive exercise at this point is to attempt to

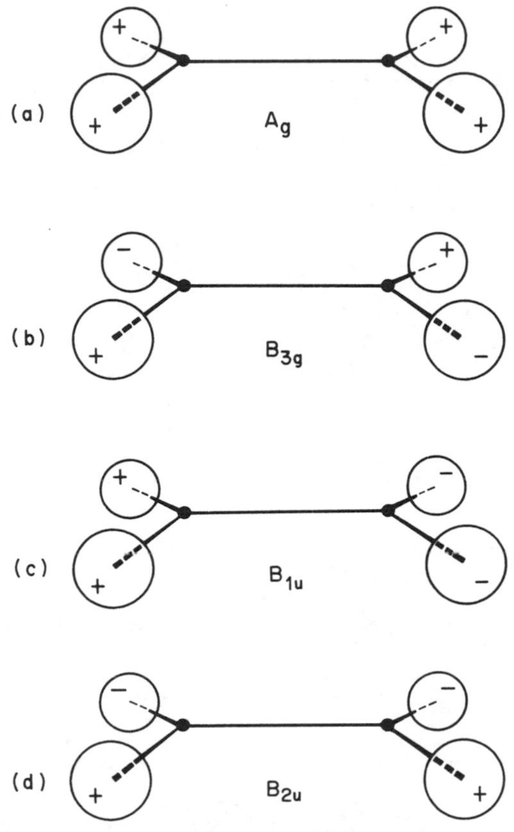


Figure 4.13 The symmetry-adapted combinations of hydrogen 1s orbitals in ethylene

generate from the data in Table 4.5 a combination transforming as an irreducible representation which we know to be absent (and so is not listed in either Table 4.6 or Figure 4.13) — e.g. $B_{1g}$. It will be found that the method we have developed is self-correcting!

**Problem 4.14** Attempt to generate a combination of 1s orbitals which does not, in fact, exist. Any of the irreducible representations $B_{1g}$, $B_{2g}$, $A_u$ or $B_{3u}$ may be chosen for this.

## 4.7 BONDING IN THE ETHYLENE MOLECULE

The symmetry-adapted linear combinations of hydrogen 1s orbitals which we have obtained are of the correct symmetries to interact with some of the carbon orbitals. Thus, the $A_g$ and $B_{1u}$ combinations interact with the carbon sp hybrids which we formed earlier and which are shown in Figure 4.7(c) and (d) respectively. The $B_{3g}$ and $B_{2u}$ are of the same symmetries as the carbon $2p_y$

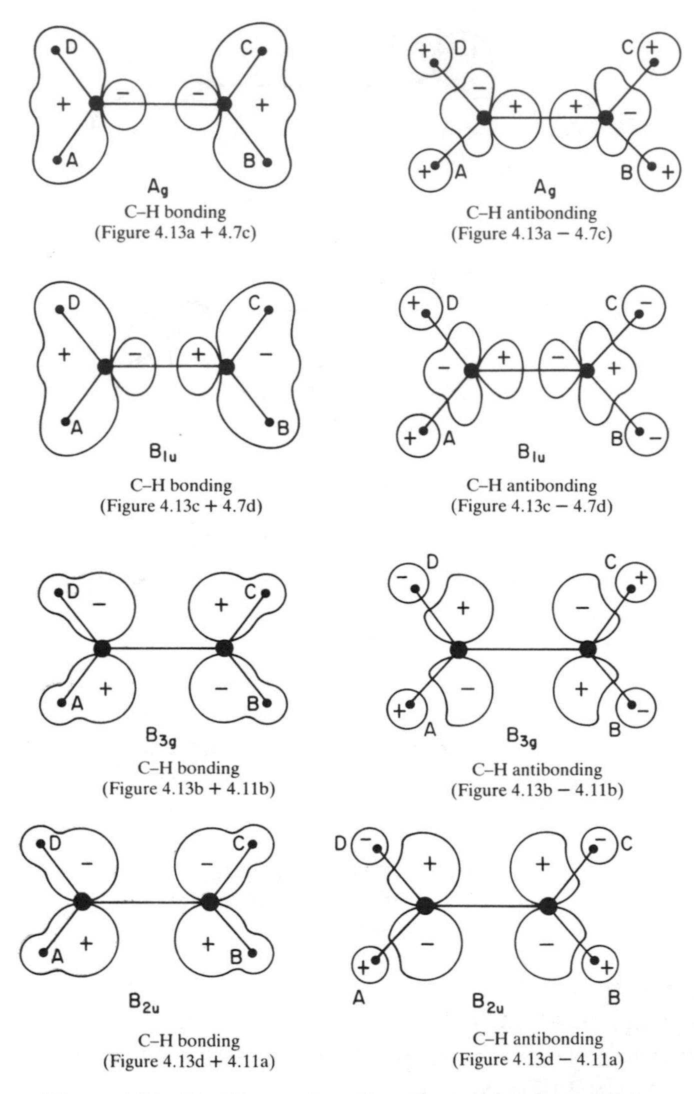


**Figure 4.14 Bonding and antibonding molecular orbitals in ethylene**

combinations (Figure 4.10b and 4.10a respectively). The resultant combinations are shown in Figure 4.14 where we indicate, qualitatively, how they are derived from the earlier figures.

**Problem 4.15** Check that the molecular orbitals shown in Figure 4.14 are correctly described by combining, qualitatively, the diagrams indicated below each molecular orbital.

There are four primarily C—H bonding molecular orbitals and four corresponding C—H antibonding orbitals (these orbitals are also either weakly C—C bonding or weakly C—C antibonding). In order to obtain even a qualitative molecular orbital energy level diagram we have to obtain some idea of the relative energies of the various C—H and the C—C $\sigma$- and $\pi$-bonding molecular orbitals. We will first look at those orbitals involved in C—H bonding; it will probably be found to be helpful to refer frequently to Figure 4.14 throughout the next few paragraphs.

There is no doubt about the most stable C—H bonding molecular orbital. This is the $A_g$ orbital. It has two features which lead to its stability. Firstly, just as the largely 2s(O)-containing molecular orbital was the most stable in $H_2O$, so too here the orbital containing an appreciable 2s(C) component is expected to be very stable. Secondly, the important interactions in which the $A_g$ orbital is involved are bonding — it is both C—H and C—C $\sigma$ bonding. Rather similar arguments hold for the $B_{1u}$ largely C—H bonding molecular orbital. It contains a 2s(C) contribution and is C—H bonding but is C—C $\sigma$ antibonding. We conclude that the $B_{1u}$ orbital is next in stability after the $A_g$.

The $B_{2u}$ and $B_{3g}$ C—H bonding molecular orbitals contain only carbon 2p orbitals so we expect them to be at higher energy than the $A_g$ and $B_{1u}$, which contain carbon 2s orbitals. Their relative energies can be related to the residual C—C bonding (which will be of $\pi$ type) associated with each. The $B_{2u}$ C—H bonding orbital is also C—C bonding but the $B_{3g}$ is C—C antibonding. We conclude that the $B_{2u}$ orbital is the more stable. In summary, then, we expect the C—H bonding molecular orbitals to decrease in stability:

$$A_g > B_{1u} > B_{2u} > B_{3g}$$

We now turn to the orbitals which are largely responsible for the carbon–carbon bonding. They are shown in Figures 4.7(a) and 4.9(a). There is no doubt that the $A_g$ largely carbon–carbon $\sigma$-bonding molecular orbital will be more stable than the $B_{3u}$ carbon–carbon $\pi$-bonding molecular orbital because one contains 2s(C) whereas the other contains $2p_z$(C). However, it is not easy to unambiguously relate their energies to those of the C—H bonding molecular orbitals. The following argument is indicative. The bond energy of a single C—C $\sigma$ bond is *ca.* 360 kJ mol$^{-1}$, although it is to be noted that this figure is appropriate to a bond length slightly longer than that found in ethylene. In contrast, the energy of an average C—H bond is *ca.* 420 kJ mol$^{-1}$. It seems reasonable, then, to anticipate that the stabilization resulting from the C—H bonding interactions should be somewhat greater than that of the $A_g$ C—C $\sigma$ interaction. This means that we would expect the C—H bonding molecular orbitals which have a carbon 2s component (those of $A_g$ and $B_{1u}$ symmetries) to be of lower energy than the carbon–carbon bonding orbital with a 2s component (that of $A_g$ symmetry). We have, then, the stability order $A_g$ (C—H bonding) > $B_{1u}$ (C—H bonding) > $A_g$ (C—C bonding). The next lowest C—H bonding orbital is $B_{2u}$ and we have to question whether its stability is sufficient to make it lower in energy than the $A_g$ (C—C bonding). If we interpret the bond energy data given above as 'the centre of gravity of the energies of the four C—H bonding interactions should be below the energy of the single C—C bonding interaction', then an order $B_{2u}$ (C—H bonding) < $A_g$ (C—C bonding) seems probable, although we cannot be certain. It seems likely that the two will be of similar energies with perhaps the $B_{2u}$ the more stable. As we shall see, this is the pattern experimentally observed.

The carbon–carbon $\pi$-bonding molecular orbital, of $B_{3u}$ symmetry, is also best placed by appeal to experiment. A great deal of spectroscopic and other information on carbon–carbon $\pi$-bonded systems can be rationalized on the assumption that it is a carbon–carbon $\pi$ orbital which is the highest occupied orbital and so we place $B_{3u}$ (C—C bonding) above $B_{3g}$ (C—H bonding). Together with our other arguments this leads to the molecular orbital energy level pattern shown in Figure 4.15. There are four valence electrons from each carbon and one from each hydrogen to be placed in these orbitals — a total of twelve. They occupy the six lowest orbitals in Figure 4.15; in this figure only one antibonding orbital, the lowest, C—C $\pi$-antibonding orbital of $B_{2g}$ symmetry, is included.

We can check Figure 4.15 in two ways. Firstly, we can appeal to detailed

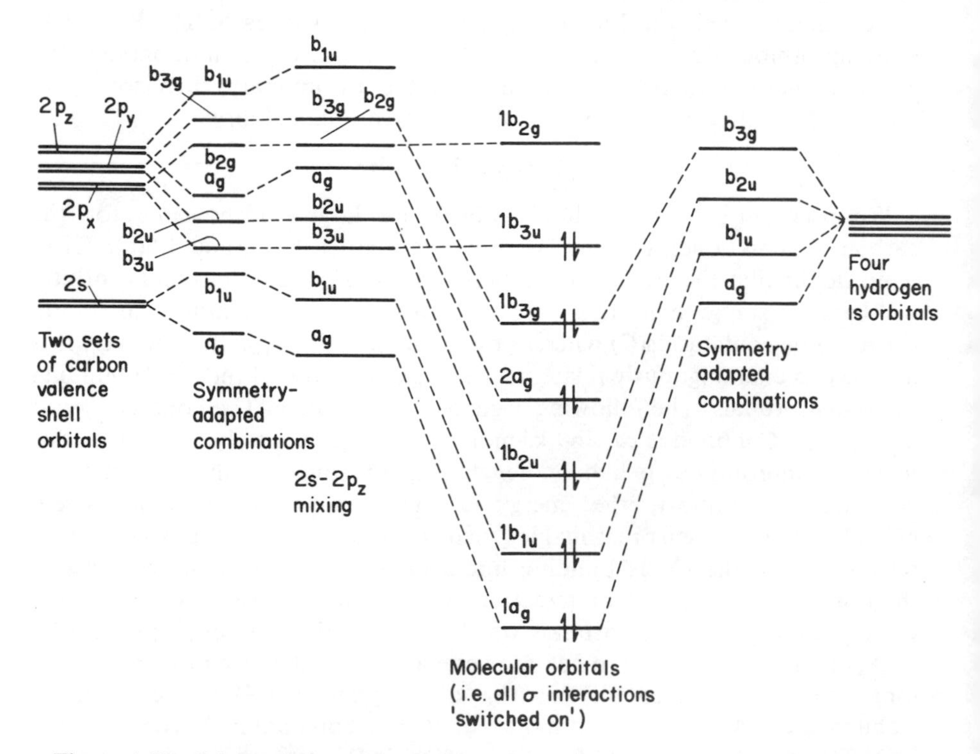


Figure 4.15 Schematic bonding orbital energy level diagram for ethylene (with the exception of $1b_{2g}$, antibonding orbitals are omitted)

accurate calculations on this molecule and, secondly, use the results of photoelectron spectroscopy. Experimental and theoretical work agree on the energy level sequence of ethylene. The results are given below, the calculated values[1] being given in brackets:

| | |
|---|---|
| $1b_{3u}$ (C—C bonding) | 10.51(10.44) eV |
| $1b_{3g}$ (C—H bonding) | 12.85(13.04) eV |
| $2a_g$ (C—C bonding) | 14.66(14.70) eV |
| $1b_{2u}$ (C—H bonding) | 15.87(16.07) eV |
| $1b_{1u}$ (C—H bonding) | 19.1 (19.44) eV |
| $1a_g$ (C—H bonding) | 23.5 (26.00) eV |

The agreement with our qualitative picture is excellent, giving some confidence in the arguments that we have used. In particular, we note that the hope that increased molecular symmetry would offset the greater molecular complexity compared with the water molecule has been justified.

## 4.8 BONDING IN THE DIBORANE MOLECULE

Our discussion of the ethylene molecule can be extended to another molecule, diborane. Diborane, $B_2H_6$, is of interest because it is the simplest of the boron hydrides (boranes). These, as a class, are often called 'electron deficient' because, whereas at least $(n-1)$ electron pairs are regarded as necessary to bond $n$ atoms they all have fewer than $2(n-1)$ electrons. Thus, there are only twelve valence shell electrons available in diborane to bond eight atoms. However, as we shall see, the term 'electron deficient' is a misnomer because the molecular structure is such that all bonding molecular orbitals are filled with electrons. Whereas diborane was such a problem for simple bonding models that it appeared necessary to give it a separate classification, a symmetry-based discussion shows that there is no need to invoke new concepts.

The structure of diborane is shown in Figure 4.16, from which it can be seen that it has four terminal hydrogen atoms and two borons which together have the same symmetry, $D_{2h}$, as ethylene (although the bond lengths and angles, of

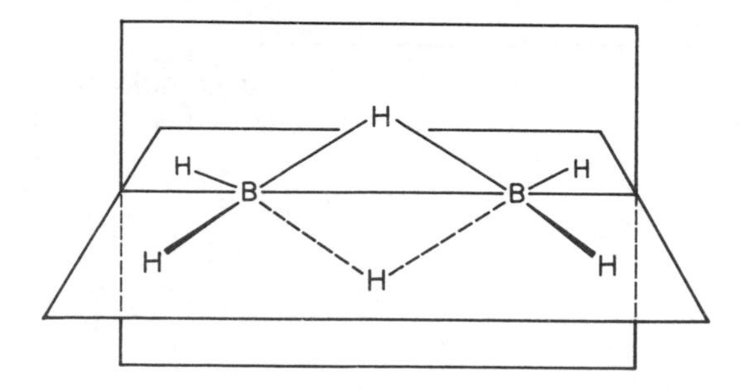


Figure 4.16 The structure of diborane, $B_2H_6$

course, are different). In addition, diborane has two hydrogen atoms out of, what is for ethylene, the molecular plane. These two hydrogens are usually called the (boron–boron) bridge hydrogen atoms. It is the presence of these bridging hydrogen atoms in place of the C—C $\pi$ bond of ethylene that plays a major part in causing diborane to have a rather different chemistry from ethylene. Figure 4.16 does not show all of the symmetry elements of diborane. Comparison with Figure 4.1 shows that the bridging hydrogens, located on the $C_2(x)$ axis of Figure 4.1, in no way diminish the $D_{2h}$ symmetry of the ethylene-like $B_2H_4$ unit. Diborane, like ethylene, has $D_{2h}$ symmetry. It follows that apart from that involving the bridging hydrogen atoms the bonding in the diborane molecule must, qualitatively, be similar to that given in the previous section for ethylene since boron, like carbon, has 2s and $2p_x$, $2p_y$ and $2p_z$ valence orbitals. We therefore expect to retain the same energy level sequence, $A_g$ (B—$H_t$ bonding) < $B_{1u}$ (B—$H_t$ bonding) < $B_{2u}$ (B—$H_t$ bonding) < $B_{3g}$ (B—$H_t$ bonding), where we have added the suffix 't' to distinguish *t*erminally bonded hydrogens from the bridging hydrogens. There is little doubt that there is a substantial difference between the carbon–carbon bonding in ethylene and the boron–boron bonding in diborane, as is shown by even a cursory study of the experimental data. The carbon–carbon bond length in ethylene is 1.34 Å whilst the boron–boron bond length in diborane is 1.77 Å. The details of the B—B bonding will also be different from the C—C bonding in ethylene because only the former has bridging hydrogens; we will start our discussion of the B—B bonding by looking at these bridge hydrogens.

The transformations of the 1s orbitals of the two bridging hydrogen atoms in diborane generate the following reducible representation:

| E | $C_2(z)$ | $C_2(y)$ | $C_2(x)$ | i | $\sigma(xy)$ | $\sigma(zx)$ | $\sigma(yz)$ |
|---|---|---|---|---|---|---|---|
| 2 | 0 | 0 | 2 | 0 | 2 | 2 | 0 |

a representation which has $A_g + B_{3u}$ components. As usual, the functions transforming as these irreducible representations are simply the sum and difference of the two 1s orbitals (which we label E and F as in Figure 4.17 and take to have the same phase). That is, they are:

| Symmetry species | Linear combination of bridge hydrogen orbitals |
|---|---|
| $A_g$ | $\frac{1}{\sqrt{2}}(E + F)$ |
| $B_{3u}$ | $\frac{1}{\sqrt{2}}(E - F)$ |

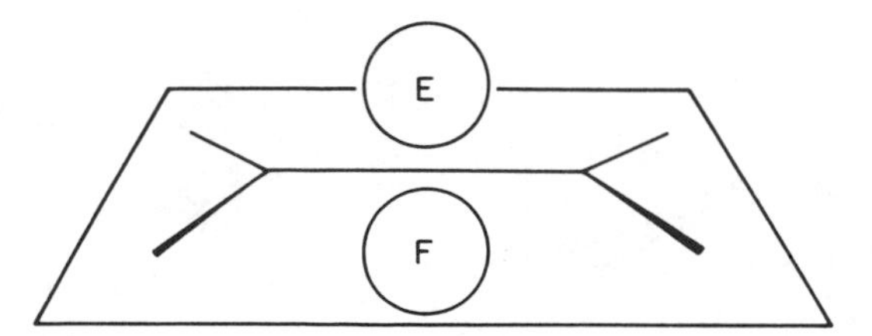


Figure 4.17 The 1s orbitals of the two bridging hydrogen atoms of diborane

**Problem 4.16** Check that the transformation of the two bridge hydrogen atoms in diborane are as given above. It is of particular importance to show that the two linear combinations of these orbitals transform as indicated.

The only orbitals shown in Figure 4.5 with which it is reasonable to expect any important interaction involving these bridge hydrogen orbitals are the boron–boron $\sigma$-bonding orbital of $A_g$ symmetry (which will be similar to that shown in Figure 4.7a but with boron atoms in place of carbon) and the boron–boron $\pi$-bonding orbital of $B_{3u}$ symmetry (which will resemble that shown in Figure 4.10a). The interactions between the bridge hydrogen orbitals and these boron–boron orbitals are shown qualitatively in Figures 4.18 and 4.19. Which of the bonding interactions shown in Figures 4.18 and 4.19 is the more important? For the $B_{3u}$ (boron–boron $\pi$ bonding) orbital the $B_2H_4$ plane

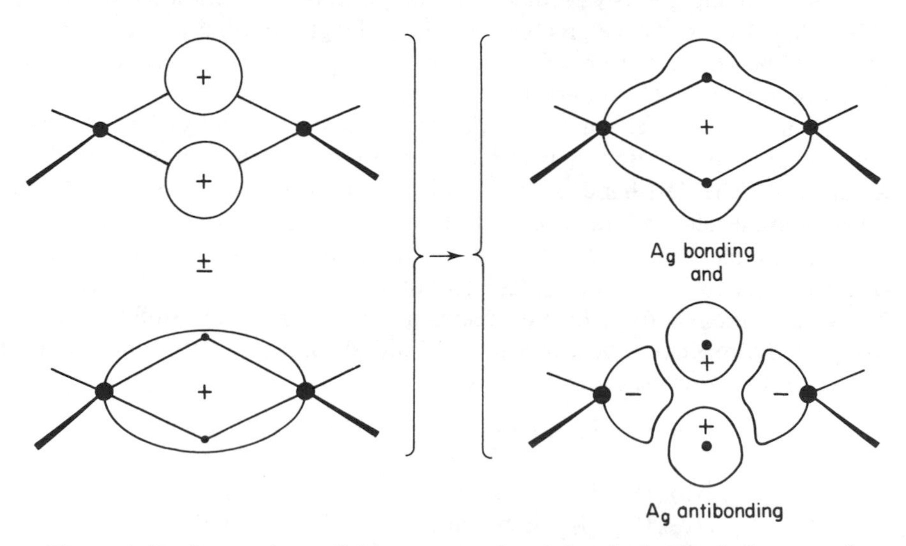


Figure 4.18 Interactions of $A_g$ symmetry involving the bridge hydrogens of diborane

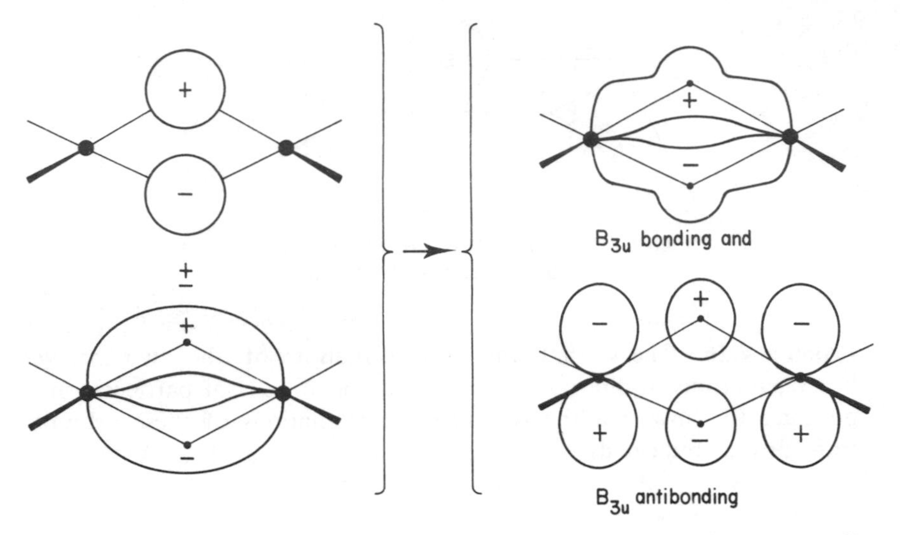


Figure 4.19 Interaction of $B_{3u}$ symmetry involving the bridge hydrogens of diborane

is a nodal plane; its maximum amplitude must be out of this plane. In contrast, the maximum amplitude of the $A_g$ (boron–boron σ-bonding) orbital is in the $B_2H_4$ plane. Because the bridging hydrogens are above and below this plane it seems probable that the interaction will be greater with the $B_{3u}$ boron combination than with the $A_g$. Whether this difference will lead to the orbital of $B_{3u}$ symmetry being beneath that of $A_g$ (in Figure 4.15 the $B_{3u}$ is above the $A_g$) we cannot unambiguously predict — in fact, it does. An additional reason for this is that because of the greater B—B bond length in diborane compared to the C—C bond in ethylene we expect that the $A_g$ B—B σ-bonding interaction is less than the C—C σ interaction in ethylene.

In Figure 4.20 we give a schematic molecular orbital energy level diagram for the diborane molecule in which we bring together all of the above arguments. The left-hand side of this diagram shows schematically the ethylene molecular orbital energy level pattern (Figure 4.15) which is then modified to take account of the bridge hydrogens. Qualitatively, the problem of the relative order of the $B_{2u}$ (B—$H_t$ bonding) and $A_g$ (B—$H_b$ bonding) orbitals, encountered for ethylene, reappears here (where the suffix 'b' is for *b*ridging hydrogens). The experimental and theoretical data[2] (the latter in brackets) are given below:

| | |
|---|---|
| $1b_{3g}$ (B—$H_t$ bonding) | 11.81(11.95) eV |
| $2a_g$ (B—$H_b$—B bonding) | 13.3 (13.12) eV |
| $1b_{2u}$ (B—$H_t$ bonding) | 13.9 (13.73) eV |
| $1b_{3u}$ (B—$H_b$—B bonding) | 14.7 (14.04) eV |
| $1b_{1u}$ (B—$H_t$ bonding) | 16.06(16.34) eV |
| $1a_g$ (B—$H_t$ bonding) | 21.4 (22.57) eV |

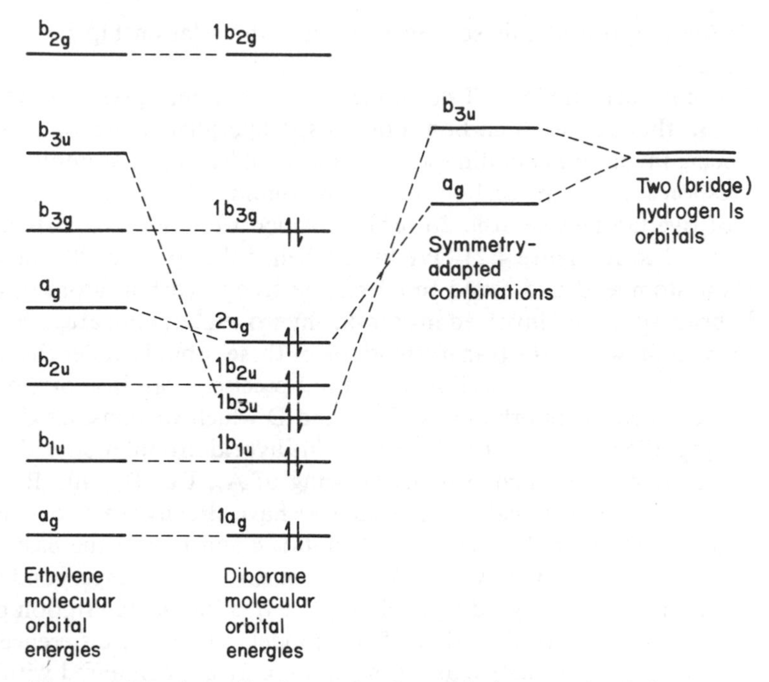


Figure 4.20 A qualitative molecular orbital energy level diagram for $B_2H_6$ and its relationship to that for $C_2H_4$

Again, we find that we have obtained an excellent qualitative prediction of the orbital energies using our simple symmetry-based model. It is interesting to note that with the sole exception of that of $B_{3u}$ symmetry, every orbital in this list is at a higher energy than its counterpart in ethylene, in accord with the higher chemical reactivity of diborane.

**Problem 4.17** The molecule $N_2H_4$, unlike $B_2H_6$ and $C_2H_4$, does not have $D_{2h}$ symmetry (it has a low-symmetry structure which may be regarded as similar to ethane with one hydrogen removed from each nitrogen atom). Use Figure 4.15 to explain why a $D_{2h}$ structure is not stable for $N_2H_4$. The discussion in the text associated with Figure 4.15 hints at the answer to this problem.

## 4.9 COMPARISON WITH OTHER MODELS

Most discussions of the electronic structures of the ethylene and diborane molecules concern themselves almost exclusively with the carbon–carbon double bond and the bridge bonding respectively. Some of these descriptions appear rather different from those which we have given in the present chapter and it is the purpose of this section to discuss the relationship between the various models.

We first consider ethylene. Two models are commonly presented for this molecule. In the first, each carbon atom is $sp^2$ hybridized, two of these $sp^2$ hybrids being involved in bonding with the terminal hydrogen atoms whilst the third is responsible for the carbon–carbon $\sigma$ bonding. A $\pi$ bond is formed as a result of overlap between the 2p orbitals which were not hybridized. This model is pictured in Figure 4.21. We have labelled the $sp^2$ hybrid orbitals on one carbon atom a, d and e and those on the second carbon atom b, c and f. The hybrids which are involved in carbon–hydrogen bonding are a, b, c and d. It is easy to show that the transformations of these orbitals under the operations of the $D_{2h}$ point group follow (or, more precisely, are isomorphous to) those of the hydrogen 1s orbitals A, B, C and D which we considered earlier in this chapter (Section 4.5). It follows that this hybrid orbital model identifies the C—H bonding molecular orbitals as being of $A_g$, $B_{3g}$, $B_{1u}$ and $B_{2u}$ symmetries, a conclusion identical to that which we have already reached. It is also straightforward to show that the hybrid orbitals e and f form the basis for a reducible representation with $A_g$ and $B_{2u}$ components, which correspond to the C—C $\sigma$-bonding and -antibonding orbitals. The qualitative description of the C—C $\sigma$ bonding is identical to that of our model. The main differences can be seen when the two orbitals e and f in Figure 4.21 are compared with their counterparts in the model we used. We regarded these orbitals as equal mixtures of the carbon 2s and $2p_z$ orbitals whereas in the $sp^2$ hybrid orbital model the s orbital contribution is only one third. However, it will be recalled that when we mixed the 2s and $2p_z$ orbitals in equal amounts we mentioned that this was an arbitrary mixing, made on grounds of simplicity. We could, accidentally, have been correct in our choice of a ratio of 1 : 1 or the $sp^2$ hybrid orbital model could have been right in its ratio of 1 : 2. Detailed calculations

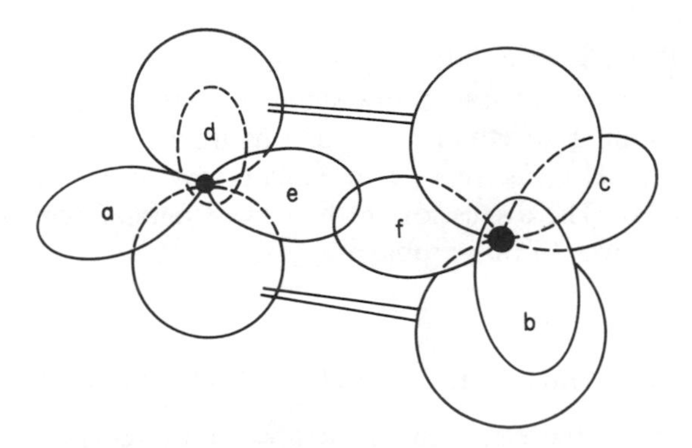


Figure 4.21 The '$sp^2 + p_\pi$' carbon atom model for the bonding in ethylene. For simplicity the hydrogen atoms are omitted

show that both are wrong — there are two $A_g$ orbitals contributing to C—C bonding, one largely involving 2s(C) and the other $2p_z$(C). The aggregate $2s:2p_z$ ratio is $1:1.3$ so, from this viewpoint, our sp model is not too bad.

A point of apparent divergence between the two approaches is to be found in the carbon–hydrogen bonding orbitals of $B_{3g}$ and $B_{2u}$ symmetries. In our description these orbitals contained no contribution from the carbon 2s orbitals. In contrast, one might expect there to be such a contribution in the hybrid orbital description since each hybrid contains a 2s component. This is not the case. If the form of the hybrid orbitals is written out explicitly and the appropriate linear combinations of them obtained using the projection operator method (these combinations are those given in Table 4.6 but with capital letters replaced by lower case letters), it will be found that carbon 2s orbital contributions also vanish in the hybrid orbital description.

**Problem 4.18** The explicit forms of the relevant carbon $sp^2$ hybrid orbitals are:

$$a = \frac{1}{\sqrt{3}}\, s(C_1) + \frac{1}{\sqrt{2}}\, p_y(C_1) - \frac{1}{\sqrt{6}}\, p_z(C_1)$$

$$d = \frac{1}{\sqrt{3}}\, s(C_1) - \frac{1}{\sqrt{2}}\, p_y(C_1) - \frac{1}{\sqrt{6}}\, p_z(C_1)$$

$$b = \frac{1}{\sqrt{3}}\, s(C_2) + \frac{1}{\sqrt{2}}\, p_y(C_2) + \frac{1}{\sqrt{6}}\, p_z(C_2)$$

$$c = \frac{1}{\sqrt{3}}\, s(C_2) - \frac{1}{\sqrt{2}}\, p_y(C_2) - \frac{1}{\sqrt{6}}\, p_z(C_2)$$

where $C_1$ and $C_2$ refer to the two carbon atoms. By substituting these in the explicit expressions for the $B_{3g}$ and $B_{2u}$ linear combinations given in Table 4.6 (but substituting the expression given above for a in place of A in Table 4.6, etc.) show that the carbon 2s orbital contributions vanish.

A model of the carbon–carbon double bond in ethylene which is historically important and which is still encountered is that in which the carbon atoms are $sp^3$ hybridized and each bond of the double bond is equivalent, as shown in Figure 4.22. However, the two carbon–carbon bonding orbitals labelled a and b in Figure 4.22 provide a basis for a reducible representation with $A_g$ and $B_{3u}$ components. These are the symmetries which we have deduced as being those of the carbon–carbon bonding orbitals. Indeed, if the projection operator method is used to obtain the $A_g$ and $B_{3u}$ combinations of a and b (they are the sum and difference of the two) then one obtains orbitals which are, essentially, identical to the carbon–carbon bonding orbitals shown in Figures 4.7 and

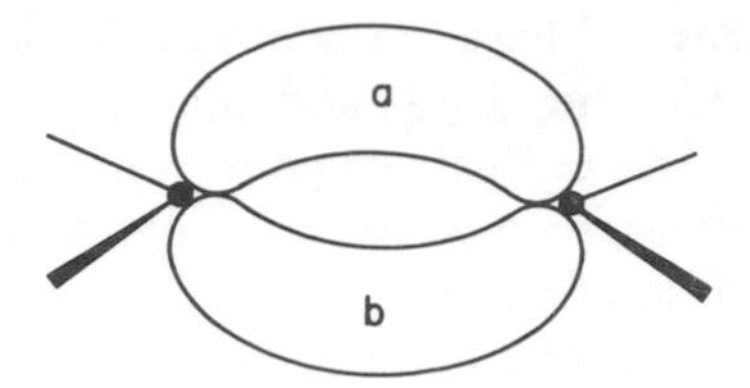


Figure 4.22 The $sp^3$ carbon atom model for the bonding in ethylene. Hydrogens are omitted and the $sp^3$ hydrids to which they bond are represented by rods. Shown in the diagram are the bonding orbitals formed by the overlap of $sp^3$ hybrids on the carbon atoms

4.10. That is, the use of $sp^3$ hybrids at each carbon atom is also consistent with the model of C—C bonding that we have derived, although such a description pictures the orbital on each carbon atom which is involved in this bonding as being one quarter composed of the carbon 2s orbital — the third value we have met! The use of $sp^3$ hybrids to explain the C—H bonding is also consistent with our symmetry-based discussion. Again C—H bonding molecular orbitals of $A_g$, $B_{1u}$, $B_{2u}$ and $B_{3g}$ symmetries are obtained when the four C—H bonding $sp^3$ hybrids are used to generate a reducible representation of the group.

Perhaps the simplest description of the bonding of the bridging hydrogen atoms in diborane is the so-called banana bond picture. These bonds are shown in Figure 4.23; the close similarity with Figure 4.22 is immediately apparent. It is not at all difficult to show that the bridge bonds in Figure 4.23 form the basis for two linear combinations, one of $A_g$ symmetry and the other of $B_{3u}$. These symmetries are the same as those of the orbitals shown in

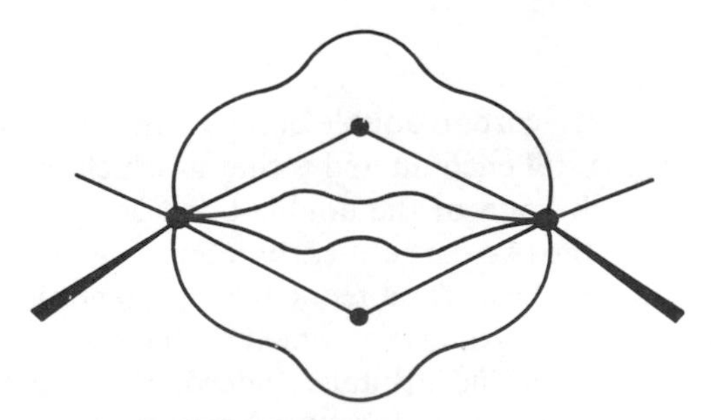

Figure 4.23 The 'banana bond' model for the bonding of the bridge hydrogens in diborane

Figure 4.20 which are responsible for the bridge bonding. The similarity between the two descriptions follows at once.

When a chemist speaks of a quantity such as 'the carbon–carbon bond' or 'the carbon–hydrogen bond' this frequently refers to quantities which do not, themselves, transform as an irreducible representation of the point group of a molecule. In such cases several other symmetry-related bonds exist and these together provide a basis set from which appropriate irreducible representations can be generated. Thus, in the present context one can say that the C—H bonds in ethylene (or B—H bonds in diborane) can be combined into combinations which transform as irreducible representations of the $D_{2h}$ point group. Localized orbitals constructed so that they are equivalent to one another in this way and which can be used to derive symmetry-adapted combinations are often referred to as equivalent orbitals. The C—H bonding molecular orbitals shown in Figure 4.14 are all different; in contrast the chemist prefers to think of equivalent (or localized) orbitals. As we recognized in the case of the water molecule (Chapter 3), and again in the present chapter for ethylene and diborane, these two pictures are usually equivalent to each other. Indeed, for all of the simple models which we have shown to be basically similar to the pictures obtained by our symmetry-based approach, the orbitals which we have transformed to obtain reducible representations are equivalent orbitals. That is, the approach developed in this chapter to the electronic structures of ethylene and diborane is, fundamentally, no different from those with which the chemist is more familiar (the same is true of the discussion at the end of Chapter 3). On the other hand, the symmetry-based approach has considerable advantages. Thus, the observation that the C—H bonds in ethylene are equivalent does not imply that the removal of any one C—H bonding electron requires the same energy as the removal of any other. The fact that there are several ionization potentials — as shown by photoelectron spectroscopy — only becomes clear in a symmetry-based description of the bonding. Despite this emphasis on symmetry it must be recognized that symmetry arguments, by themselves, tell us nothing about energy levels. It is only when we elaborate these arguments by including additional concepts, such as nodality, orbital composition and relative magnitudes of interactions that relative energies emerge.

There is one final point. In Section 4.2 we saw that $\sigma_v$ operations are equivalent to $C_2$ followed by i. This is a characteristic of all 'improper rotation' operations — they correspond to a proper rotation combined with i. We shall meet other examples later in this book.

## 4.10 SUMMARY

In this chapter we have seen that point groups may be related to each other. When a point group is the direct product of two smaller groups (the jargon is to refer to such smaller groups as 'invariant subgroups' (page 62) of the larger group) then the multiplication tables of the larger group may be derived from those of the smaller groups (page 62) — as may its symmetry operations (page 62), character table (page 62) and (usually) labels for its irreducible representations (page 63).

The technique of using projection operators to obtain linear combinations of a particular symmetry is most important (page 73).

As in the previous chapter, symmetry-based models led to qualitative predictions of electronic structure which were both in accord with the results of photoelectron spectroscopy (page 79, 82) and consistent with more traditional bonding models (page 83).

## REFERENCES

1. W. Von Niessen, G. H. F. Diercksen, L. S. Cederbaum and W. Domcke, *Chem. Phys.*, **18**, 469 (1976).
2. D. R. Lloyd, N. Lynaugh, P. J. Roberts and M. F. Guest, *J. Chem. Soc.* (Far. II), **71**, 1382 (1975).

CHAPTER 5

# *The electronic structure of bromine pentafluoride, $BrF_5$*

Although an object of this chapter is to discuss the electronic structure of bromine pentafluoride, this topic represents only about a third of its contents. We must first generalize the group theoretical methods that wc have developed in the previous chapters to a point which will enable us to discuss almost any molecule, irrespective of its symmetry; this generalization is the major purpose of this chapter and takes up most of it.

The structure of the bromine pentafluoride molecule is shown in Figure 5.1; the bromine is surrounded by four fluorines at the corners of a square and by a fifth, unique, apical, fluorine situated so that the five fluorines form a square-based pyramid around the bromine atom. Perhaps surprisingly, the bromine is slightly beneath the plane defined by the four coplanar fluorines and therefore is below the centre of gravity of the five fluorines. A valence electron count shows that there are two non-bonding electrons on the bromine atom and the electron pair repulsion (Sidgwick–Powell–Nyholm–Gillespie) model (Chapter 1) suggests that lone pair–bond pair repulsion will have a greater effect on the four coplanar fluorine atoms than will the repulsion between these bromine–fluorine bonds and the apical one. The consequence of this inequality will be that the coplanar B—F bonds will be bent towards the apical fluorine atom, giving the observed geometry of the molecule. As a simplifying

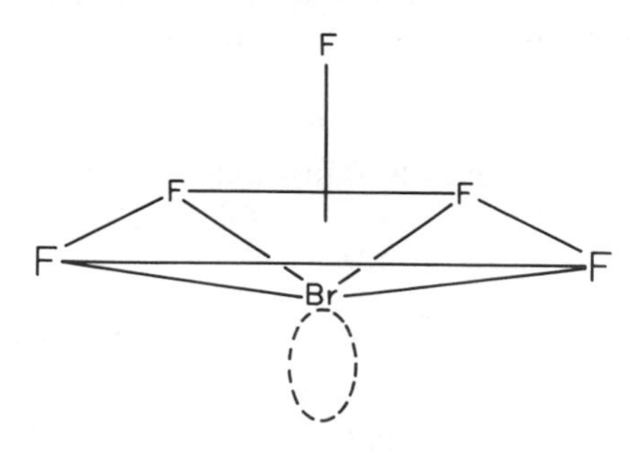


Figure 5.1 The structure of the $BrF_5$ molecule

assumption, however, we shall in this chapter assume that the central bromine is coplanar with the surrounding fluorine atoms.

We shall proceed by presenting the character table for the $C_{4v}$ group because, as we shall see, the bromine pentafluoride molecule has this symmetry. However, there are many differences between this character table and those which we have previously used. Much of this chapter will be occupied by an exploration of these differences — this study is most important because it will lead us to the generalization of group theoretical concepts and techniques referred to above. We shall find that there may be more than one symmetry operation corresponding to a single symmetry element and, correspondingly, that character tables may contain numbers other than 1 and $-1$. The most important of our generalizations will be the generalization of the orthonormality theorems. We shall use this generalization to generate the $C_{4v}$ character table, given in Table 5.1.

The reader will notice that, confusingly, E appears in the list of irreducible representation labels as well as in the list of operations. In this new usage it labels an irreducible representation which describes the transformation of two things simultaneously. It is sometimes called a 'doubly degenerate' irreducible representation. The reason for this will become evident later in this chapter.

Table 5.1

| $C_{4v}$ | E | $2C_4$ | $C_2$ | $2\sigma_v$ | $2\sigma_v'$ |
|---|---|---|---|---|---|
| $A_1$ | 1 | 1 | 1 | 1 | 1 |
| $A_2$ | 1 | 1 | 1 | $-1$ | $-1$ |
| $B_1$ | 1 | $-1$ | 1 | 1 | $-1$ |
| $B_2$ | 1 | $-1$ | 1 | $-1$ | 1 |
| E | 2 | 0 | $-2$ | 0 | 0 |

**Problem 5.1** Both the $D_{2h}$ and $C_{4v}$ groups are of order eight — a total of eight operations is listed at the top of each table (compare Tables 4.1 and 5.1). However, their structures are rather different. Make a list of the qualitative differences between the two tables.

We shall start our discussion by looking in detail at the symmetry operations of the bromine pentafluoride molecule — of the $C_{4v}$ group.

## 5.1 SYMMETRY OPERATIONS OF THE $C_{4v}$ GROUP

The perspective shown in Figure 5.1 is not the best to show the symmetry of the $BrF_5$ molecule. This is most readily recognized by viewing the molecule along the bromine–axial fluorine bond, as shown in Figure 5.2, from which it is clear that the four other fluorine atoms lie at the corners of a square.

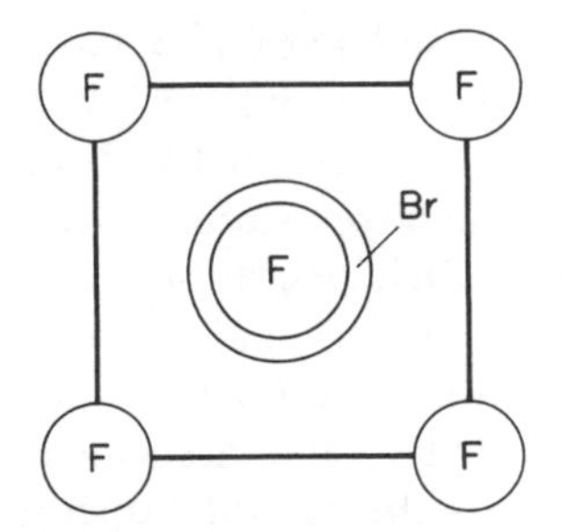


Figure 5.2 A view of the $BrF_5$ molecule looking down the apical (axial) F—Br bond

Evidently, the bromine–axial fluorine bond coincides with a fourfold rotation axis (i.e. $C_4$) of the molecule. In all of the symmetries we have previously considered there has always been a single symmetry operation associated with each symmetry element of a molecule. Further, we have chosen to use the same symbol for operation and element, leaving it to the context to make it clear which was the subject of discussion. Although we shall persist with the latter convention we must now recognize that there is not always a one-to-one correspondence between symmetry elements and symmetry operations. Thus, in the present case, although there is just one fourfold rotation axis in the $BrF_5$ molecule there are two corresponding symmetry operations. The molecule is turned into itself by a rotation of 90° in either a clockwise or an anticlockwise direction about the fourfold axis. These two operations have the effect of interchanging the fluorine atoms of $BrF_5$ in different ways and so are distinct operations. The clockwise and anticlockwise $C_4$ rotation operations associated with the $C_4$ axis are inseparable — one cannot have one without the other. Usually we can group these operations together as $2C_4$, thus recognizing both their distinction and similarity, and it will be seen that they are written this way in the $C_{4v}$ character table (Table 5.1). Operations paired and written in this way are said to be 'members of the same class'. The concept of class is an important one and is dealt with more formally in Appendix 1. Because a fourfold axis exists it follows that a rotation of 180° about this axis turns the molecule into itself. However, this operation is *not* a $C_4$ rotation but a $C_2$ one. We conclude that, strictly, we should regard there as being a $C_2$ axis as being coincident with the $C_4$ axis. A high rotational symmetry may, automatically, imply the simultaneous existence of coincident axes of lower symmetry. A $C_2$ rotation operation may, of course, be regarded as a $C_4$ rotation operation carried out twice in succession (in the same clockwise or anticlockwise sense). Symbolically one can write

$$C_4 \times C_4 = C_4{}^2 \equiv C_2$$

where, following the discussion of Chapter 2, we have multiplied $C_4$ by $C_4$ to obtain $C_2$.

In the same way it is easy to see that $C_4^3 \equiv C_4^{-1}$ (i.e. carrying out three $C_4$ rotations in one sense — clockwise or anticlockwise — is equivalent to a single $C_4$ rotation in the opposite sense) and that $C_4^4 = E$. This is another point of difference from the groups which we met in previous chapters. For all of these groups we found that any of their operations carried out twice in succession gave the identity, regenerating the original arrangement. For the $C_4$ operation it takes four steps (and for a general $C_n$ rotation it takes $n$ steps).

The other symmetry elements (and associated operations) of $BrF_5$ are fairly evident. In addition to the identity operation, the two $C_4$ rotation operations and the associated $C_2$ rotation operation (which, it should be noted, comprises a class of its own), there are four mirror planes which are indicated in Figure 5.3. It can be seen from this figure that these mirror planes are of two types. Firstly, there are those which we have labelled $\sigma_v(1)$ and $\sigma_v(2)$, in each of which lie the bromine and three fluorine atoms. It is impossible to have one of these operations without the other because of the $C_4$ axis. A $C_4$ operation rotates one of these $\sigma_v$ mirror planes into the other. They are therefore inextricably paired together. These operations therefore comprise a class which is written as $2\sigma_v$ and it appears in this form in the character table. Secondly, there are those mirror planes which we have labelled $\sigma_v'(1)$ and $\sigma_v'(2)$ in Figure 5.3. Each contains the bromine and the axial fluorine atom and again are interrelated by a $C_4$ operation. They comprise the class $2\sigma_v'$.

Several comments are relevant at this point. Firstly, all four of the mirror planes which we have discussed contain the $C_4$ axis and so are $\sigma_v$ mirror planes, as we have labelled them. Secondly, many authors prefer to give the mirror planes which we have called $\sigma_v'(1)$ and $\sigma_v'(2)$ the labels $\sigma_d(1)$ and $\sigma_d(2)$, or, as a class, $2\sigma_d$. This is because in a closely related group — that of the symmetry operations of a square — they carry this label. Strictly, however, the loss in symmetry in going from this group to $C_{4v}$ forbids the use of the $\sigma_d$ symbol (as we shall see in Section 7.1 this symbol has a rather precise meaning). Thirdly,

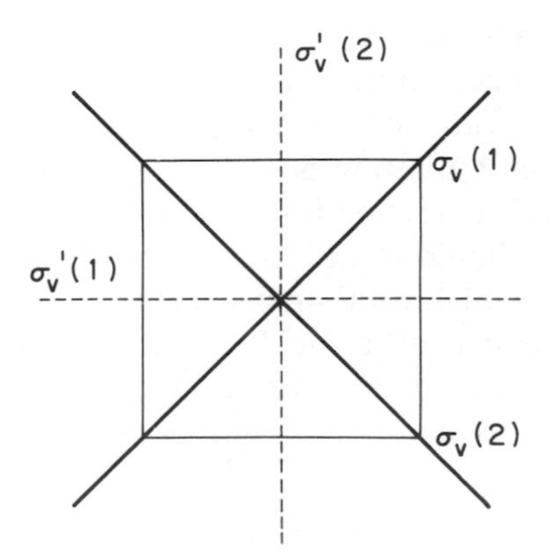


Figure 5.3 Mirror planes of symmetry in Figure 5.2

we have seen that $\sigma_v(1)$ and $\sigma_v(2)$ are interconverted by a $C_4$ rotation, as also are $\sigma_v'(1)$ and $\sigma_v'(2)$. When symmetry elements are interconverted by another operation of the group, it is a sure sign that the corresponding operations fall into the same class. Finally, we note that it is the presence of the $C_4$ axis, together with the vertical mirror planes, that give rise to the shorthand symbol for the group, $C_{4v}$.

Collecting all of the symmetry operations of the $C_{4v}$ group together we have

$$E \qquad 2C_4 \qquad C_2 \qquad 2\sigma_v \qquad 2\sigma_v'$$

and it is these operations that head the character table (Table 5.1). Although our next major task is to derive this character table, it is convenient first to consider a problem which we shall encounter when using it.

## 5.2 PROBLEMS IN USING THE $C_{4v}$ GROUP

When we are considering the transformation of something — an orbital or set of orbitals perhaps — what should we do when there are two operations in a class? How do we generate a character in such a case? Although the formal answer to this is unattractive, the practical answer is simple. Formally, the correct procedure is to consider the transformations under each of the individual operations in the class and to take the average of characters generated. However, it is invariably the case that each of the symmetry operations in a class always gives the same character. This means that, in practice, all that we have to do is to:

> Select a single symmetry operation from a class (and it quite often happens that it is possible to set up the problem in such a way that there is one operation with which it is particularly easy to work) and take the character generated by this operation.

There is yet one more problem which it is as well to consider before turning to the $C_{4v}$ character table. As we have seen, the axis of highest rotational symmetry is conventionally chosen as the z axis so that the $C_4$ axis of $BrF_5$ is clearly to be taken as the z axis. However, we are left with the problem of where to place the x and y axes. Perhaps the most evident choice of directions is that shown in Figure 5.4(a), in which the four fluorines are taken to define the x and y axes. However, what is wrong with the alternative choice given in Figure 5.5(a)? The solution to this problem becomes clearer when we note that the x and y axes, just like the $\sigma_v$ mirror planes in which they lie, are interchanged by the $C_4$ operations, irrespective of whether we choose the orientation of Figure 5.4 or Figure 5.5. The orientations for x and y axes in these figures have an obvious attraction — they are choices which place the axes in mirror planes. A less attractive choice, such as that shown in Figure 5.6, retains the property that x and y are interrelated by a $C_4$ rotation. Clearly, we have to treat the x and y axes as a pair, just like the two $\sigma_v$ and also the two $\sigma_v'$ mirror

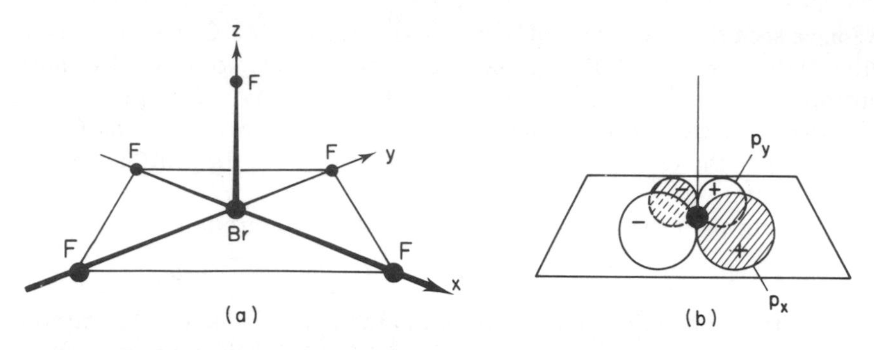


Figure 5.4 One choice of direction for x and y axes in $BrF_5$ (and consequent directions for the bromine $4p_x$ and $4p_y$ orbitals)

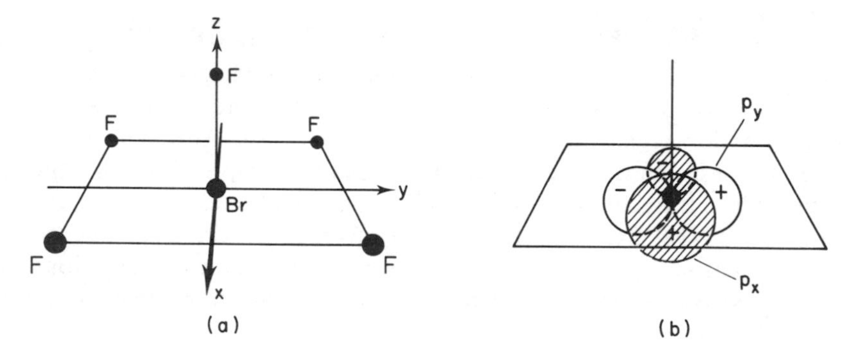


Figure 5.5 An alternative choice of direction for x and y axes in $BrF_5$ (and consequent directions for the bromine $4p_x$ and $4p_y$ orbitals)

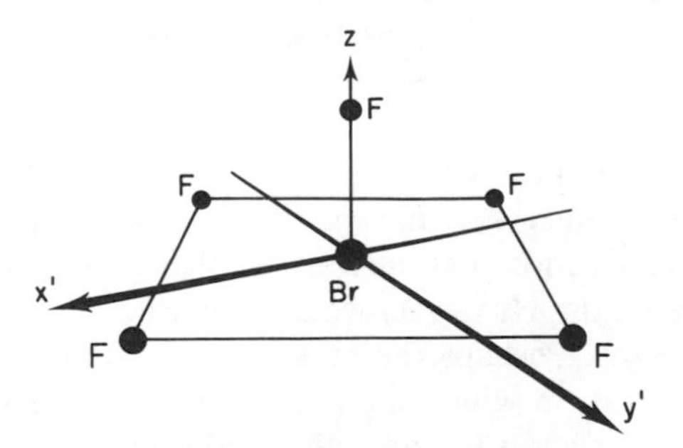


Figure 5.6 A third choice of direction for x and y axes in $BrF_5$

planes. This intimate pairing of x and y axes is basic to the difference between the $C_{4v}$ character table and the Abelian character tables of earlier chapters in this book. The x and y axes are said to 'transform as a pair'. We shall see later that this statement is manifest in that the x and y axes transform, together, as the E irreducible representation in Table 5.1. The choice of x and y axis directions — Figures 5.4, 5.5 and 5.6 — is ultimately unimportant; after all, no physical property can in any way depend on the way we choose to place axes. This is discussed in more detal in Appendix 2.

## 5.3 ORTHONORMALITY RELATIONSHIPS

We now return to the problem of generating the character table of the $C_{4v}$ point group. We could follow the procedure of Chapter 3 and use the transformation of atomic orbitals of the bromine atom to generate irreducible representations of this group. Unfortunately, however, we could not obtain a complete set of irreducible representations even if we included f orbitals on the bromine atom — although g orbitals would suffice! If this is the only method available for the compilation of a character table, one would be led to wonder whether a study of yet higher orbitals might uncover further, previously unrecognized, irreducible representations. Fortunately there are systematic methods available for the generation of character tables. We shall now describe one of these methods, one which relies on the existence of the orthonormality theorems which we have already used in a simple form in earlier chapters. The general proof of these theorems is rather mathematical and so we simply present the results and hope that the reader will be prepared to accept them as reasonable extrapolations from the cases we have already discussed in detail. The derivation and mathematial form of the theorems is given in Appendix 2. Of these theorems, numbers 2, 3, 4 and 5 are those that are commonly called the orthonormality theorems.

**Theorem 1** In every character table there exists a totally symmetric irreducible representation.

*Comment:* The totally symmetric irreducible representation is the first given in any character table and has a character of 1 associated with each class of operation. It describes the symmetry properties of something which is turned into itself under every operation of the group. This really is a rather trivial theorem, introduced here for convenience. By definition, a molecule is turned into itself by every operation of the point group used to describe it. A totally symmetric irreducible representation must therefore exist for every point group.

**Theorem 2** Take each element of any row of a character table (i.e. the characters of any irreducible representation), square each, multiply by the number of operations of the class to which the character belongs and add the answers together. The number that results is an integer which is equal to the order of the group (i.e. equal to the total number of symmetry operations in the group).

*Comment:* We met this theorem when, in Chapter 3, we obtained a systematic method of reducing a reducible representation into its irreducible components. Because we were then working with an Abelian group the number of operations in each class was one, so there was no need to include the step of multiplying by the number of operations in a class. For non-Abelian groups, for which there are invariably several classes containing more than one operation, we must take care to include this additional step, otherwise the number obtained is not the order of the group.

**Problem 5.2** Apply Theorem 2 to each of the irreducible representations of the $C_{4v}$ point group (Table 5.1). The order of this group is eight.

**Theorem 3** Take any two different rows of a character table (i.e. any two irreducible representations) and multiply together the two characters associated with each class. Then in each case multiply the product by the number of operations in the class. Finally, add the answers together. The result is always zero.

*Comment:* This theorem, again, we met when reducing reducible representations. Again, too, because the number of operations in each class was one we did not need to explicitly include multiplication by the number of operations in the class. In general, however, this step must be included.

**Problem 5.3** Apply Theorem 3 to at least five pairs of irreducible representations of the $C_{4v}$ point group (Table 5.1). The E irreducible representation should be included in at least two cases.

As we shall see Theorems 2 and 3 are at the heart of the method used to reduce reducible representations into their irreducible components. They are sometimes loosely referred to as 'the orthornomality relationships'.

**Problem 5.4** Look back at Section 3.1 and read the discussion on orthornomality given there. Why are Theorems 2 and 3 referred to as 'orthonormality relationships'?

The fourth and fifth theorems are similar to Theorems 2 and 3 but relate to the columns of a character table instead of the rows. They are new to the reader but it can readily be checked that they are correct when applied to all of the character tables we have so far met.

| **Theorem 4** Consider any class (column) of a character table and square each of the elements in it; sum the squares and multiply the answer by the number of operations in the class. The answer is always equal to the order of the group. |
|---|

**Problem 5.5** Apply Theorem 4 to the columns of the $C_{4v}$ character table.

| **Theorem 5** Consider any two different classes (columns) of the character table. This selects two characters of each irreducible representation. Multiply these pairs of characters of the same irreducible representation together and sum the results. The answer is zero. |
|---|

*Comment:* In this case we have not made explicit allowance for the number of operations in a class. This is because multiplying by any factor which is common to all contributions to the sum would not change the final answer — it would still be zero.

**Problem 5.6** Apply Theorem 5 to at least five pairs of columns of the $C_{4v}$ character table.

| **Theorem 6** This states that a character table is always square — it has the same number of columns as it has rows. There are as many irreducible representations as there are classes of symmetry operations. |
|---|

*Comment:* Yet again, it is easy to see that this theorem holds for all the character tables that we have so far encountered.

**Problem 5.7** As indicated at the end of Section 2.4, some character tables contain complex numbers. Sometimes, authors of introductory texts attempt to protect their readers from such horrors by manipulation of the character table. The character table for the group $C_4$ taken from one such text is given below:

| $C_4$ | E | $2C_4$ | $C_2$ |
|---|---|---|---|
| A | 1 | 1 | 1 |
| B | 1 | −1 | 1 |
| E | 2 | 0 | −2 |

Show that this character table does not obey Theorems 2, 4 and 5. (The correct character table will be discussed in detail in Chapter 11 and is given in Table 11.1.)

## 5.4 THE DERIVATION OF THE $C_{4v}$ CHARACTER TABLE USING THE ORTHONORMALITY THEOREMS

We start our study of the $C_{4v}$ character table by using Theorem 6. The total number of symmetry operations in the $C_{4v}$ group is eight and we have seen that these fall into five classes. Because the table is square it follows from Theorem 6 that there are just five irreducible representations in the character table. Now, from Theorem 4, we know that the sum of squares of characters lying in the column corresponding to the identity operation is eight (the order of the group). We therefore have to find five integers, the squares of which total eight. Further, because of the nature of the identity operation, none of these integers can be negative or zero. The only set of integers which satisfies these conditions is the set 1, 1, 1, 1 and 2 ($1^2+1^2+1^2+1^2+2^2=8$). Including the totally symmetric irreducible representation (Theorem 1) we can write down the skeleton character table shown in Table 5.2, where the quantities a → p have yet to be determined.

Because an irreducible representation which describes the behaviour of a single object has characters which can only be plus or minus one* (the object always goes into itself or minus itself, never into a different object under a symmetry operation) the entries a to l in Table 5.2 all have values of either plus one or minus one.

Consider now the column corresponding to the $C_2$ rotation operation.

Table 5.2

| E | $2C_4$ | $C_2$ | $2\sigma_v$ | $2\sigma_v'$ |
|---|---|---|---|---|
| 1 | 1 | 1 | 1 | 1 |
| 1 | a | b | c | d |
| 1 | e | f | g | h |
| 1 | i | j | k | l |
| 2 | m | n | o | p |

*More strictly, are always of modulus unity.

Again, by Theorem 4, the sum of characters in this column has to equal eight and since the squares of b, f and j are each +1 it follows that $n^2$ must be 4 so that $n$ is either +2 or −2. Remembering that entries a to l when squared all give the number 1 it follows, again by Theorem 4, that because each of the $2C_4$, $2\sigma_v$ and $2\sigma_v'$ classes have two operations in them the elements m, o and p must each be equal to zero. If they had any other value the sum of squares of elements in each column when multiplied by two, the order of the class, would give a number of greater than eight. We are thus led to Table 5.3 in which all ± signs are to be regarded as independent of each other.

Table 5.3

| E | $2C_4$ | $C_2$ | $2\sigma_v$ | $2\sigma_v'$ |
|---|---|---|---|---|
| 1 | 1 | 1 | 1 | 1 |
| 1 | ±1 | ±1 | ±1 | ±1 |
| 1 | ±1 | ±1 | ±1 | ±1 |
| 1 | ±1 | ±1 | ±1 | ±1 |
| 2 | 0 | ±2 | 0 | 0 |

Consider the identity column in Table 5.3 together with one of the columns corresponding to any class of order two. In order that Theorem 5 be satisfied (i.e. zero is obtained when the products of corresponding characters are summed) the three characters listed for each class as ±1 must, in fact, contain one +1 and two −1's. Since at this point in our argument the middle three rows of Table 5.3 are identical we may arbitrarily choose two negative elements for any one class of order two. This we shall do for the $2C_4$ class. This is a convenient point at which to apply Theorem 3 simultaneously to the double degenerate irreducible representation given in Table 5.3 and the first (totally symmetric) irreducible representation. The sum of the products of characters multiplied by the number of elements in each class must be zero. This is only the case when the character for the $C_2$ class of the doubly degenerate irreducible representation is −2. We have thus completely generated this particular irreducible representation. These results are summarized in Table 5.4.

We now apply Theorem 5 to the columns in Table 5.4 headed by the E and $C_2$ operations (the E and $C_2$ classes). The sum of products of elements of the

Table 5.4

| E | $2C_4$ | $C_2$ | $2\sigma_v$ | $2\sigma_v'$ |
|---|---|---|---|---|
| 1 | 1 | 1 | 1 | 1 |
| 1 | 1 | ±1 | ±1 | ±1 |
| 1 | −1 | ±1 | ±1 | ±1 |
| 1 | −1 | ±1 | ±1 | ±1 |
| 2 | 0 | −2 | 0 | 0 |

Table 5.5

| E | $2C_4$ | $C_2$ | $2\sigma_v$ | $2\sigma_v'$ |
|---|---|---|---|---|
| 1 | 1 | 1 | 1 | 1 |
| 1 | 1 | 1 | $-1$ | $-1$ |
| 1 | $-1$ | 1 | $\pm 1$ | $\pm 1$ |
| 1 | $-1$ | 1 | $\pm 1$ | $\pm 1$ |
| 2 | 0 | $-2$ | 0 | 0 |

two classes must be zero. This can only happen if all of the characters in the $C_2$ class are $+1$. Remembering this result, consider the first two rows (irreducible representations) of Table 5.4 and apply Theorem 3. The only way in which a sum of zero can be obtained from products of elements in the two rows is if negative signs are taken for the two characters of undetermined sign in the second irreducible representation. These results are summarized in Table 5.5.

The most evident thing about Table 5.5 is that the characters associated with the third and fourth irreducible representations are, so far, indistinguishable. The application of either Theorem 3 or Theorem 5 readily shows that the four characters of undetermined sign in Table 5.5 must be either

$$1 \qquad -1 \qquad -1 \qquad 1$$

*or*

$$-1 \qquad 1 \qquad 1 \qquad -1$$

Substitution of these sets of numbers alternately into Table 5.5 readily establishes that they generate the same two irreducible representations; the alternatives merely differ in the order in which the irreducible representations are listed.

**Problem 5.8** The derivation of the $C_{4v}$ character table has been explained in some detail. It is important that each step is followed closely because this will give valuable practice in the use of the orthonormality theorems. If it has not already been done in reading this section, carefully check each step in the derivation of the $C_{4v}$ character table.

The final character table is given in Table 5.6, where we have also given the commonly adopted symbols for the irreducible representations. Note the difference between irreducible representations labelled B and those labelled A. Both are singly degenerate but the B's are antisymmetric with respect to a rotation about the axis of highest symmetry ($C_4$) whereas the A's are symmetric.

This may be contrasted with the discussion in Section 2.4, where the distinction between A's and B's in the $C_{2v}$ point group related to their behaviour under the $C_2$ rotation. The generalization is clear — for a group for which the highest rotational axis is $C_n$, A's are symmetric with respect to this operation whereas B's are antisymmetric.

**Problem 5.9** A fragment of the $C_{8v}$ character table is shown below. Complete this fragment.

| $C_{8v}$ | E | $2C_8$ | $2C_4$ | $2C_8{}^3$ | ... |
|---|---|---|---|---|---|
| $A_1$ | 1 | | | | ... |
| $A_2$ | 1 | | | | ... |
| $B_1$ | 1 | | | | ... |
| $B_2$ | 1 | | | | ... |
| . | . | . | . | . | ... |
| . | . | . | . | . | ... |

(*Hint:* within this fragment there is no distinction apparent between $A_1$ and $A_2$ or between $B_1$ and $B_2$. Consider first the behaviour under $C_8$; the other entries follow because $C_4$ and $C_8{}^3$ are multiples of $C_8$ and the characters under these operations must be consistent with that for $C_8$.) This problem illustrates yet another approach to the compilation of character tables and the sort of relationships that exist within them. A discussion which parallels that required to answer this problem is to be found at the end of Section 11.5.

**Problem 5.10** Use the second part of the hint in Problem 5.9 to explain why there are no B irreducible representations in the character table of the $C_{7v}$ group (or, indeed, any group containing a $C_n$ axis when $n$ is odd).

We are now in a better position to discuss Table 5.6 than we were when we first met it as Table 5.1. There are five aspects of it on which we wish to comment, all associated with the E irreducible representation. First is the label E itself. This is identical to the label used to describe the identity operation. Although this appears confusing, in practice it is not. This is because the contexts in which the two labels are used are always quite different; the context

Table 5.6

| $C_{4v}$ | E | $2C_4$ | $C_2$ | $2\sigma_v$ | $2\sigma_v{}'$ | |
|---|---|---|---|---|---|---|
| $A_1$ | 1 | 1 | 1 | 1 | 1 | $z, z^2, x^2+y^2$ |
| $A_2$ | 1 | 1 | 1 | −1 | −1 | |
| $B_1$ | 1 | −1 | 1 | 1 | −1 | $x^2-y^2$ |
| $B_2$ | 1 | −1 | 1 | −1 | 1 | $xy$ |
| E | 2 | 0 | −2 | 0 | 0 | $(x, y)(zx, yz)$ |

tells which is intended. In some texts, however, the ambiguity is avoided by a difference in typeface or, more simply, by the use of the label I for the *I*dentity operation.

Secondly, the occurrence of the characters 2 and 0 in this irreducible representation is something new and requires comment. The appearance of the character 2 for the identity, leave-alone, operation tells us that the E irreducible representation describes the behaviour of two objects simultaneously. The x and y axes of $BrF_5$ which we described earlier is such a pair. Another, closely related, example which we shall consider in detail shortly are the valence shell $p_x$ and $p_y$ orbitals of the bromine atom in $BrF_5$. The character 2, then, means the same as when we met it in Chapter 3 in the 'two things are left alone' sense. The same thing is not true for the 0. Previously, we obtained this character when every object under consideration moved as a result of a symmetry operation. There is another way in which it can appear. This is when each object which remains unchanged is matched by one which changes its sign. The sum of 1 and $-1$ is, of course, 0. We shall meet an example when we consider the $p_x$ and $p_y$ orbitals of the bromine atom in $BrF_5$.

Thirdly, we note the way in which we indicate the members of a set of functions, the transformations of which generate the E representation. The x and y axes are such a pair and are written (x, y) in contrast to functions which, separately and independently, provide a basis for a repesentation. The functions z and $z^2$ form a basis for the $A_1$ irreducible representation; the way that either can do this independently of the other is indicated by the way they are written: z, $z^2$.

Fourthly, this is a convenient point at which to formally introduce a piece of useful jargon (which we have already used!). Irreducible representations which describe the transformation of two objects simultaneously are said to be 'doubly degenerate' whereas those describing the transformation of one are said to be 'singly degenerate'. There are fundamental reasons for this notation but it is simplest here to note that if the objects which the irreducible representations describe are orbitals then for E irreducible representations there *must* be two orbitals with exactly the same energy. Were they not the same, the act of carrying out, for example, a $C_4$ rotation would have the effect of changing energies; the energy of an orbital would depend on whether or not we choose to do a $C_4$ operation and this clearly is ridiculous.

Fifthly, if we were to construct a group multiplication table for the $C_{4v}$ group we would find that only the characters of the various A and B irreducible representations could be substituted for their corresponding operations to give an arithmetically correct numerical multiplication table. The substitution fails for the E irreducible representation. The reason is that we should really use $2 \times 2$ matrices to describe the E irreducible representation, not a simple number. When these matrices are substituted for the corresponding characters and multiplied by the laws of matrix multiplication then a correct multiplication table is obtained. For the reader unfamiliar with matrices and their multiplication this is explained in more detail in Appendix 2. At this point we would merely make the comment that ordinary numbers may be regarded as $1 \times 1$ matrices (whereupon the laws of matrix multiplication reduce to the ordinary laws of numerical multiplication) so that those irreducible representations containing only characters of $+1$ or $-1$ may also be regarded as involving matrices.

We have now encountered two quite different methods of generating character tables, that of using the transformations of suitable basis functions and that of the use of various theorems relating to character tables. From this point on in the text we shall not systematically derive character tables, although we shall frequently comment on various aspects of those that we meet. Rather, we shall, effectively, adopt what is a more usual attitude — to simply refer to a collection of character tables such as that given in Appendix 3 and copy out the one we need.

## 5.5 THE BONDING IN THE $BrF_5$ MOLECULE

We now return to the problem of the bonding in the $BrF_5$ molecule. The discussion will be simplified by considering only $\sigma$ interactions between the fluorine and bromine atoms. Further, we shall ignore the possibility that d orbitals on the bromine may be involved in the bonding. These are reasonable simplifications but it is well to anticipate their consequences. Firstly, each fluorine atom will have six valence shell non-bonding electrons, a total of thirty. There will be peaks arising from these electrons in the photoelectron spectrum of the molecule which may make it difficult for us to test our model. Secondly, the neglect of bromine d orbitals will mean that we will find, at most, three bonding molecular orbitals responsible for the $\sigma$ bonding of the four coplanar fluorines to the bromine atom — of the bromine's four valence shell orbitals one, $4p_z$, has a node in the plane containing the four fluorines and so cannot be involved in this bonding.

We start by considering the transformational properties of the bromine valence shell orbitals, 4s, $4p_x$, $4p_y$ and $4p_z$ (for simplicity, we shall drop the prefix 4 in the following discussion). It is simple to show that the bromine s and $p_z$ orbitals separately transform as $A_1$.

**Problem 5.11** Show that the bromine s and $p_z$ orbitals do indeed transform as $A_1$. (*Hint:* the z coordinate axis is shown in Figure 5.4. Viewing the orbitals from the view of Figure 5.2 should prove helpful.)

As we have indicated earlier, $p_x$ and $p_y$ transform together as E but we have both to show that this is so and that the result is independent of the choice of orientation of x and y axes (although our demonstration of this latter point will be incomplete because we shall only consider the alternative axis sets of Figures 5.4 and 5.5).

The transformations of the $p_x$ and $p_y$ orbitals of the bromine atom are detailed in Table 5.7 for the two choices of x and y axes and this table should be worked through carefully. Note, in particular, the different behaviour of the two sets of p orbitals under the mirror plane reflections depending on the choice of x and y axis directions. Despite these differences, the character resulting from these transformations is the same for either choice. Similarly, the sum of characters generated by $p_x$ and $p_y$ is the same for all operations in any one class. The — consensus — characters generated by the $p_x$ and $p_y$ orbitals under the operations of the $C_{4v}$ point group are given at the bottom of Table 5.7. The argument that we have used does not show that the choice of x and y axes of Figure 5.6 (or any other arbitrary choice) would lead to the same set of characters as those given in Table 5.7. However, in Appendix 2 we show that this is the case. Comparison with Table 5.6 shows that the representation which we have generated using $p_x$ and $p_y$ as bases is the E irreducible representation of the $C_{4v}$ group. Because the x and y axes transform similarly to $p_x$ and $p_y$ (just drop the p's in Table 5.7 to obtain the transformation of the axes) it follows that these also transform as E, as we asserted earlier in this chapter.

We shall not attempt to specify in detail the composition of the orbital on each fluorine which is involved in the $\sigma$ bonding with the bromine. It will be a mixture of s and p orbitals, but the participation of each is not symmetry-determined and, in any case, does not affect our qualitative conclusions. For simplicity, in the diagrams this hybrid orbital will be drawn as a sphere (in contrast, they are drawn as pure p orbitals in Appendix 4). We must now consider the transformational properties of these fluorine hybrid orbitals. The axial fluorine lies on all of the symmetry elements of the $C_{4v}$ group; all of the corresponding operations turn the orbital into itself. It therefore transforms as the totally symmetric irreducible representation of the $C_{4v}$ point group ($A_1$). The four symmetry-related fluorine atoms transform as a set and form a basis for a reducible representation which must be decomposed into its irreducible components. Following the usual procedure it is readily found that this reducible representation is

| $E$ | $2C_4$ | $C_2$ | $2\sigma_v$ | $2\sigma_v'$ |
|---|---|---|---|---|
| 4 | 0 | 0 | 2 | 0 |

which has $A_1 + B_1 + E$ components. Labelling the fluorine orbitals as indicated in Figure 5.7 and proceeding as in Section 4.6 we find that the normalized form of the linear combinations of fluorine orbitals which transform as the $A_1$ and $B_1$ irreducible representations — the symmetry-adapted combinations — are:

| Symmetry species | Symmetry-adapted combinations of fluorine and orbitals |
|---|---|
| $A_1$ | $\frac{1}{2}(a + b + c + d)$ |
| $B_1$ | $\frac{1}{2}(a - b + c - d)$ |

Table 5.7 The transformations of the bromine $p_x$ and $p_y$ orbitals in $BrF_5$. The table shows the orbital obtained when each operation operates on $p_x$ and $p_y$. In brackets after each orbital is given its contribution to the aggregate character

| | E | $C_4$ (clockwise) | $C_4$ (anticlockwise) | $C_2$ | $\sigma_v(1)$ | $\sigma_v(2)$ | $\sigma_v'(1)$ | $\sigma_v'(2)$ |
|---|---|---|---|---|---|---|---|---|
| $p_x$ (Figure 5.4) becomes | $p_x(1)$ | $-p_y(0)$ | $p_y(0)$ | $-p_x(-1)$ | $-p_x(-1)$ | $p_x(1)$ | $p_y(0)$ | $-p_y(0)$ |
| $p_y$ (Figure 5.4) becomes | $p_y(1)$ | $p_x(0)$ | $-p_x(0)$ | $-p_y(-1)$ | $p_y(1)$ | $-p_y(-1)$ | $p_x(0)$ | $-p_x(0)$ |
| $p_x$ and $p_y$ together (Figure 5.4) | 2 | 0 | 0 | $-2$ | 0 | 0 | 0 | 0 |
| $p_x$ (Figure 5.5) becomes | $p_x(1)$ | $-p_y(0)$ | $p_y(0)$ | $-p_x(-1)$ | $-p_y(0)$ | $p_y(0)$ | $-p_x(-1)$ | $p_x(1)$ |
| $p_y$ (Figure 5.5) becomes | $p_y(1)$ | $p_x(0)$ | $-p_x(0)$ | $P_y(-1)$ | $-p_x(0)$ | $p_x(0)$ | $p_y(1)$ | $-p_y(-1)$ |
| $p_x$ and $p_y$ together (Figure 5.5) | 2 | 0 | 0 | $-2$ | 0 | 0 | 0 | 0 |

| $C_{4v}$ | E | $2C_4$ | $C_2$ | $2\sigma_v$ | $2\sigma_v'$ |
|---|---|---|---|---|---|
| Representation generated by $p_x$ and $p_y$ together | 2 | 0 | $-2$ | 0 | 0 |

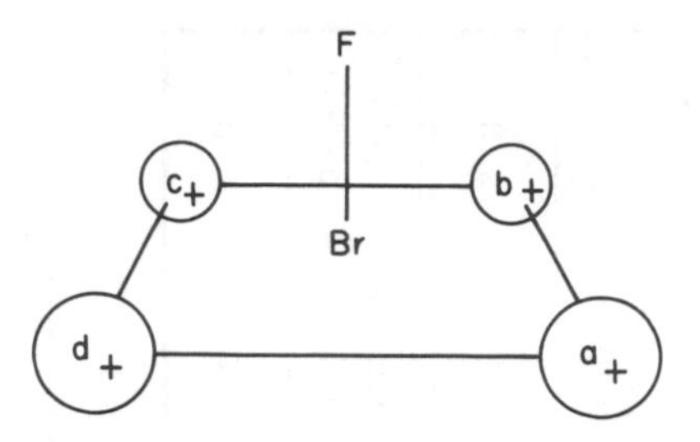


Figure 5.7 The labelling and phases of the $\sigma$ hybrid orbitals of the four coplanar fluorines. For simplicity these hybrid orbitals are drawn as circles

**Problem 5.12** Generate the above reducible representation, reduce it to its irreducible components and use the projection operator method to generate the $A_1$ and $B_1$ functions given above. (*Warning:* this is not a trivial problem. It will be necessary to slightly modify the method previously used to reduce a reducible representation (Section 4.6) because there is more than one operation in each class.) (*Hint:* use Theorems 3 and 4 of Section 5.3. The answer to this problem is given at the end of the chapter.)

The generation of the two combinations which transform as E is a more difficult problem and we shall consider it in some detail. As for the $A_1$ and $B_1$ combinations, we use the projection operator method described in the last chapter. Using the fluorine hybrid orbital labelled a in Figure 5.7 as the generating element and mirror plane operations labelled as in Figure 5.3 we find the following transformations:

| Operation | E | $C_4$ (clockwise) | $C_4$ (anticlockwise) | $C_2$ | $\sigma_v(1)$ | $\sigma_v(2)$ | $\sigma_v'(1)$ | $\sigma_v'(2)$ |
|---|---|---|---|---|---|---|---|---|
| Under the operation orbital a becomes | a | d | b | c | c | a | b | d |
| The E irreducible represent-ation | 2 | 0 | 0 | −2 | 0 | 0 | 0 | 0 |
| Multiply | 2a | 0 | 0 | −2c | 0 | 0 | 0 | 0 |

The sum of products is 2a – 2c which gives, on normalization, $(1/\sqrt{2})(a - c)$ as one of the E functions. Note in particular that we have listed each operation separately, so that when there is a class comprising two symmetry operations, underneath *each* operation we give the corresponding character of the E irreducible representation. The wavefunction which we obtain, $(1/\sqrt{2})(a-c)$, is one member of the pair of functions transforming as E. How may we obtain its partner?

In the function which we have generated there is no contribution from the orbitals b and d; we might reasonably expect them to contribute to the orbital we are seeking. If we consider the transformations of either of these orbitals and follow the projection operator technique used above it is a simple task to show that the function $(1/\sqrt{2})(b - d)$ is generated. This is the second function for which we have been looking.

**Problem 5.13** Generate the second E function, $(1/\sqrt{2})(b - d)$.

The functions $(1/\sqrt{2})(a - c)$ and $(1/\sqrt{2})(b - d)$ transform as a pair under the E irreducible representation of the $C_{4v}$ group and are shown in Figure 5.8. Our method of obtaining the second member of the degenerate pair was based on an enlightened guess. In the next chapter we will meet a more systematic approach. One final word on these combinations. Whereas in the $A_1$ combination adjacent fluorine $\sigma$ orbitals have the same phase — and so any interaction between them is bonding — in the $B_1$ they are always of opposite phase. Any interaction between them is antibonding. In each of the E combinations there is no interaction between adjacent $\sigma$ orbitals (only trans orbitals appear in any one combination). This argument leads us to expect a relative energy order $A_1 < E < B_1$, an order which will be reflected in the way we construct the molecular orbital energy level diagram for $BrF_5$ (Figure 5.12).

We are now almost ready to consider the interaction between bromine and fluorine $\sigma$ orbitals. Firstly, however, we recall that the s and $p_z$ bromine

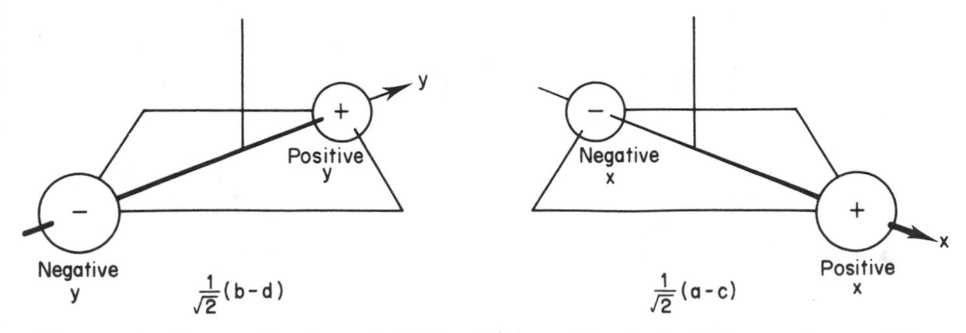


Figure 5.8 The two fluorine $\sigma$ hybrid orbital combinations which transform as E in the $C_{4v}$ point group and which are derived in the text

orbitals separately transform as the $A_1$ irreducible representation. We have encountered analogous situations previously and we have then combined the corresponding orbitals to obtain two mixed orbitals of the form $(1/\sqrt{2})(s \pm p_z)$. We shall also follow this simplifying procedure in the present case. One of the mixed orbitals is orientated in a way that should give good overlap with the $\sigma$ orbital of the apical fluorine atom, an orbital which, as we have seen, is also of $A_1$ symmetry; these steps are shown schematically in Figures 5.9 and 5.10. The remaining $s - p_z$ mixed orbital on the bromine atom points in the direction indicated by dotted lines in Figure 5.1 and therefore might be regarded as the orbital which (in the electron pair repulsion model) causes the distortion of the molecule that we noted at the beginning of this chapter. Unfortunately, as we shall see, reality is perhaps more complicated than this. The complication arises from the fact that there is also a combination of $\sigma$ orbitals from the planar fluorines which has $A_1$ symmetry. Clearly, it can interact with $A_1$ orbitals of the bromine. However, one of these latter $A_1$ orbitals is $p_z$ and this has a nodal plane in which the fluorines lie (in our simplified geometry of coplanar bromine and fluorines). The fluorine $A_1$ combination interactions will therefore be almost entirely with the bromine *s* orbital. Correspondingly, it seems probable that the bromine *s* orbital involvement with the axial fluorine and in the basal lone pair will be rather less than we have assumed.

The only other valence orbitals on the bromine atom are $p_x$ and $p_y$ which, together, are of E symmetry. They are shown in Figure 5.4(b). They interact with the two fluorine $\sigma$ orbital combinations of E symmetry which we generated earlier and which are shown in Figure 5.8. Provided the p orbitals

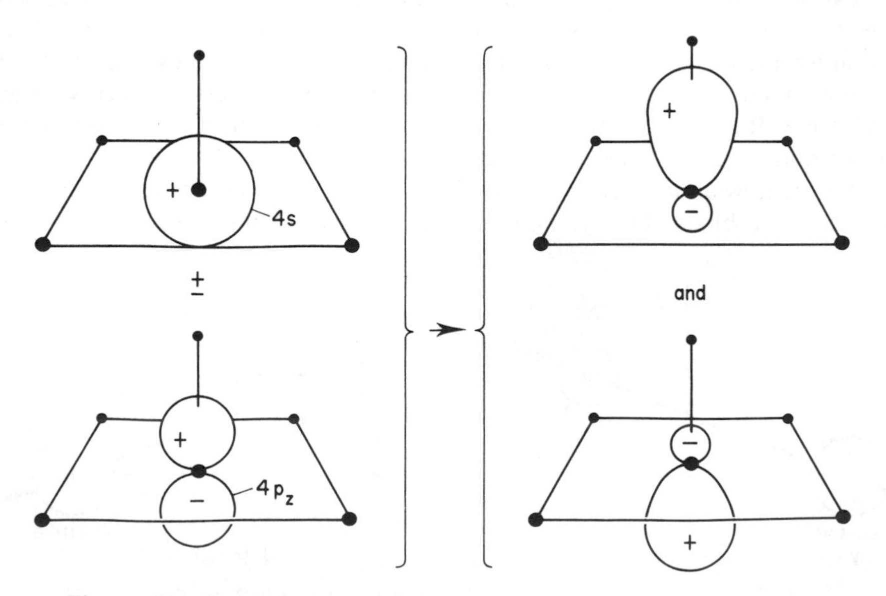


Figure 5.9 Hybrid orbitals derived from bromine atomic orbitals of $A_1$ symmetry in $BrF_5$

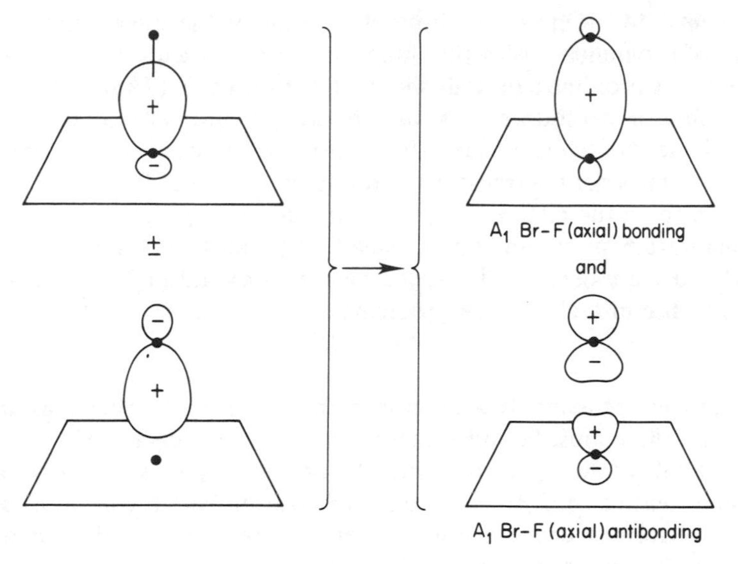


Figure 5.10 Bonding of the axial fluorine to the bromine atom in $BrF_5$

and the fluorine $\sigma$ orbital combinations are properly chosen — and this means that the same set of coordinate axes is used for each — then each p orbital only interacts with one $\sigma$ orbital combination. The orbitals in Figures 5.4(b) and 5.8 are properly chosen and the result of their interactions gives what we have represented as sum (bonding) and difference (antibonding) combinations, in Figure 5.11.

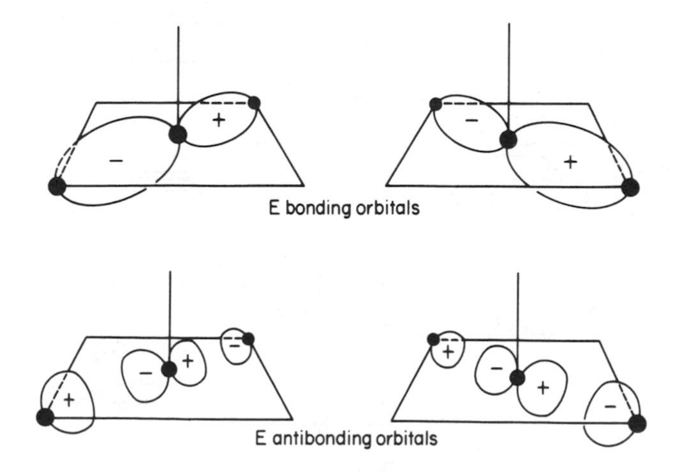


Figure 5.11 The two bonding orbitals and the two antibonding orbitals of E symmetry in $BrF_5$ arising from interactions of the four coplanar fluorines with the central bromine atom

**Problem 5.14** Repeat the above discussion of the interactions of orbitals of E symmetry using the choice of coordinate axes shown in Figure 5.5 (use the bromine p orbitals shown in Figure 5.5b). (*Hint:* the projection operator method which we used in the text automatically selected the coordinate axis choice shown in Figure 5.8 because of our (implicit) choice to consider the transformation of an individual fluorine $\sigma$ orbital. We can force the method to give combinations appropriate to the axes of Figure 5.5 by considering, instead, the transformation of pairs of neighbouring $\sigma$ orbitals. Thus, the pairs (a + b) and (a + d) are suitable pairs to use in tackling this problem.)

The above discussion is summarized in Figure 5.12, where, as we have recognized, there must be some uncertainty about the details of the positions of the orbitals of $A_1$ symmetry. Into the orbital pattern shown in this figure we have to place a total of twelve electrons (seven from the bromine and one from each fluorine $\sigma$ orbital). It will be remembered that this diagram does not include the fluorine non-bonding electrons.

There are some interesting consequences of Figure 5.12 and of our discussion of the bonding of the $BrF_5$ molecule. We have suggested that one molecular orbital (of $A_1$ symmetry) is primarily involved in the bonding of the axial fluorine to the bromine. If this view is correct then this bromine-fluorine bond involves two electrons. In contrast, in the picture we have developed, the strongly bonding molecular orbitals involving the planar fluorine atoms are of E symmetry, although there will be a smaller contribution from an $A_1$ orbital. That is, the four coplanar fluorine atoms are bonded to the central bromine

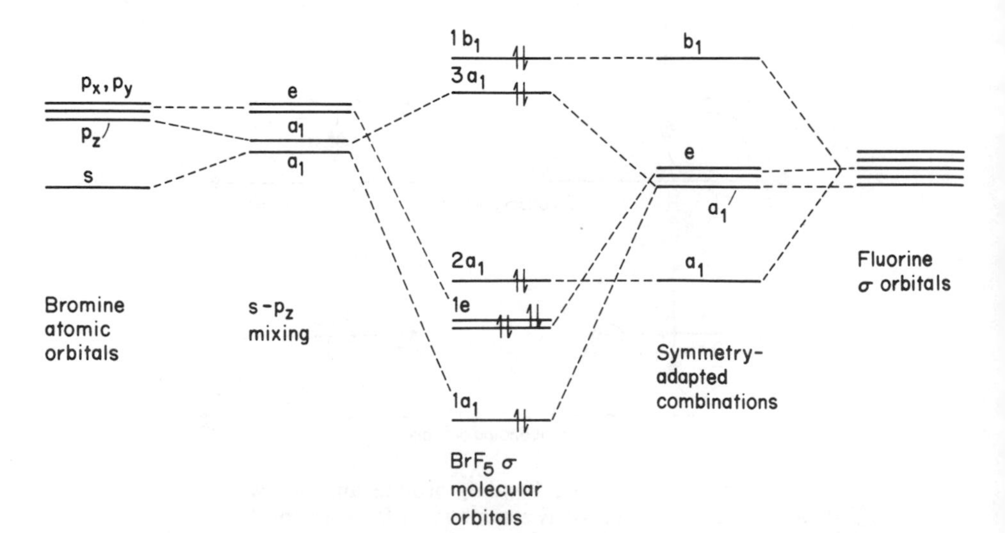


Figure 5.12 Schematic molecular orbital energy level diagram for $BrF_5$

atom by little more than two molecular orbitals. If this conclusion is correct, it suggests that the bonding between the bromine and each of the four coplanar fluorines is rather weak, a view supported by the fact that bromine pentafluoride is an extremely powerful fluorinating agent. Some further support for this difference between axial and planar fluorines is to be found in molecular structure determinations which show Br—F bond lengths of 1.68 Å (axial) and 1.78 Å (equatorial).

What light can photoelectron spectroscopy or theoretical calculations shed on this problem? Unfortunately, rather little. The photoelectron spectrum of $BrF_5$ has been reported[1] but its interpretation is ambiguous. It seems that the highest lying orbital, at 13.2 eV, corresponds to ionization from the lone pair of electrons on the bromine atom. Then come at least three peaks corresponding to ionization of the fluorine 2p non-bonding electrons. Peaks at *ca.* 17 and 20 eV probably correspond to ionization from Br—F $\sigma$-bonding molecular orbitals. There do not appear to have been any detailed calculations on $BrF_5$, but some exist for $ClF_5$[2] (for which there are no photoionization data!). Here the bonding pattern obtained is similar to ours and indicates that there are Cl—F $\sigma$-bonding molecular orbitals of $A_1$ symmetry at *ca.* 16 and 22 eV and of E symmetry at *ca.* 22 eV. The fluorine $\sigma$ orbital non-bonding combination of $B_1$ symmetry is at *ca.* 16 eV.

**Problem 5.15** In our discussion of the bonding in $BrF_5$ we have ignored the presence of 4d orbitals on the bromine. The justification for this is that the 4d orbitals of the isolated bromine atom are so large and diffuse that they cannot overlap effectively with a valence shell atomic orbital of any other atom unless there is something which causes them to contract. Something may exist in $BrF_5$ because the polarity of each of the Br—F bonds will be such that there will presumably be a significant build-up of positive charge on the bromine atom. One effect of this would be to lower the energy and decrease the size of the bromine 4d orbitals and thus perhaps make them available for chemical bonding. If this occurs we should have included the d orbitals in our discussion. This is an attractive hypothesis but is one that is extremely difficult to test, even by

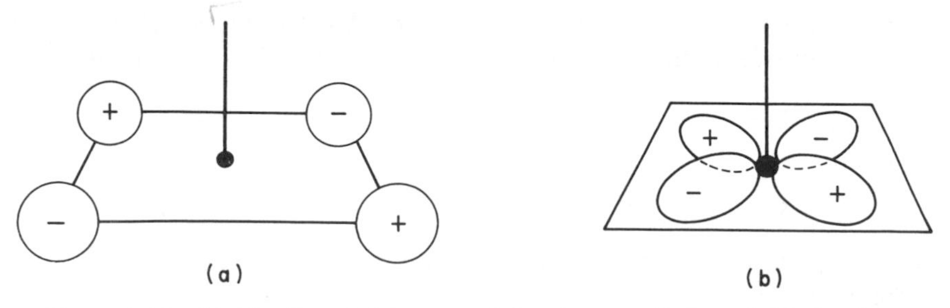


Figure 5.13 (a) The $B_1$ combination of fluorine $\sigma$ orbitals and (b) the $d_{x^2-y^2}$ ($B_1$) orbital of bromine

detailed calculations. Show that the bromine $d_{z^2}$ orbital has $A_1$ symmetry and that its $d_{x^2-y^2}$ orbital has $B_1$ symmetry.

This latter orbital is shown (contracted!) in Figure 5.13 together with the $B_1$ combination of fluorine $\sigma$ orbitals with which it potentially interacts. Show that both of the labels $B_1$ and $d_{x^2-y^2}$ would have to be changed if the coordinate axis set of Figure 5.5 were used in the discussion.

## 5.6 SUMMARY

In this chapter we have found that operations may be divided into classes (page 91) and that when some classes contain more than one operation the character table contains a degenerate representation (page 90). The presence of a degenerate representation enabled the orthonormality relationships to be presented in a more general form (page 95). The procedures previously used to reduce a reducible representation has to be modified in the more general case (Problem 5.12) although the projection operator technique is unchanged (page 106). Application of these techniques to the problem of the bonding in $BrF_5$ suggests both a reason for its existence — polar Br—F bonds possibly enabling participation of bromine d orbitals in the bonding (Problem 5.14) — and for its reactivity — the four coplanar fluorines are not strongly bonded (page 110).

## ANSWER TO PROBLEM 5.12

An object is only turned into itself under a symmetry operation when it is located on the corresponding symmetry element. Since the four fluorine $\sigma$ orbitals lie on the $\sigma_v$ mirror planes, two on each plane, it is only under the E and $2\sigma_v$ operations that non-zero characters are generated by their transformation and we have

| E | $2C_4$ | $C_2$ | $2\sigma_v$ | $2\sigma_v'$ |
|---|---|---|---|---|
| 4 | 0 | 0 | 2 | 0 |

To reduce this it is convenient to first multiply by the number of operations in each class (see Theorems 3 and 4 of Section 5.3). That is we multiply by

| | | | | |
|---|---|---|---|---|
| 1 | 2 | 1 | 2 | 2 |

and obtain

| | | | | |
|---|---|---|---|---|
| 4 | 0 | 0 | 4 | 0 |

As an example we test for the $B_1$ irreducible representation; i.e. we multiply by the characters

| | | | | |
|---|---|---|---|---|
| 1 | −1 | 1 | 1 | −1 |

to obtain

| 4 | 0 | 0 | 4 | 0 |
|---|---|---|---|---|

Addition and division by the order of the group (8) gives $8/8 = 1$, i.e. the $B_1$ irreducible representation occurs just once in the reducible representation. Explicit $A_1$ and $B_1$ functions are obtained by the method described in the text immediately following Problem 5.12 with the substitution of the characters of the $A_1$ or $B_1$ irreducible representation for those of the E irreducible representation.

## REFERENCES

1. R. L. De Kock, B. R. Higginson and D. R. Lloyd, *Faraday Disc. Chem. Soc.*, **54**, 84 (1972).
2. M. B. Hall, *Faraday Disc. Chem. Soc.*, **54**, 97 (1972).

CHAPTER 6

# *The electronic structure of the ammonia molecule*

In the first chapter of this book we discussed in outline four different qualitative descriptions of the bonding in the ammonia molecule. We have now developed our symmetry-based approach to a point at which we may reconsider this problem in more detail. At the same time we shall meet again a problem which we encountered in the last chapter — that of the choice of directions of x and y axes. The form in which this problem appears is one which will lead us to a general solution, a solution which will enable us to tackle molecules of high symmetry such as those which will be the subject of Chapter 7.

## 6.1 THE SYMMETRY OF THE AMMONIA MOLECULE

The structure of the ammonia molecule is given in Figure 6.1 which also shows the symmetry elements possessed by this molecule. The axis of highest rotational symmetry (which we shall therefore take as the z axis) is a $C_3$ rotation axis and has associated with it clockwise and anticlockwise rotation operations (which, to help the reader remember this distinction, we shall call $C_3^+$ and $C_3^-$; in a more general notation they would be called $C_3$ and $C_3^2$). In addition, there are three mirror planes, each of which contains the threefold axis (and are vertical with respect to it, i.e. they are $\sigma_v$ mirror planes), with one hydrogen atom lying in each mirror plane. The symmetry operations which turn the ammonia molecule into itself are, therefore,

$$E \qquad C_3^+ \qquad C_3^- \qquad \sigma_v(1) \qquad \sigma_v(2) \qquad \sigma_v(3)$$

This group is called the $C_{3v}$ point group, the shorthand symbol $C_{3v}$ indicating the coexistence of the $C_3$ axis and the *v*ertical mirror planes.

**Problem 6.1** Show that this set of operations comprises a group. (*Hint:* it will be found helpful to refer back to Problem 4.3. The group multiplication table for the $C_{3v}$ group is given in Table 8.2.)

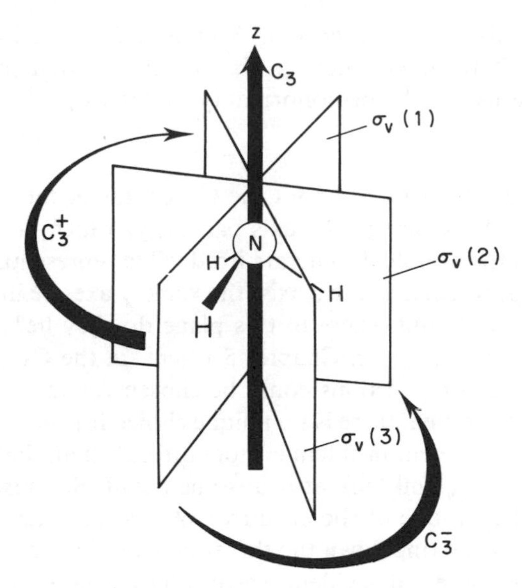


Figure 6.1 The symmetry elements of the ammonia molecule

The class structure of the symmetry operations of the $C_{3v}$ group is suggested from the similarities between the various operations and is

$$E \qquad 2C_3 \qquad 3\sigma_v$$

Alternatively, the formal methods described in Appendix 1 may be used to deduce this class structure (the $2C_3$ class is given as a worked example in Appendix 1).

The character table of the $C_{3v}$ point group is given in Table 6.1. In this table we have followed Table 5.6 and moved a step further towards the usual presentation of character tables. On the right-hand side of the table are shown functions which are a basis for a particular representation. Thus the z axis, chosen following the convention which locates it along the $C_3$ axis, transforms as $A_1$, and the x and y axes, together, are a basis for the E irreducible representation.

Table 6.1

| $C_{3v}$ | E | $2C_3$ | $3\sigma_v$ | |
|---|---|---|---|---|
| $A_1$ | 1 | 1 | 1 | $z, z^2, x^2+y^2$ |
| $A_2$ | 1 | 1 | $-1$ | |
| E | 2 | $-1$ | 0 | $(x, y)(zx, yz)$ $(xy, x^2-y^2)$ |

**Problem 6.2** Use the theorems of Section 5.3 to derive Table 6.1. (*Note:* this is a relatively short problem but one that gives excellent practice in the use of the orthonormality theorems.)

There is one particular point about the $C_{3v}$ character table which we must discuss in detail. This concerns the axis pair (x, y) which, as shown in Table 6.1, transform as the doubly degenerate irreducible representation E. Because they must be perpendicular to the z axis, the x and y axes lie in a plane perpendicular to the $C_3$ axis. But where in this plane do they lie? This problem is similar to one which we met in Chapter 5 where, in the $C_{4v}$ point group, we found that a variety of directions could be chosen for the x and y axes. So, too, in the present problem there is no unique choice for the x and y axis directions. However, the present problem is more difficult than that encountered in the $C_{4v}$ case and so we will look at it in some detail. Suppose we choose the x axis so that it lies in one of the $\sigma_v$ mirror planes, as is done in Figure 6.2, which shows a view looking down the threefold axis. Two related problems at once arise. Firstly, there is no evident reason why we should select a particular mirror plane rather than one of the others. Secondly, the choice which we have made for the x axis means that the y axis is forced to be quite differently orientated in space. However, having made a choice we will stay with it and move on to the next problem, that of the effect of a $C_3$ rotation operation on these x and y axes, shown in Figure 6.3. It is seen from this figure that the x axis is rotated so that it lies along one of the directions which we could have originally taken as the x axis but chose not to. Similarly, the y axis is rotated into a direction appropriate to this second choice of x axis. In Figure 6.3 we have indicated the alternative x and y axes by primes (so that x is rotated into x′ and y into y′). This is a quite new situation. So far in this book symmetry operations have turned objects into themselves or interchanged them. Here, a symmetry operation has generated something which did not previously exist, or so it seems. The truth is that the x′ and y′ axes did previously exist — they

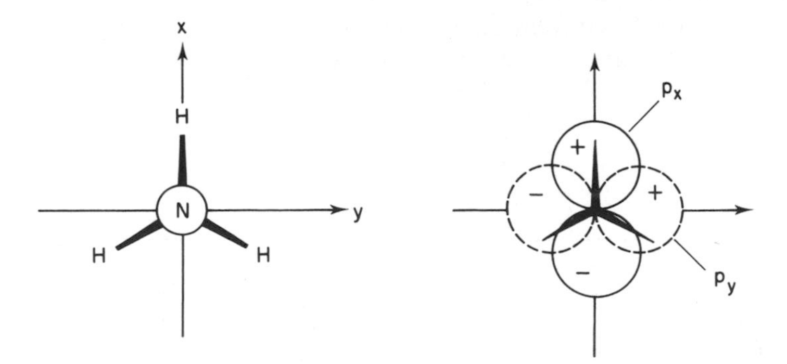


Figure 6.2 The choice of direction of x and y axes discussed in the text and consequent orientation of the $p_x$ and $p_y$ orbitals

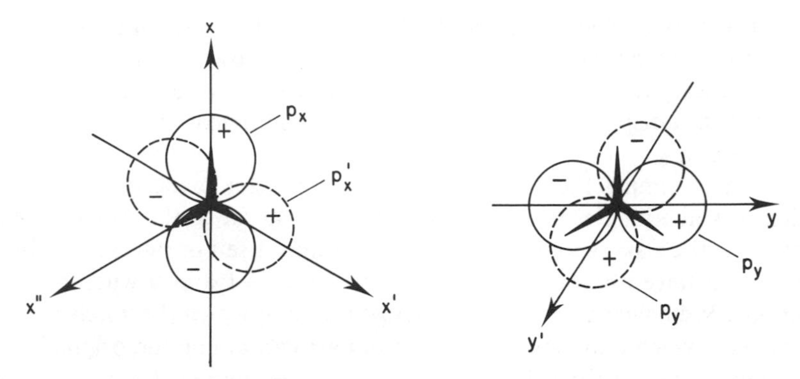


Figure 6.3 The x and y axes of Figure 6.2 together with an alternative set (x′ and y′) produced by a $C_3$ rotation of x and y. In both cases the corresponding p orbitals are also shown. The x″ axis will be referred to later in the text (following Problem 6.6)

were just not revealed. A little thought shows that this must be the case. As is clear from Table 6.1, the $C_3$ rotation acting on the x, y axis pair, and which converts them into the x′, y′ axis pair, is associated with a character of $-1$ (this is the character of the E irreducible representation under $C_3$ rotations). In some way or other the x′, y′ set is $-1$ times the x, y set. How?

We tackle this problem by investigating the relationship between two axis sets (x, y) and (x′, y′) related by a rotation of an angle $\alpha$ (later we shall take $\alpha = 120^\circ$, as appropriate to the $C_{3v}$ point group). If an object were to start at the origin of coordinates in Figure 6.4 and be displaced along the x′ axis it is

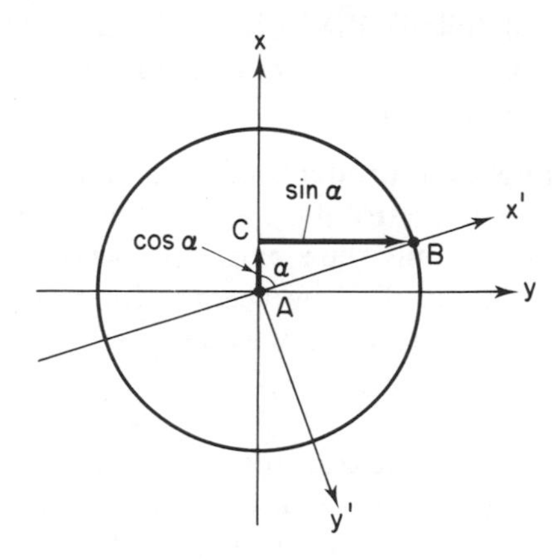


Figure 6.4 In this figure the circle is taken to be of unit radius. It follows that the unit displacement AB is the sum of the displacements AC and CB which, respectively, have magnitudes of cos $\alpha$ and sin $\alpha$

evident that this displacement could, alternatively, be represented as a sum of displacements along the original x and y axes. As shown in this figure, for an angle $\alpha$ relating the x and x′ axes, a unit displacement along the x′ axis is equivalent to a displacement of $\cos\alpha$ along x combined with a displacement of $\sin\alpha$ along y.

When determining the contribution to a character made by the transformation of something such as an x axis we have so far in this book asked the question 'is the x axis turned into itself, into minus itself or into something different', and have associated the characters of 1, $-1$ and 0 with these three situations. We have now encountered a situation in which the x axis is rotated into an axis which may be described as in part containing the original x axis. Accordingly, we must modify our question to the simpler, but more general, form, 'To what extent is the old axis contained in the new?' As is evident from Figure 6.4 and the discussion above, the numerical answer to this question is $\cos\alpha$, where $\alpha$ is the angle of rotation. An axis which is left unchanged by a rotation corresponds to $\alpha = 0$, so $\cos\alpha = 1$ — the character we have associated with this situation. Similarly, for a rotation of 180°, $\cos\alpha = -1$. For $\alpha = 90°$ (when the x is rotated so that it becomes the y axis), $\cos\alpha = 0$.

> When an axis is rotated by an angle $\alpha$ by a symmetry operation its contribution to the character for that operation is $\cos\alpha$.

*Comment:* This statement holds for axes; for products of axes it has to be modified. Thus, because

x contributes $\cos\alpha$
it follows that $x^2$ contributes $\cos^2\alpha$
and $x^3$ contributes $\cos^3\alpha$
and so on.

Note that this rule applies to products of axes which are perpendicular to the axis of rotation. Thus, if the rotation axis is the z axis then the function xz will vary as $\cos\alpha$ because only the x axis is perpendicular to the rotation axis — the z axis is left unchanged. However, the function xy will vary as $\cos^2\alpha$ because both x and y separately vary as $\cos\alpha$.

**Problem 6.3** Show that the transformation of $x^2$ under a rotation of $\alpha$ about the z axis is given by the factor $\cos^2\alpha$. (*Hint:* it is sufficient to check that this relationship holds for particular values of $\alpha$; $\alpha = 0°, 90°, 180°, 270°$ and 360° are particularly convenient.)

**Problem 6.4**
(a) In the $C_{5v}$ point group a pair of functions transforming as the doubly degenerate irreducible representation $E_1$ have a character of $2\cos 72°$

under a $C_5$ rotation. Suggest a pair of functions which might form a basis for this irreducible representation.

(b) Repeat this problem for the $E_2$ irreducible representation, for which the character is 2 cos 144°. Solution of this problem requires a small extension of the argument developed above.

Solutions to both these problems are given in Appendix 3.

Returning to to the case of the $C_{3v}$ point group, we conclude that the x and y axes each make a contribution of $\cos 120° = -\frac{1}{2}$ to the character under the $C_3$ rotation operations. The sum of these two, $-1$, is, indeed, the character of the E irreducible representation under this operation.

It was in Section 3.2 that we first mentioned that two quantities, such as axes or orbitals, can be mixed by the operations of a group. We are now able to understand just what this means. The effect of a $C_3$ rotation on the original x and y axes is to rotate them to give new axes, each of which is a mixture of the original axes. In such cases the contribution that each axis makes to the character is always fractional.

Everything that we have said about the x and y axes also holds for the $2p_x$ and $2p_y$ orbitals of the nitrogen atom in ammonia because the transformations of $2p_x$(N) is isomorphous to that of x, as is that of $2p_y$(N) to y (Figure 6.3). We have already anticipated this parallel by taking the molecular x and y axes to pass through the nitrogen atom — although they could be chosen to pass through any point along the $C_3$ axis — so that the above argument could be used as a basis for a discussion of the bonding in the ammonia molecule without need to redefine axes.

## 6.2 THE BONDING IN THE AMMONIA MOLECULE

We now complete this chapter by a discussion of the bonding in the ammonia molecule. As is evident from the discussion above, the transformation properties of the nitrogen valence shell 2p orbitals follow those of the coordinate axes given in Table 6.1, the $2p_z$ has $A_1$ symmetry and $2p_x$ and $2p_y$, as a pair, have E symmetry; it is a trivial exercise to show that the nitrogen 2s orbital is totally symmetric (this orbital is spherical and lies on all symmetry elements; it therefore transforms as $A_1$). The transformation of the three hydrogen 1s orbitals under the operations of the group gives rise to the reducible representation

| E | $2C_3$ | $3\sigma_v$ |
|---|---|---|
| 3 | 0 | 1 |

which is a linear sum of the irreducible representations $A_1$ and E.

**Problem 6.5** Reduce the above reducible representation into its irreducible components. (*Hint:* if this problem is found to be difficult refer to Problem 5.12 — which is similar and to which an explicit solution has been given at the end of Chapter 5.)

Much of the discussion so far in this chapter has developed from the fact that the operation of rotation by 120° has the effect of mixing functions which provide a basis for the E irreducible representation. This same problem reappears again when we try to determine the symmetry-adapted combinations of hydrogen 1s orbitals in the ammonia molecule which transform as the E irreducible representation, a problem which we now consider in detail. Labelling the hydrogen 1s orbitals as indicated in Figure 6.5 and considering the transformation of the orbital labelled a we find that the six symmetry operations of the group lead to the following transformations:

| E | $C_3^+$ | $C_3^-$ | $\sigma_v(1)$ | $\sigma_v(2)$ | $\sigma_v(3)$ |
|---|---|---|---|---|---|
| a | b | c | a | c | b |

Application of the projection operator technique described in Section 4.6 shows the $A_1$ function to be

$$\frac{1}{\sqrt{3}}(a + b + c)$$

There is no difficulty in obtaining one of the E functions. The steps involved are shown in Table 6.2 and lead to the function

$$\frac{1}{\sqrt{6}}(2a - b - c)$$

A problem arises when we try to obtain the second E function. We met a similar problem in Section 5.5 when discussing the four $\sigma$ orbitals of the

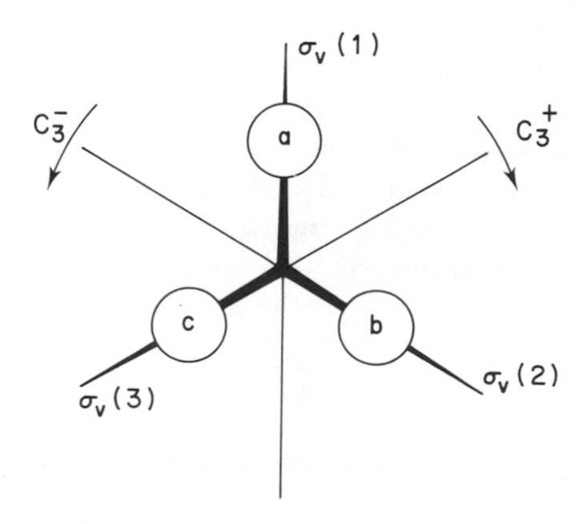


Figure 6.5 The labels used in the text for the hydrogen 1s orbitals of the ammonia molecule

Table 6.2

| Operation | E | $C_3^+$ | $C_3^-$ | $\sigma_v(1)$ | $\sigma_v(2)$ | $\sigma_v(3)$ |
|---|---|---|---|---|---|---|
| The operation turns a into: | a | b | c | a | c | b |
| Characters of the E irreducible representation | 2 | −1 | −1 | 0 | 0 | 0 |
| Multiply | 2a | −b | −c | 0 | 0 | 0 |
| Sum | | 2a − b − c | | | | |
| Normalize | | $\frac{1}{\sqrt{6}}(2a - b - c)$ | | | | |

coplanar fluorine atoms in $BrF_5$. In that case the problem was relatively simple because the first E function contained contributions from only two of the $\sigma$ orbitals; the projection operator technique applied to one of the other $\sigma$ orbitals immediately gave the second E function. There is no such simple solution to the present problem; all three hydrogen 1s orbitals appear in the E function that we have generated, albeit with unequal weight. Following the procedure described in Chapter 5 we could use the transformations of either the hydrogen 1s orbital b or c as a basis for the projection operation method — but which? If we use b then we obtain the combination

$$\frac{1}{\sqrt{6}}(2b - c - a)$$

whilst if we use c we obtain

$$\frac{1}{\sqrt{6}}(2c - a - b)$$

**Problem 6.6** Show, by constructing tables analogous to Table 6.2, that the transformation of the hydrogen 1s orbitals b and c lead to the E functions

$$\frac{1}{\sqrt{6}}(2b - c - a) \quad \text{and} \quad \frac{1}{\sqrt{6}}(2c - a - b)$$

respectively.

We have, apparently, obtained three quite different functions transforming as E — yet we know that, for a doubly degenerate irreducible representation, there can only be two. As indicated above, this problem is closely related to the three possible choices for the x axis that we discussed earlier and the solution to the problem is also similar. What we have, in fact, done in using a, b,

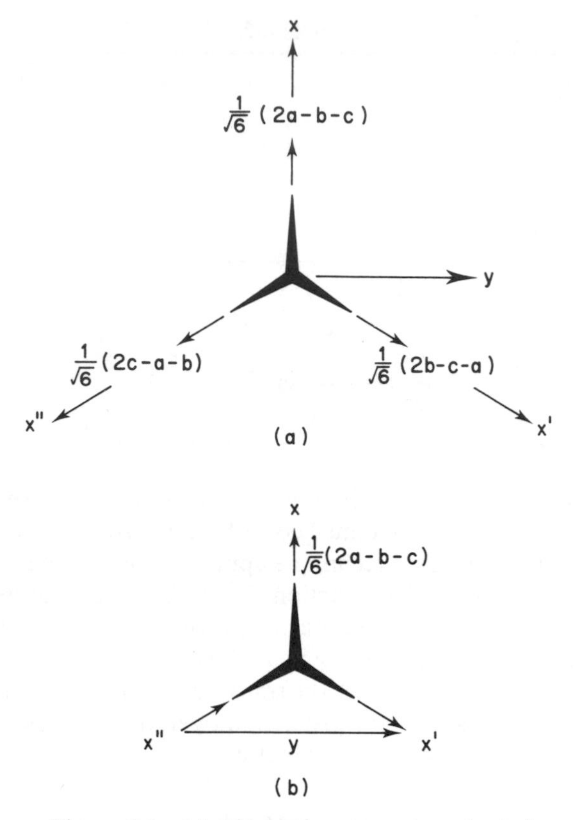


Figure 6.6 (a) Alternative symmetry-adapted combinations of hydrogen 1s orbitals in $NH_3$ corresponding to the axes x, x′ and x″ of Figure 6.3. Just as one of these axes has to be selected so, too, does one of the three symmetry-adapted combinations. (b) The (vector) sum of displacements along −x″ and x′ is a displacement along y

and c separately to generate an E function is to generate functions appropriate to the x, x′ and x″ axes, respectively, of Figure 6.3, as indicated in Figure 6.6(a). The functions corresponding to the x′ and x″ axes, like these axes themselves, are mixtures of the functions appropriate to the original x and y axes. It is the latter that we are seeking.

The simplest way of obtaining the second E function is to exploit the fact that if the first function corresponds to the x axis then the second corresponds to the y axis. This vector (axis-like) property of our functions is indicated by the arrows in Figure 6.6(a). If, as shown in Figure 6.6(b), we reverse the direction of the vector pointing in the direction x″ and add it to that pointing in the x′ direction we obtain a vector pointing in the y direction. We now have to repeat these steps using functions rather than vectors.

The negative of the function associated with x″ is

$$-\frac{1}{\sqrt{6}}(2c - a - b)$$

and adding it to the function associated with x′

$$\frac{1}{\sqrt{6}}(2b - c - a)$$

gives

$$\frac{1}{\sqrt{6}}(-2c + a + b + 2b - c - a) = \frac{1}{\sqrt{6}}(3b - 3c)$$

That is, the second E function is of the form

$$(b - c)$$

or, normalized,

$$\frac{1}{\sqrt{2}}(b - c)$$

**Problem 6.7** The sum of vectors pointing along x′ and x″ of Figure 6.6(a) is the negative of a vector pointing along x. Show that an analogous statement is true for the corresponding E functions.

**Problem 6.8** The fact that the two E functions which we have just obtained have quite different mathematical forms tends to be received with suspicion. Show that their forms are such that the orbitals a, b and c make equal total contributions to the E functions. (*Hint:* sum the squares of coefficients in the normalized E functions.)

The symmetry-adapted combinations of hydrogen 1s orbitals which we have just generated are shown in Figure 6.7, together with the nitrogen orbitals of the same symmetry with which they interact. Notice, in this figure, that the unequal contribution of a, b and c to each of the two symmetry-adapted combinations of E symmetry is reflected in the diagramatic representation of the orbitals. It will also be noted that in Figure 6.7 we have followed the approximate procedure of taking a combination of nitrogen 2s and $2p_z$ orbitals as the nitrogen orbital which interacts with the $A_1$ combination of hydrogen 1s orbitals. A schematic molecular energy level diagram of ammonia is shown in Figure 6.8. There are eight valence electrons which have to be allocated to these orbitals (five from the nitrogen and one from each of the three hydrogens) and they are accommodated in the lowest molecular orbitals of $A_1$ and E symmetry, all of which are M—H bonding and the second $A_1$ orbital,

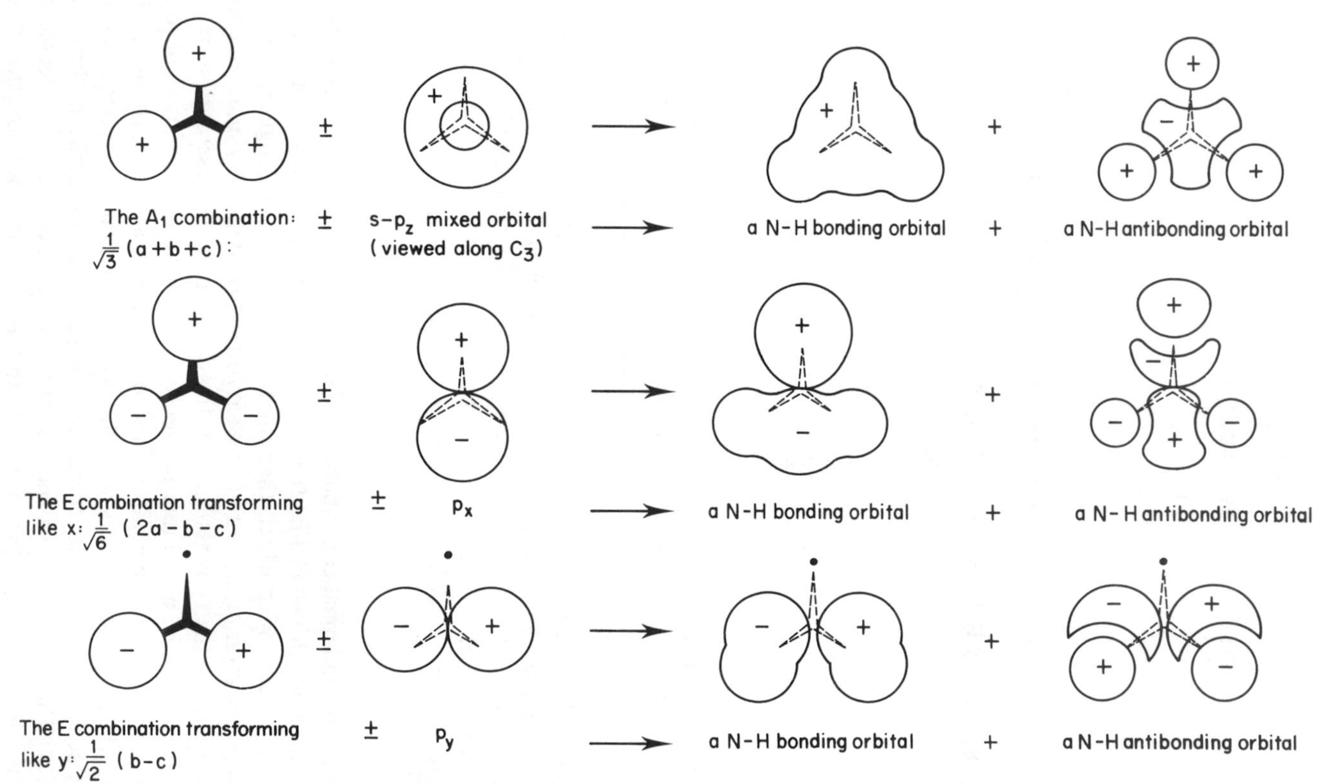


Figure 6.7 The bonding and antibonding molecular orbitals of $A_1$ and E symmetry in $NH_3$

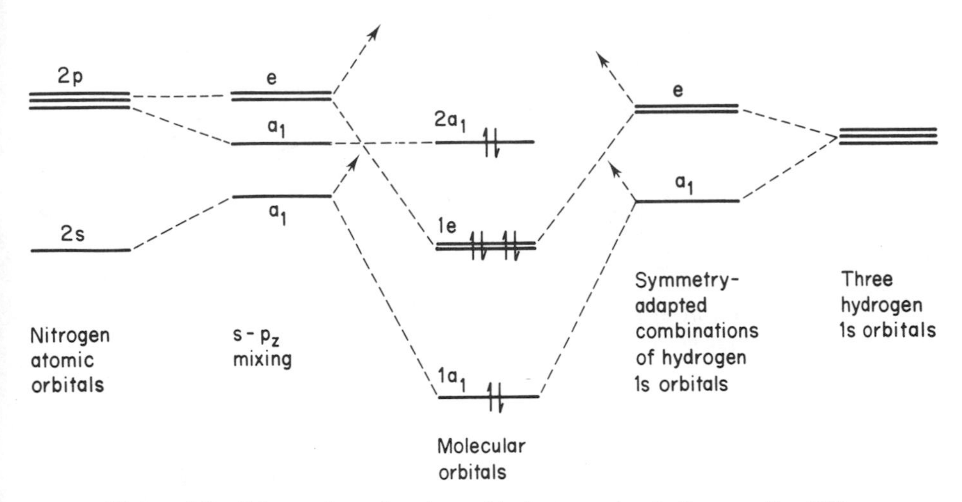


Figure 6.8 Schematic molecular orbital energy level diagram for $NH_3$

which is, essentially, the nitrogen lone pair orbital. These qualitative conclusions are to be compared with the results of detailed calculations and with the results of photoelectron spectroscopy.

Calculations[1] show that the $A_1$ orbitals of the ammonia molecule have energies of *ca.* −11.6 and −31.3 eV and that of E symmetry *ca.* −17.1 eV. Of these, the more stable of the $A_1$ orbitals has, as we have come to expect, a major contribution from the nitrogen 2s atomic orbital. These data are in general agreement with the photoelectron spectroscopic results[2] which give energies of *ca.* 10.2, 27.0 and 15.0 eV for these levels respectively. Again, we obtain the encouraging result that our model is in good qualitative agreement both with detailed calculations and with experiment.

Ammonia is a molecule for which, like the water molecule, it is a simple matter to describe the angles at which the various contributions to the molecular bonding maximize. Using arguments entirely similar to those of Chapter 3 for the water molecule, we conclude that the bonding interactions of $A_1$ symmetry involving the nitrogen $2p_z$ orbital (Figure 6.9a) maximizes at small bond angles, whereas the interactions between orbitals of E symmetry maximize for the planar molecule (Figure 6.9b). These angular variations are conveniently summarized in a Walsh diagram, just as for the water molecule in Chapter 3. This diagram is given, qualitatively, in Figure 6.10.

**Problem 6.9** Check that Figure 6.10 does, indeed, summarize the discussion of the above paragraph. What can you conclude about the non-bonding nature of the highest $A_1$ orbital from this diagram?

As we indicated in Chapter 1, calculations show that the total bonding in

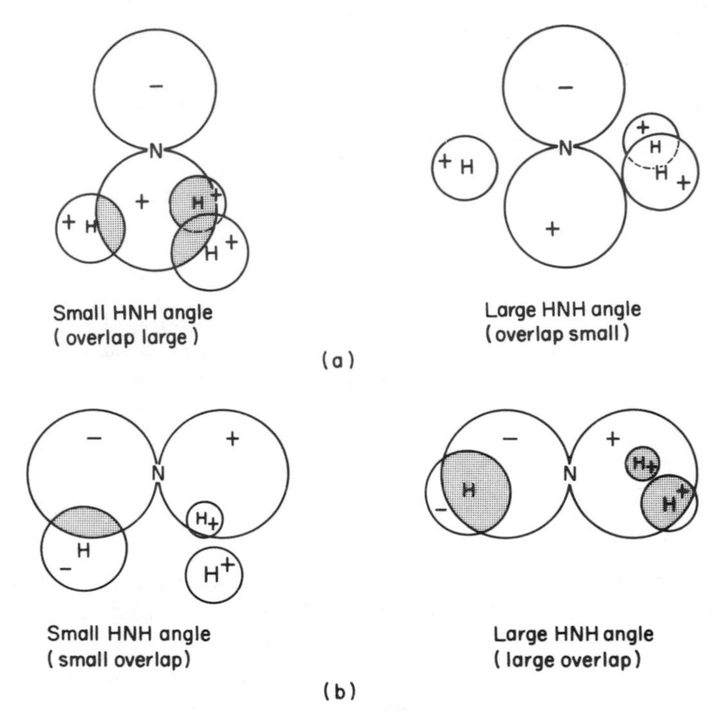


Figure 6.9 (a) The overlap between the $A_1$ symmetry-adapted hydrogen 1s combination and the nitrogen $2p_z$ orbital decreases as the HNH bond angle increases. (b) The overlap between an E symmetry-adapted hydrogen 1s combination and a nitrogen $2p_x$ orbital increases with the HNH bond angle

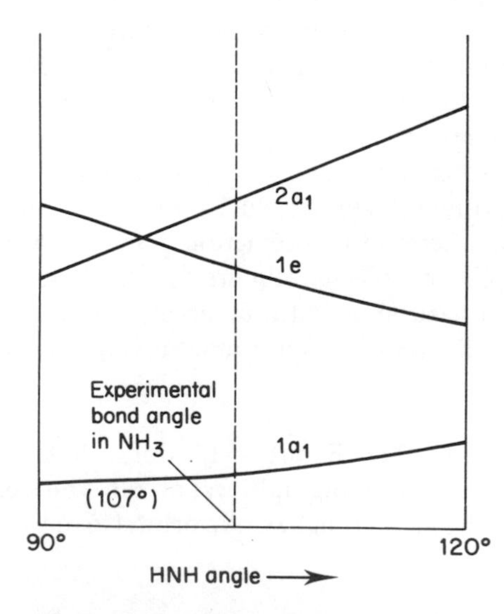


Figure 6.10 Walsh diagram for $NH_3$

the ammonia molecule is a maximum when the molecule is planar so we conclude that the E interactions dominate. However, this argument neglects the effects of repulsive forces on the molecular geometry and, as we stated in Chapter 1, the same calculations show that it is these that lead (only just) to the molecule adopting a pyramidal shape. At the observed bond angle there are both $A_1$ and E contributions to the bonding. Were we to remove an electron from the highest $A_1$ molecular orbital, which contains a large nitrogen $2p_z$ contribution and which, despite our simplified discussion, makes a contribution to the molecular bonding, we might expect that a more nearly planar molecule would result. Experiment, indeed, indicates that in its ground state $NH_3^+$ is a planar molecule.

Although we have not discussed the bonding in planar $NH_3$ or $NH_3^+$ it is, nonetheless, of interest to consider a related planar species. This is the molecule trisilylamine, $(SiH_3)_3N$. In this molecule the $Si_3—N$ framework is planar (unlike the $C_3—N$ skeleton in trimethylamine $(CH_3)_3N$, which is pyramidal like ammonia). The question of why trisilylamine should be planar has been widely discussed and is frequently associated with Si—N $\pi$ bonding in which an empty 3d orbital of silicon accepts electrons from the lone pair on nitrogen (which, in this geometry, occupy a pure 2p orbital) (Figure 6.11). Our discussion has indicated that the planarity of this molecule could arise if the delicate balance between bonding and repulsive forces which we found for ammonia — and which appears to occur for many such molecules — is such as to favour the planar form of trisilylamine. This argument, of course, does not require the existence of any Si—N $\pi$ bonding. It could be, of course, that the presence of a small amount of $\pi$ bonding is decisive in tipping a delicate balance. Equally, such an interaction might be important, not for any, small, $\pi$-bonding stabilization which results but because the resulting more diffuse electron distribution leads to a reduction in the destabilization

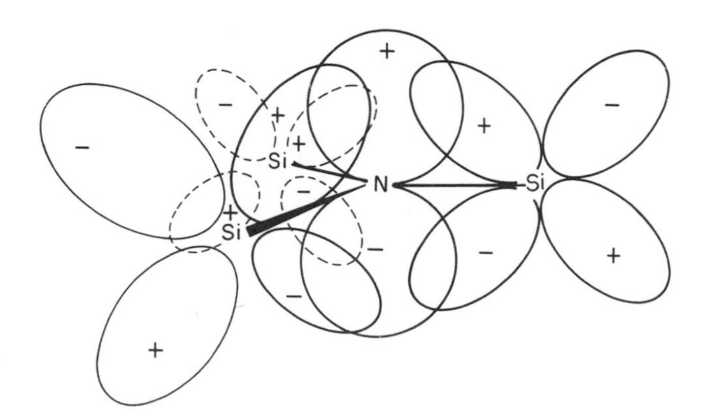


Figure 6.11 Postulated $p_\pi / d_\pi$ bonding in the planar molecule trisilylamine, $N(SiH_3)_3$

resulting from electron repulsion. However, it is important to recognize that the observed planar geometry of trisilylamine does not of itself prove the existence of significant d–p $\pi$ bonding in this molecule.

## 6.3 SUMMARY

In this chapter we have discussed the problem of the transformation of functions which form the basis for a degenerate reducible representation but which appear to be differently oriented with respect to the symmetry elements and may, indeed, have different mathematical forms (page 123). Despite these superficial differences, the fact that they are mixed or interchanged by some operations of the group is sufficient to ensure their ultimate equivalence (page 116).

## REFERENCES

1. C. D. Ritchie and H. F. King, *J. Chem. Phys.*, **47**, 564 (1967).
2. A. W. Potts and W. C. Price, *Proc. Roy. Soc.*, **326**, 181 (1972).

CHAPTER 7

# *The electronic structures of some cubic molecules*

The methods which we have developed so far in this book will be exploited to the full in the present chapter, where we shall consider in detail the electronic structure of the octahedral molecule $SF_6$. This is quite a large molecule but its symmetry is also considerable — enough to enable us to consider not only the bonding between sulphur and fluorine but also the non-bonding electrons on the fluorines. There are several shortcuts which can be used in symmetry arguments and the present discussion will enable us to introduce several of them.

Throughout the book we have met new symmetry operations in each chapter. This is also true for the present chapter; the operations we meet complete the list of the types encountered in point groups and so we shall include a general review of point group classification. This will prepare the way for the next chapter, in which we look in some detail at the relationships between point groups.

Although the molecule $SF_6$ is the major subject of this chapter we shall extend the discussion to include transition metal complexes of octahedral symmetry. Both tetrahedral and octahedral molecules have x, y and z axes equivalent to each other and the chapter also contains a discussion of both methane and of tetrahedral transition metal complexes.

In Figure 7.1 we show a cube, an octahedron and a tetrahedron. An octahedron is closely related to a cube. If the mid-points of faces of a cube are joined together the figure that is generated is an octahedron. The octahedron has eight faces but what is of more concern to us is the fact that it has six apices — corners — because when these apices are occupied by six atoms around an atom at the centre of the figure an octahedral molecule results (Figure 7.2).

In the majority of octahedral $ML_6$ compounds the central atom is a metal ion whilst the surrounding atoms or ions are usually those of an electronegative element and are called ligands. Such species are referred to as 'octahedral complexes' and are of major importance. Although it is not convenient to start our discussion with such molecules we shall look at them in more detail later in the chapter.

A tetrahedron (Figure 7.1) is also derived from a cube, as we recognized in

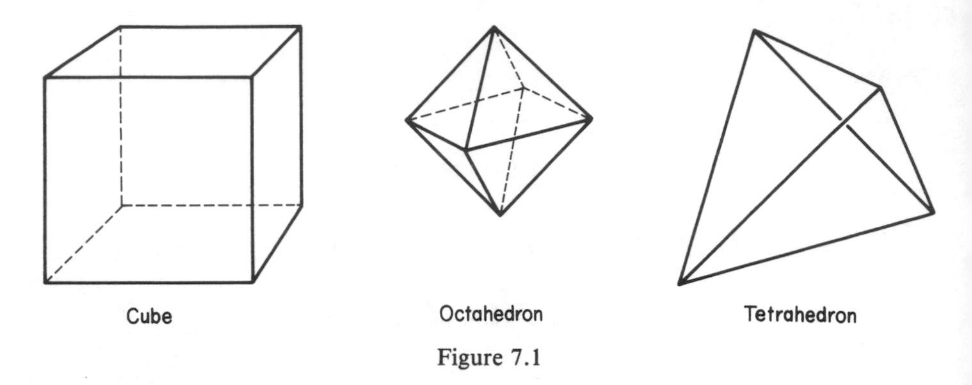


Figure 7.1

Chapter 1 (see Figure 1.4 and the discussion in Section 1.2.4). The fact that both the octahedron and tetrahedron are related to the cube means that it is possible to give a common discussion of the electronic structure of octahedral and tetrahedral transition metal complexes. In the present book we shall not embark on this discussion although we shall indicate the starting point. First, however, we must look at the symmetry of the octahedron in more detail.

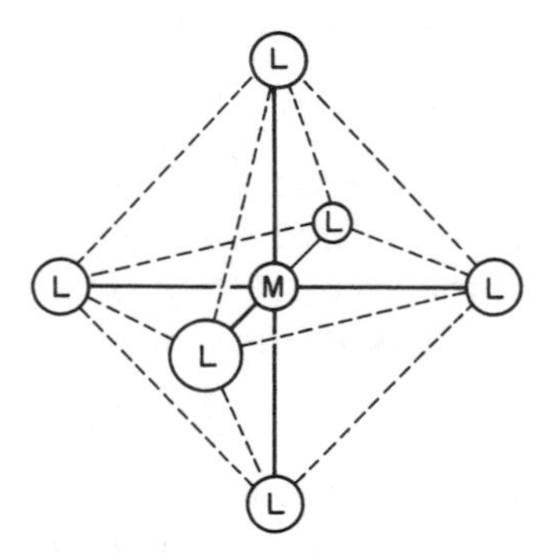


Figure 7.2 An octahedral molecule $ML_6$

## 7.1 SYMMETRY OPERATIONS OF THE OCTAHEDRON

In Figure 7.3(a) we show those pure rotational symmetry operations which turn an octahedral $ML_6$ molecule into itself. The octahedron contains three fourfold rotation axes and, of necessity, three coincident twofold rotation axes. There are also six twofold axes which are quite distinct from those that are coincident with the fourfold axes. Figure 7.3(a) shows a rather bewildering array of symmetry axes but there is a simple way of reducing the complexity; this is by identifying symmetry elements with geometrical features. Thus, each $C_3$ axis passes through the mid-points of a pair of equilateral triangular faces on opposite sides of the octahedron. There are eight faces and so four pairs of opposite faces. It follows that there are four different $C_3$ axes. Similarly,

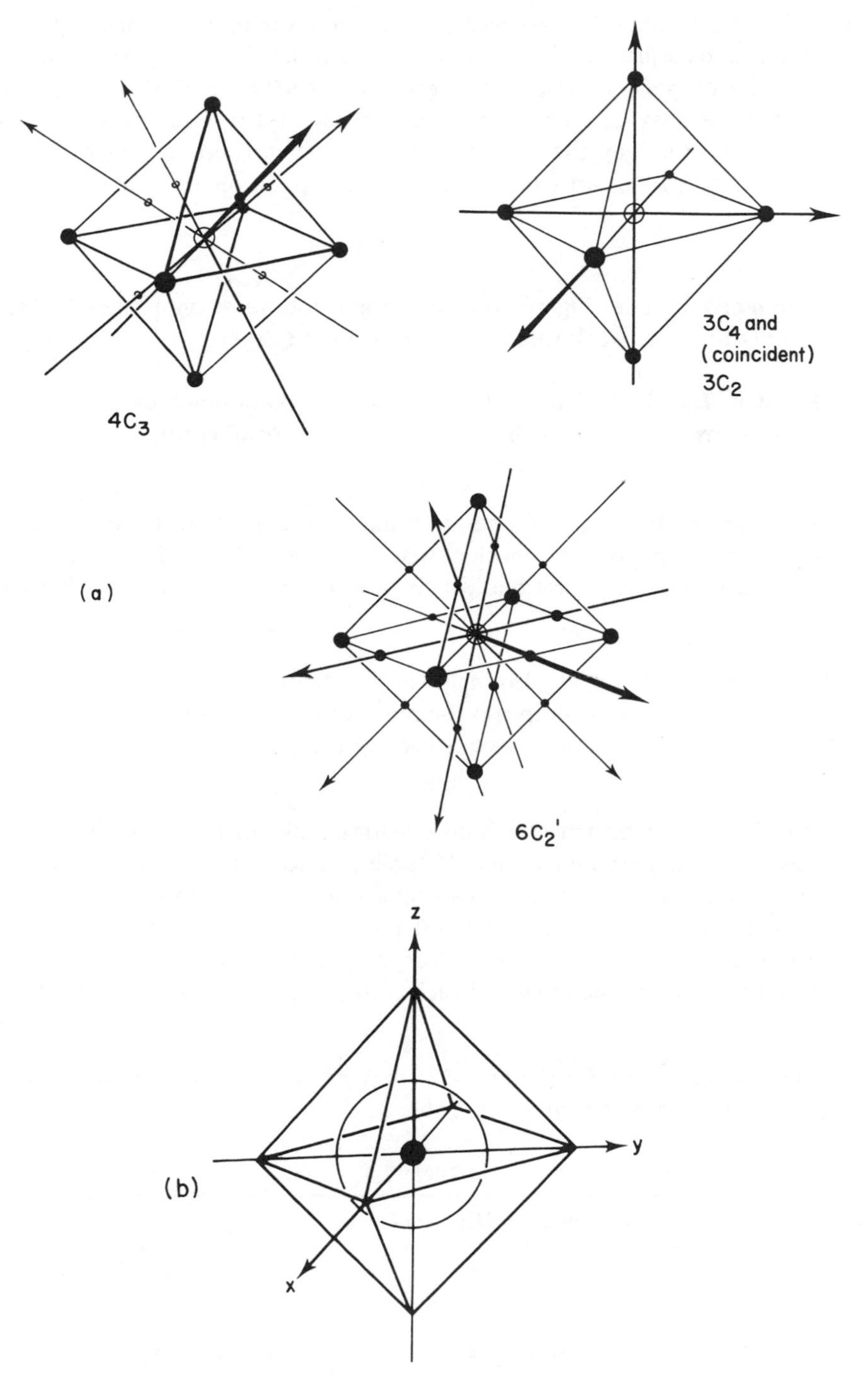


Figure 7.3 (a) Rotational symmetry elements of an octahedron. (b) The conventional choice of coordinate axes for an octahedron

the $C_4$ and coincident $C_2$ axes pass through opposite pairs of apices; there are six apices and so just three $C_4$'s and $C_2$'s. The other $C_2'$ axes pass through the mid-points of pairs of opposite edges. Because the octahedron has twelve edges there are six $C_2'$ axes. The fact that each set of axes forms a class is evident from the way that members of each set are interchanged by other operations of the group (for instance, a $C_3$ operation interchanges the $C_4$ axes).

**Problem 7.1** Use Figure 7.1 to obtain the rotational axes of an octahedron (i.e. work through the above argument).

**Problem 7.2** Use Figure 7.1 to obtain the rotational axes of a cube. Compare your answer with that found for an octahedron.

For each of the $C_3$ and $C_4$ rotation axes there are two distinct symmetry operations — those of rotation clockwise and anticlockwise. It follows that the rotational symmetry operations which turn an $ML_6$ molecule into itself are

$$E, \quad 8C_3, \quad 6C_4, \quad 3C_2 \quad \text{and} \quad 6C_2'$$

where we have, of course, also included the identity operation. This group of twenty-four operations comprises the point group O. The fact that it is a complete group may be shown by constructing the group multiplication table.

**Problem 7.3** Construct the multiplication table for the group O. (*Note:* this means constructing a $24 \times 24$ table and so will take some time.) A good model (perhaps made of cardboard) is almost essential. Follow the transformations of a general point (i.e. one not lying on a symmetry element). (*Hint:* it is invariably true that each operation appears once, and once only, in each row and each column of the multiplication table.)

The character table for the group O may be derived using the theorems we met in Chapter 5 and is given in Table 7.1.

Table 7.1

| O | E | $8C_3$ | $6C_4$ | $3C_2$ | $6C_2'$ | |
|---|---|---|---|---|---|---|
| $A_1$ | 1 | 1 | 1 | 1 | 1 | $x^2+y^2+z^2$ |
| $A_2$ | 1 | 1 | $-1$ | 1 | $-1$ | |
| E | 2 | $-1$ | 0 | 2 | 0 | $\left[\frac{1}{\sqrt{3}}(2z^2-x^2-y^2),\ x^2-y^2\right]$ |
| $T_1$ | 3 | 0 | 1 | $-1$ | $-1$ | $(x, y, z)$ |
| $T_2$ | 3 | 0 | $-1$ | $-1$ | 1 | $(xy, yz, zx)$ |

**Problem 7.4** Derive the character table for the group O using the theorems of Section 5.3. (*Hint:* it may help to look again at the solution to Problem 6.2.)

There are several aspects of Table 7.1 which call for comment. For the first time we encounter triply degenerate irreducible representations; they are labelled T. Their existence was implied earlier in this chapter when we commented that 'octahedral molecules have x, y and z axes equivalent to each other'. These axes provide the basis for either a reducible or an irreducible representation. In the event, it is irreducible and, as indicated by the basis functions given to the right-hand side of Table 7.1, they actually form a basis for the $T_1$ irreducible representation.

**Problem 7.5** Show that the x, y and z axes, as a set, form a basis for the $T_1$ irreducible representation of the point group O. (*Hint:* take the Cartesian axes to coincide with the $C_4$ axes as in Figure 7.3(b). For each class of operation select that operation which makes the transformation simple to follow.)

The answer to this problem will be detailed — in an equivalent form — at the beginning of Section 7.2.

On the right-hand side of Table 7.1 are shown more basis functions than we have previously met. The reason is that when we discuss transition metal complexes later in this chapter we shall need to know how the d orbitals of the transition metal transform. Table 7.1 shows that the d orbitals $d_{xy}$, $d_{yz}$ and $d_{zx}$ are degenerate and transform as $T_2$ whilst $d_{z^2}$ (or, more accurately, $d_{(1/\sqrt{3})(2z^2-x^2-y^2)}$) and $d_{x^2-y^2}$ are degenerate and transform as E. The function $x^2+y^2+z^2$, which like an s orbital has spherical symmetry, transforms as $A_1$.

It is evident from Figures 7.1 and 7.2 that an octahedron — and a cube — contain symmetry elements in addition to the rotations that we have discussed so far. It contains a centre of symmetry, i, $\sigma_h$ and $\sigma_d$ mirror planes, and some rotation–reflection axes which are denoted $S_n$. The octahedron contains $S_4$ and $S_6$ rotation–reflection axes. Such elements are not easy to fully appreciate and we shall look at them in detail shortly. All are shown in Figure 7.4. Of these, the i and $\sigma_h$ (a mirror plane horizontal with respect to an axis of highest symmetry, here $C_4$) have been met in Chapter 4. The $\sigma_d$ mirror plane is something new. Mirror planes that bisect the angle between a pair of twofold axes are called $\sigma_d$ mirror planes, the suffix 'd' being the first letter of the word *d*ihedral (the same word which gives its initial letter to groups such as $D_2$, $D_{2h}$ and $D_{3h}$, groups which have, respectively, two, two and three twofold axes perpendicular to the axis of highest symmetry). In the octahedron there are six $\sigma_d$ mirror planes. Although they, indeed, bisect the angles between the $C_2$ axes it is easier to count them by noting that each $\sigma_d$ mirror plane cuts opposite

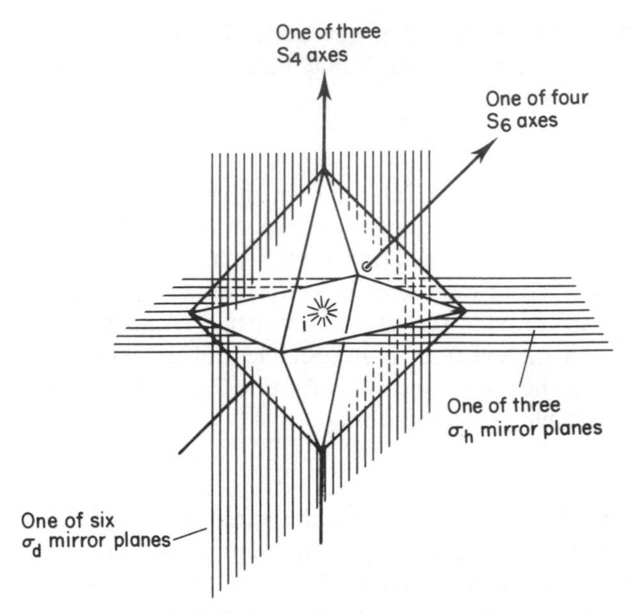


Figure 7.4 Some of the symmetry elements associated with improper rotation operations of an octahedron

edges of the octahedron, just like the $C_2'$ axes. There are six such pairs of edges and so six $\sigma_d$ mirror planes.

**Problem 7.6** Start with the definition of $\sigma_d$ mirror planes as those that bisect the angle between pairs of $C_2$ axes, and thus show that there are six $\sigma_d$ mirror planes. (*Hint:* how many pairs of $C_2$ axes are there?)

Note that the mirror planes which we have labelled $\sigma_h$ bisect the angles between pairs of $C_2'$ axes. These mirror planes could have been labelled as $\sigma_d'$. However, convention dictates that the label $\sigma_h$ takes precedence over $\sigma_d$ whenever both are applicable.

Operations such as $S_6$ and $S_4$ are interesting because, as we shall see, they are two-part operations, conventionally taken as a rotation part and a reflection part. Hence they are called rotation–reflection operations. We have seen that the cube and octahedron have the same rotational symmetry (Problem 7.2) and they also have the same additional operations. It follows that both have $S_6$ and $S_4$ axes. The $S_4$ axes are easier to see for the cube and are illustrated in Figure 7.5(a). As this figure shows, the operation consists of a rotation by 90° (clockwise and anticlockwise rotations being associated with different $S_4$ operations) followed by reflection in a mirror plane perpendicular to the axis about which the 90° rotation was made. It is clear that this opera-

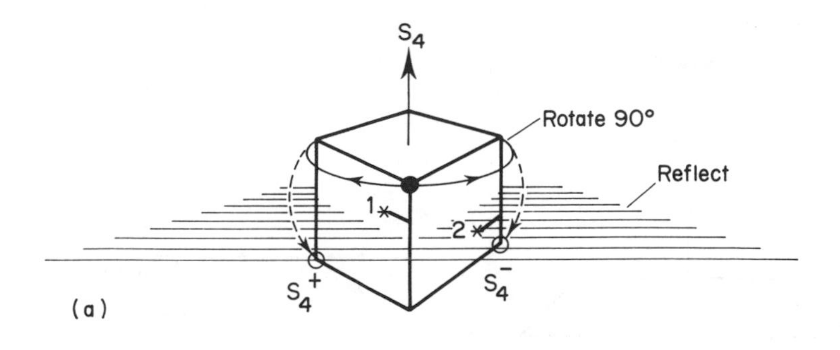


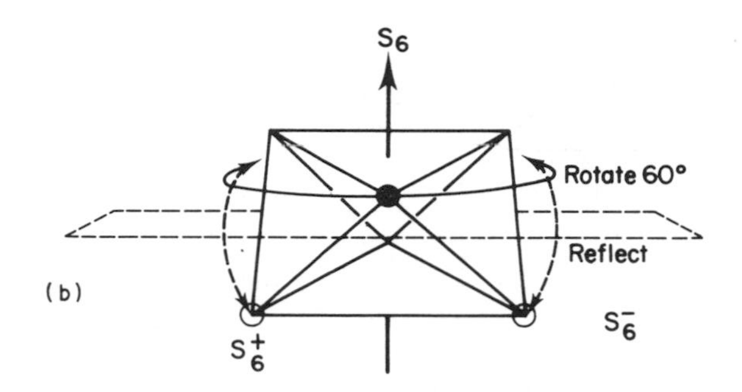


Figure 7.5 (a) An $S_4$ symmetry operation. (b) An $S_6$ symmetry operation

tion interconnects corners of the cube; what is not so clear is that it is necessary — for the pairs of corners connected by the $S_4$ operations in Figure 7.5(a) are also connected by $C_2$ operations (the $C_2$ axes emerging through mid-points of the cube faces on the right- and left-hand sides of Figure 7.5a). The difference between the $S_4$ and $C_2$ operations is shown by the stars in Figure 7.5(a). The star labelled 1 moves to the position occupied by star 2 under the $S_4^-$ operation, but these two points are not interconnected by a $C_2$ rotation. The $S_6$ operation (rotate by 60° and then reflect in a perpendicular mirror plane) is most readily seen for an octahedron standing on a face and is illustrated in Figure 7.5(b). In the case of the $S_4$ operations both the 90° rotation and reflection have a separate existence as $C_4$ and $\sigma_h$ operations respectively. In the case of the $S_6$ operations the rotation and reflection do not exist in their own right as symmetry operations in the $O_h$ group.

The $S_n$ operations seem rather strange because the operation involves two operations, $C_n$ and $\sigma_h$, which may or may not have an independent existence. This apparently paradoxical situation may be made more acceptable by returning for the moment to the $C_{2v}$ point group, discussed in Chapter 2. In Figure 7.6 we show the water molecule and the $C_2$ operation which interrelates the two hydrogen atoms. As is seen from this figure, completely equivalent to this single $C_2$ operation is the combined operation of inversion through *any* point

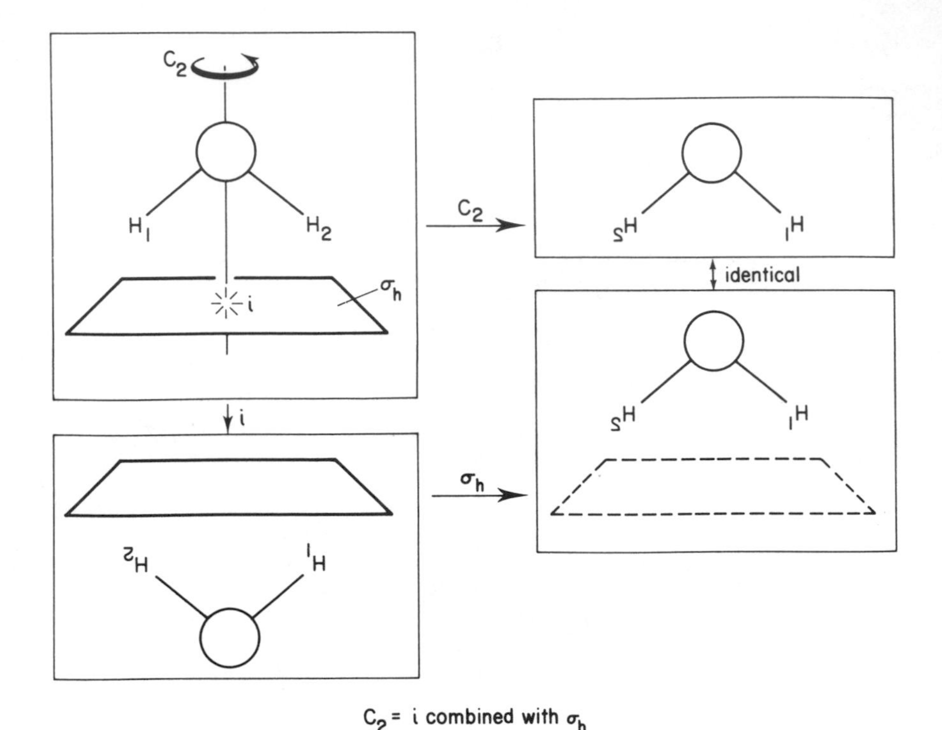


Figure 7.6 The two water molecules at the right-hand side of this diagram are the same, showing that the $C_2$ operation of the $H_2O$ molecule (shown at the top) is equivalent to an inversion at some point along this axis followed by reflection in a $\sigma_h$ mirror plane containing this inversion centre (shown at the left and bottom)

along the $C_2$ axis followed by reflection in a mirror plane perpendicular to the $C_2$ axis and containing the inversion centre. Neither the operations i or $\sigma_h$ (or their infinity of counterparts) are operations of the $C_{2v}$ point group, yet their combination is.

**Problem 7.7** Determine what combinations of two independently non-existent operations are equivalent to the (real) $\sigma_v$ and $\sigma_v'$ operations of the $C_{2v}$ point group. (*Hint:* the three operations $C_2$, $\sigma$ and i form a complementary trio. Any one can be represented in terms of the other two, provided that the other two are correctly oriented.)

It is necessary to count the $S_4$ and $S_6$ operations. The number of each follows from their correspondence with $C_4$ and $C_3$ operations; there are six $S_4$ and eight $S_6$.

**Problem 7.8** Show that the operation $S_4$ carried out twice is equivalent to $C_2$:

$$S_4^2 = C_2$$

that $S_6$ carried out twice is equivalent to $C_3$:

$$S_6^2 = C_3$$

and that $S_6$ carried out thrice is equivalent to i:

$$S_6^3 = i$$

We conclude that the complete list of symmetry operations of the octahedron is

$$E \quad 8C_3 \quad 6C_4 \quad 3C_2 \quad 6C_2' \quad i \quad 8S_6 \quad 6S_4 \quad 3\sigma_h \quad 6\sigma_d$$

The shorthand symbol for this set of operations is $O_h$ (pronounced 'ooh-aiche'). The character table of the $O_h$ group is given in Table 7.2. With some considerable effort a group multiplication table for the $O_h$ group may be constructed (it is a $48 \times 48$ table); the character table may be derived using the methods of Section 5.3 but fortunately an easier method exists. This arises from the fact that the $O_h$ group is the direct product of the groups O and $C_i$ (the group containing E and i). This means that the character table of Table 7.2 is the direct product of Table 7.1 (the character table for O) and Table 4.4 (the character table for $C_i$). That this is so is evident from the way Table 7.2 is set out: it consists of four blocks containing the characters of Table 7.1 modulated by the signs of the four characters of Table 4.4.

**Problem 7.9** Show that the labels used for the irreducible representations in Table 7.2 may be derived immediately from those of the character tables of Table 7.1 and Table 4.4.

Because of the relationship between the groups O and $O_h$ it is often possible to pretend that the symmetry of an octahedral molecule is O and then determine the g or u nature of the irreducible representations by simply considering the effect of the inversion operation, i. Thus, a p orbital is ungerade — undergoes a change of phase — under the i operation. This, together with the knowledge that a set of p orbitals transforms as $T_1$ in the group O (Table 7.1), is sufficient to establish that they transform as $T_1$ in $O_h$.

The relationship between the groups $O_h$, O and $C_i$ means that those operations possessed by $O_h$ which are not present in O may be written in such a way that each is equivalent to some operation of O together with the operation i. The operations of O are proper (or pure) rotation operations; the additional

Table 7.2

| $O_h$ | $E$ | $8C_3$ | $6C_4$ | $3C_2$ | $6C_2'$ | $i$ | $8S_6$ | $6S_4$ | $3\sigma_h$ | $6\sigma_d$ | |
|---|---|---|---|---|---|---|---|---|---|---|---|
| $A_{1g}$ | 1 | 1 | 1 | 1 | 1 | 1 | 1 | 1 | 1 | 1 | $x^2+y^2+z^2$ |
| $A_{2g}$ | 1 | 1 | −1 | 1 | −1 | 1 | 1 | −1 | 1 | −1 | |
| $E_g$ | 2 | −1 | 0 | 2 | 0 | 2 | −1 | 0 | 2 | 0 | $\left[\frac{1}{\sqrt{3}}(2z^2-x^2-y^2),\ x^2-y^2\right]$ |
| $T_{1g}$ | 3 | 0 | 1 | −1 | −1 | 3 | 0 | 1 | −1 | −1 | |
| $T_{2g}$ | 3 | 0 | −1 | −1 | 1 | 3 | 0 | −1 | −1 | 1 | $(yz,\ zx,\ xy)$ |
| $A_{1u}$ | 1 | 1 | 1 | 1 | 1 | −1 | −1 | −1 | −1 | −1 | |
| $A_{2u}$ | 1 | 1 | −1 | 1 | −1 | −1 | −1 | 1 | −1 | 1 | |
| $E_u$ | 2 | −1 | 0 | 2 | 0 | −2 | 1 | 0 | −2 | 0 | |
| $T_{1u}$ | 3 | 0 | 1 | −1 | −1 | −3 | 0 | −1 | 1 | 1 | $(x,\ y,\ z)$ |
| $T_{2u}$ | 3 | 0 | −1 | −1 | 1 | −3 | 0 | 1 | 1 | −1 | |

operations are improper rotations. As we saw at the end of Chapter 4, this name is used to denote any point group operation which is not a pure rotation. The precise correspondence between proper and improper rotations in the $O_h$ group is evident from the way that we have written Table 7.2 and is

E followed by i gives i
$C_3$ followed by i gives $S_6$
$C_4$ followed by i gives $S_4$
$C_2$ followed by i gives $\sigma_h$
$C_2'$ followed by i gives $\sigma_d$

**Problem 7.10** The above discussion indicates a different definition of $S_n$ operations from that which we have used. They may be defined as 'rotation inversion' operations and, indeed, this is the way that they are described by crystallographers: 'Rotate by $360/n^\circ$ and follow by inversion in a centre of symmetry.' $C_n$ and i may, or may not, exist in their own right as operations in a group containing $S_n$.

Show that the operation $S_4^-$ (defined as a rotation–reflection operation) has the same effect as the operation $S_4^+$ (defined as a rotation–inversion operation). (*Hint:* use Figure 7.5a; the + and − define the sense of the rotation component.)

## 7.2 THE BONDING IN THE $SF_6$ MOLECULE

We now turn to a discussion of the bonding in the $SF_6$ molecule. We shall take the valence shell atomic orbitals of the sulphur atom to be 3s and 3p, ignoring the 3d. The behaviour of d orbitals in octahedral molecules is of major importance in transition metal chemistry and this is the context in which we shall discuss them later in this chapter. Throughout our discussion of $SF_6$ it will be convenient to work in the point group O rather than the correct group $O_h$. The reason for this lies in the structure of the $O_h$ group. It is twice as large as O (48 operations compared with 24) and so is more cumbersome to handle. However, as we have seen, $O_h$ is the direct product of O with $C_i$, so that the only additional information that $O_h$ has compared with O is that of behaviour under the additional operation introduced by the group $C_i$ — i.e. behaviour under the operation i. It is easier to ask of a basis function 'how does it transform under i?' and to add either g (*gerade* = symmetric) or u (*ungerade* = antisymmetric) as a suffix to the irreducible representation of O than to plough through the whole set of $O_h$ operations.

It is easy to show that the sulphur 3s orbital, shown in Figure 7.3(b), transforms as the totally symmetric, $A_1$, irreducible representation of the point group O. In $O_h$, of course, it has $A_{1g}$ symmetry. In Figure 7.3(b) we also show the axis system we shall use in our discussion, although in many of the following figures the octahedron is drawn from a different viewpoint from that shown

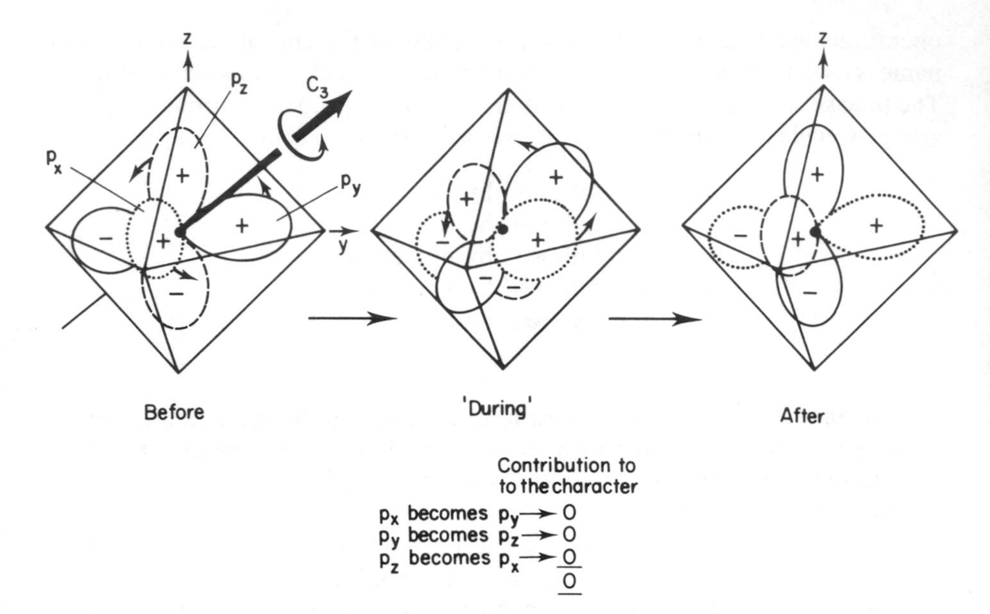


Figure 7.7 The transformation of a set of p orbitals of a central atom of an octahedral molecule under a $C_3$ rotation operation

in Figure 7.3. Like the coordinate axes of Problem 7.5, the sulphur 3p orbitals transform together as the $T_1$ irreducible representation ($T_{1u}$ in $O_h$). We shall look at this particular problem in some detail because it illustrates how to handle the sometimes bewildering task of working with several equivalent objects in a high symmetry environment. Our discussion is largely diagrammatic

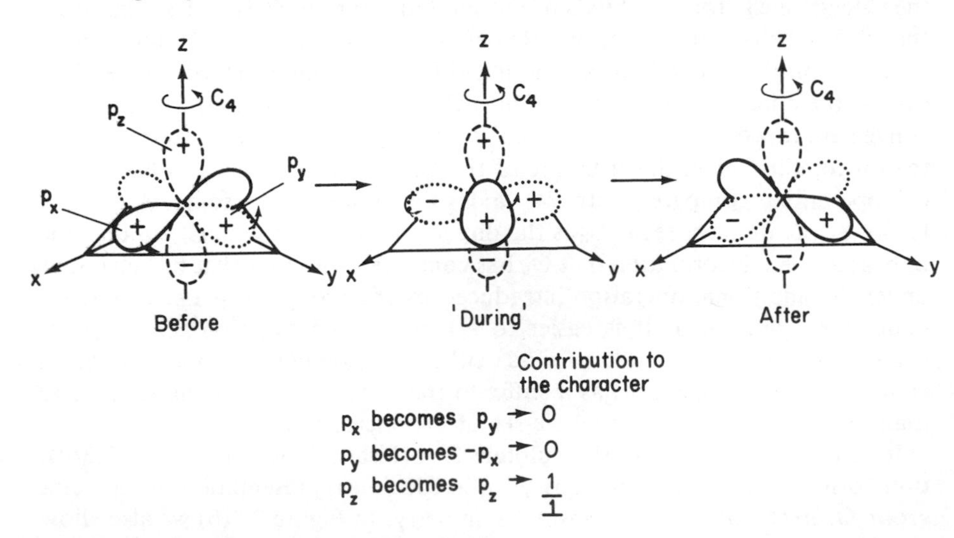


Figure 7.8 The transformation of a set of p orbitals of a central atom of an octahedral molecule under a $C_4$ rotation operation

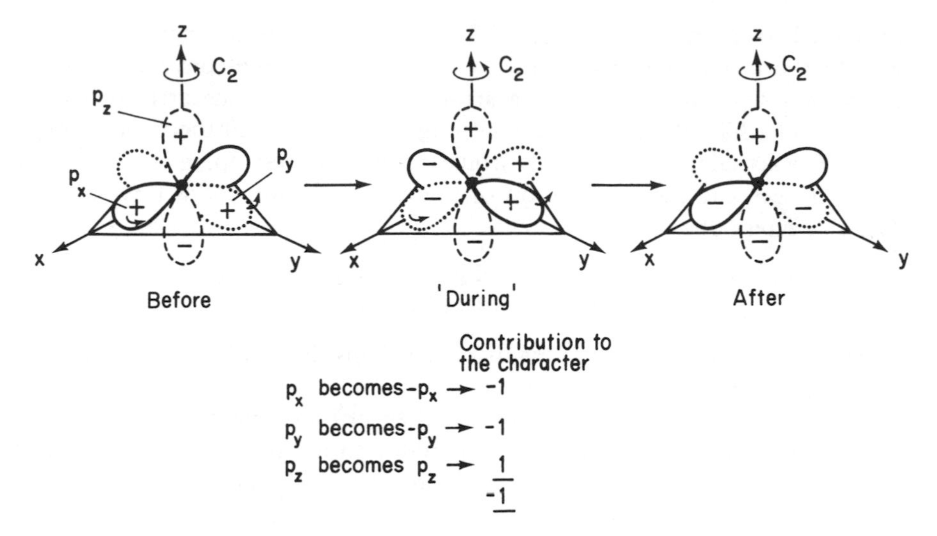


Figure 7.9 The transformation of a set of p orbitals of a central atom of an octahedral molecule under a $C_2$ rotation operation

because good diagrams — perhaps more than a good model — are very important. It is therefore essential that each figure is studied carefully and the transformations that it shows are followed in detail. Figures 7.7 to 7.10 illustrate the transformations of the set of 3p orbitals under a representative operation of each class of the O point group. For clarity of presentation we have represented the lobes of the p orbitals more as ellipses than the circles we

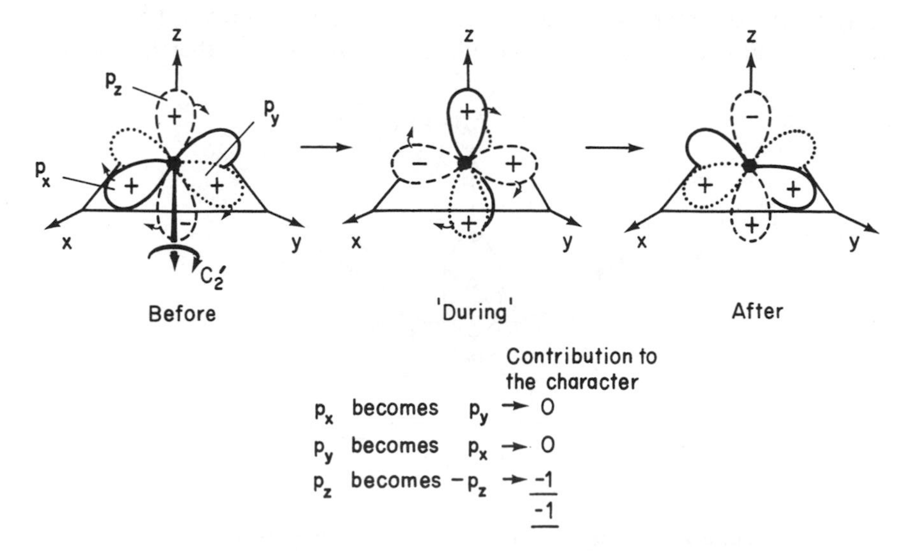


Figure 7.10 The transformation of a set of p orbitals of a central atom of an octahedral molecule under a $C_2'$ rotation operation

have used so far in this book; the different p orbitals are distinguished by the way they are outlined. As an aid to visualizing the interconversions, these figures not only show the starting arrangement but also the arrangement at some point whilst the operation is in progress as well as in the final arrangement. The actual transformations brought about by these operations are listed in the figures and the compilation of the corresponding characters detailed. The characters obtained are:

| E | $8C_3$ | $6C_4$ | $3C_2$ | $6C_2'$ |
|---|---|---|---|---|
| 3 | 0 | 1 | −1 | −1 |

Comparison with Table 7.1 immediately confirms that this set of characters is the $T_1$ irreducible representation.

When in this chapter we illustrate the effects of a $C_3$ rotation operation we shall always choose the particular $C_3$ operation shown in Figure 7.7. The effect of this choice is that coordinate axes, and thus labels, permute as follows:

$$
\begin{array}{ccc} & x & \\ \nearrow & & \searrow \\ z & \longleftarrow & y \end{array}
$$

These permutations apply equally to products of axes. We shall see later in Figure 7.24 that this $C_3$ operation turns $d_{x^2-y^2}$ into $d_{y^2-z^2}$, which is just what the permutation gives.

The s and p orbitals of the sulphur atom in $SF_6$ bond with those fluorine orbitals that point towards the sulphur atoms; without defining their composition further we shall call them the fluorine $\sigma$ orbitals. This set of orbitals is shown schematically in Figure 7.11 and, because all are symmetry-related, we consider the transformations of the six as a set. This is not a difficult task, but some mental gymnastics can be avoided by remembering that it is a general

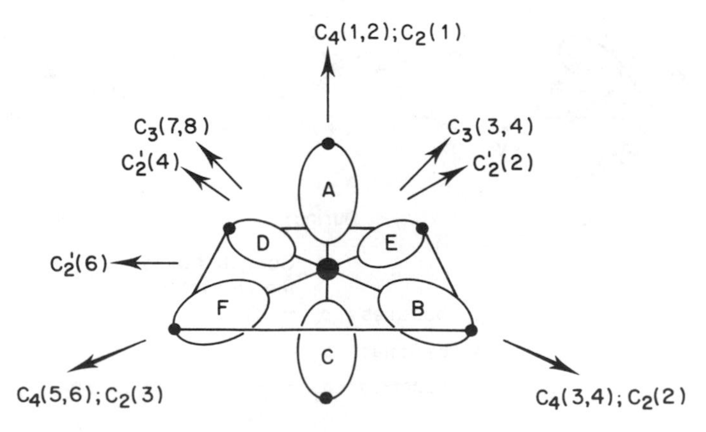


Figure 7.11 The six fluorine S—F $\sigma$-bonding hybrid orbitals of $SF_6$ together with the labels used for them in the text. Some of the axes used in obtaining Table 7.3 are indicated

principle that only if an object lies on a symmetry element can it give a non-zero contribution to the character associated with the corresponding operation. It follows, therefore, that because the fluorine atoms are located on the fourfold axes of the octahedron it is only under the fourfold and corresponding twofold rotation operations (and, of course, the identity operation) that any of the fluorine $\sigma$ orbitals can be left unchanged. Remembering that there are two flurorines on each $C_4$ axis, we conclude that the reducible representation generated by the transformation of the fluorine $\sigma$ orbitals is:

| E | $8C_3$ | $6C_4$ | $3C_2$ | $6C_2'$ |
|---|---|---|---|---|
| 6 | 0 | 2 | 2 | 0 |

This is a sum (it is sometimes called a *direct* sum) of the $A_1 + E + T_1$ irreducible representations.

**Problem 7.11** Use Table 7.1 to show that the above reducible representation has $A_1$, E and $T_1$ components. (*Hint:* this is similar to, but more difficult than, Problems 5.12 and 6.5.)

There are several ways in which we could proceed to obtain the g and u nature of the $A_1$, E and $T_1$ combinations. The most obvious way is to generate them again but using the full $O_h$ group. However, we should still be without the explicit forms of the combination of $\sigma$ orbitals which transform as each irreducible representation. We shall find it simpler to use the projection operator method to obtain these combinations (working in the point group O) and subsequently to ask which are g and which are u in nature. To some extent we may anticipate the result. The fluorine $\sigma$ orbitals which we used to generate the reducible representation were neither inherently symmetric or antisymmetric with respect to inversion in the centre of symmetry; because they were not located at this centre they were always permuted by the inversion operation. In such a situation this indifference is reflected by the generation of an equal number of combinations of g and u symmetries. That is, in the present case we must have three linear combinations of fluorine orbitals which are g and three which are u. There are, then, two possibilities: either we have $A_{1g} + E_g + T_{1u}$ in $O_h$ or, alernatively, $A_{1u} + E_u + T_{1g}$. Physically, only the first choice makes any sense, because the irreducible representations generated by the transformation of the sulphur valence shell s and p orbitals are included in this set whereas they are not in the second. That is, if the first set is correct then there can be interactions between the fluorine $\sigma$ orbitals and the sulphur orbitals — and so the existence of the molecule is explained — whereas for the second set there would be no interactions and the molecule $SF_6$ would not exist!

In order to obtain the fluorine $\sigma$ orbital combinations transforming as the $A_1$, E and $T_1$ irreducible representations we shall use the projection operator method which we have met before. Because the present case provides a particularly good example of the general method we shall give it in detail, bringing together the techniques developed in previous chapters.

First, we give each ligand $\sigma$ orbital a label, A to F as shown in Figure 7.11. We then consider in more detail the transformation of one of these orbitals under the operation of the group. In Table 7.3 are listed the twenty-four operations of the group O, and beneath each is the ligand $\sigma$ orbital into which A is transformed by the particular operation. Within each set of operations, $8C_3$ for example, the order in which the operations are considered is unimportant; what matters is that all are included.

**Problem 7.12** Use Figures 7.3(a) and 7.11 to obtain Table 7.3. (*Hint:* good diagrams are important — it may well be necessary to sketch out parts of Figure 7.3(a) several times to retain clarity in distinguishing the different effects of the various operations.) Note that because there are just six fluorine $\sigma$ orbitals but twenty-four operations, each orbital label appears four times in Table 7.3. To help with this problem as many axes as would be consistent with graphic clarity have been indicated in Figure 7.11.

We now multiply the orbitals in Table 7.3 by characters appropriate to the listed operations. The products are then added together. The sum is either the required *ligand group orbital* or is simply related to it.

For the $A_1$ group orbital, multiplying each of the orbitals by one (the value of each of the $A_1$ characters) and adding the products together gives

$$4A + 4B + 4C + 4D + 4E + 4F$$

Normalizing, the $A_1$ orbital is obtained:

$$\psi_{a_1} = \frac{1}{\sqrt{6}}(A + B + C + D + E + F)$$

Turning to the E orbitals, the sum obtained after multiplication is

$$4A + 4C - 2B - 2D - 2E - 2F$$

Table 7.3

| | | | | | | |
|---|---|---|---|---|---|---|
| E | $C_4(1)$ | $C_4(2)$ | $C_4(3)$ | $C_4(4)$ | $C_4(5)$ | $C_4(6)$ |
| A | A | A | F | E | B | D |
| $C_2(1)$ | $C_2(2)$ | $C_2(3)$ | $C_3(1)$ | $C_3(2)$ | $C_3(3)$ | $C_3(4)$ |
| A | C | C | D | E | E | B |
| $C_3(5)$ | $C_3(6)$ | $C_3(7)$ | $C_3(8)$ | $C_2'(1)$ | $C_2'(2)$ | $C_2'(3)$ |
| B | F | D | F | D | B | E |
| $C_2'(4)$ | $C_2'(5)$ | $C_2'(6)$ | | | | |
| F | C | C | | | | |

which after normalizing is

$$\psi_e(1) = \frac{1}{\sqrt{12}}(2A + 2C - B - D - E - F)$$

**Problem 7.13** Derive $\psi_e(1)$. (*Hint:* this problem is quite similar to that solved in Table 6.2 and the associated discussion.)

We now have to obtain the second E function. Were we to use B or E as the generating orbital in Table 7.3 we would have obtained the (unnormalized) combinations:

From B: $4B + 4D - 2A - 2E - 2C - 2F$
From E: $4E + 4F - 2A - 2B - 2C - 2D$

Neither of these can be the second E function for the choice between them is arbitrary; they cannot both be correct and we cannot have three different E functions. The method described in Section 6.2 may be used to systematically obtain the second function. Study of the results obtained there suggests that we take the difference between the functions given above which were derived from B and E. This difference is

$$6B + 6D - 6E - 6F$$

which on normalizing gives the second function

$$\psi_e(2) = \tfrac{1}{2}(B + D - E - F)$$

**Problem 7.14** Show by squaring and adding the coefficients with which the fluorine $\sigma$ orbitals appear in $\psi_e(1)$ and $\psi_e(2)$ that each orbital makes an equal contribution to the E set. (*Hint:* this problem resembles Problem 6.7.) Note that the sum of squares of coefficients is equal to the ratio

$$\frac{\text{Number of E functions}}{\text{Total number of } \sigma \text{ orbitals}} = \frac{2}{6}$$

In Section 6.2 the argument that we used which led to the generation of a second E function depended on the fact that the functions which we were seeking to generate had vector-like properties. Those that we have just obtained do not behave like axes (as the basis functions given at the right-hand side of Table 7.1 show, they transform like sums of products of axes). The method deduced in Section 6.2 clearly has wider generality than we could have anticipated.

The $T_1$ functions are readily obtained. The transformation of A (or C) yields $4A - 4C$ which, normalized, gives

$$\psi_{t_1}(1) = \frac{1}{\sqrt{2}}(A - C)$$

Similarly, the transformations of B (or D) and E (or F) give

$$\psi_{t_1}(2) = \frac{1}{\sqrt{2}}\,(B - D)$$

and

$$\psi_{t_1}(3) = \frac{1}{\sqrt{2}}\,(E - F)$$

respectively. Because these three functions each involve different fluorine $\sigma$ orbitals they are clearly independent of each other.

**Problem 7.15** Derive the three $T_1$ functions listed above.

The complete list of ligand group orbitals is given in Table 7.4. In order to determine their symmetries in $O_h$ we have to investigate the behaviour of these functions under the operation of inversion in the centre of symmetry. Because this operation interchanges the fluorine $\sigma$ orbitals as follows:

$$\begin{aligned} A &\longleftrightarrow C \\ B &\longleftrightarrow D \\ E &\longleftrightarrow F \end{aligned}$$

we obtain the effect of this operation by making these substitutions (A for C, C for A, etc.) in the functions given in Table 7.4. The $A_1$ and Es are left unchanged but each $T_1$ function changes sign. We conclude that they transform in $O_h$ as $A_{1g}$, $E_g$ and $T_{1u}$ respectively. We have indicated this in Table 7.4. This means that symmetries of the sulphur 3s and 3p orbitals ($A_{1g}$ and $T_{1u}$ respectively) are matched within the fluorine $\sigma$ orbital symmetries. The bonding molecular orbital of $A_{1g}$ symmetry is shown in Figure 7.12(a) and a representative $T_{1u}$ bonding molecular orbital in Figure 7.12(b). One of the fluorine or-

Table 7.4

| Symmetry | | |
|---|---|---|
| In the group O | In the group $O_h$ | Symmetry-adapted function (ligand group orbitals) |
| $A_1$ | $A_{1g}$ | $(1/\sqrt{6})(A + B + C + D + E + F)$ |
| E | $E_g$ | $(1/\sqrt{12})(2A + 2C - B - D - E - F)$ <br> $1/2(B + D - E - F)$ |
| $T_1$ | $T_{1u}$ | $(1/\sqrt{2})(A - C)$ <br> $(1/\sqrt{2})(B - D)$ <br> $(1/\sqrt{2})(E - F)$ |

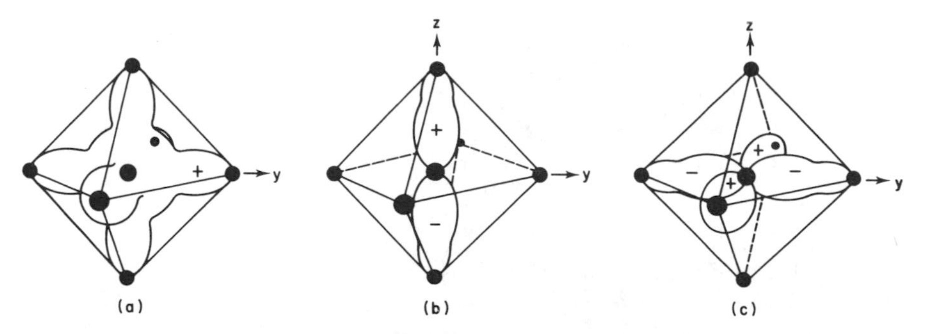


Figure 7.12 (a) The $A_{1g}$ bonding molecular orbital; (b) one of the $T_{1u}$ bonding molecular orbitals (that involving the sulphur $p_z$ orbital); (c) one of the fluorine $\sigma$ orbital symmetry-adapted combinations of $E_g$ symmetry

bital combinations — the second that we generated — is shown in Figure 7.12(c). There is no doubt that the $\sigma$-bonding orbitals have bonding energies which increase in the order

$$A_{1g} > T_{1u} > E_g$$

This order of orbital energies is also the order in terms of the number of nodes. The $A_{1g}$ bonding molecular orbital is nodeless, the $T_{1u}$ has one nodal plane and the $E_g$ two. These nodal patterns are implicit in the expressions given in Table 7.4 and are also evident in Figure 7.12.

This discussion has assumed that only $\sigma$ bonding is involved in the interaction between the central sulphur atom and the surrounding fluorines. Although this is quite a good approximation for $SF_6$ we will, nonetheless, extend our discussion to include those $p_\pi$ orbitals on the fluorine atoms which we have so far ignored. This is because they are of relatively high energy and will be seen in the photoelectron spectrum. These $p_\pi$ orbitals transform as a degenerate pair of E symmetry under the local $C_{4v}$ symmetry of each fluorine atom. It follows from our discussion of this symmetry in Chapter 5 that there is no unique specification of the direction of the local x and y axes, but the choice and notation in Figure 7.13 prove to be convenient in practice.

**Problem 7.16** Figure 7.13 appears to be rather complicated. Show that it has an internal consistency in that all $p_x$ orbitals are all oriented in the same x direction, the $p_y$ in the same y direction and the $p_z$ in the same z direction.

The next step is to determine the reducible representation generated by the transformation of these twelve $p_\pi$ orbitals. As for the case of the $\sigma$ orbitals, it is only possible to obtain non-zero characters for the identity operation, for

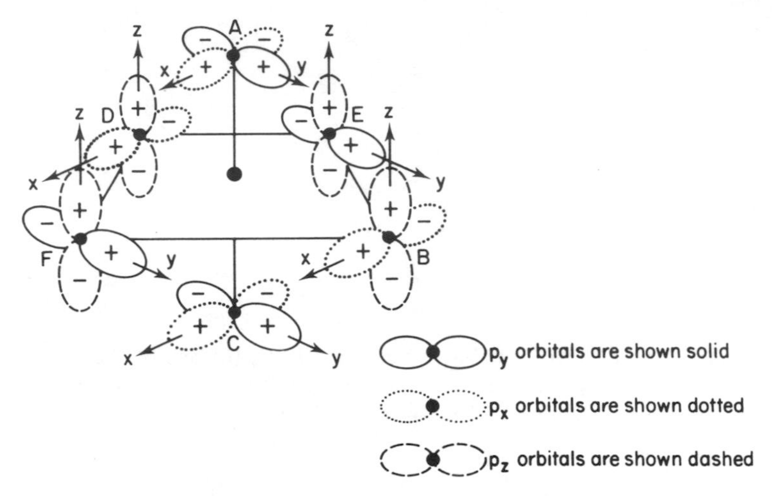


Figure 7.13 Fluorine $p_\pi$ orbitals in $SF_6$

the $C_4$ rotation operations and the corresponding $C_2$ rotation operation (because the fluorine atoms lie on the fourfold axes of the octahedron). Of these, the character for the $C_4$ rotation operation is zero (because, for those fluorine atoms left unshifted by the operation, the $p_x$ and $p_y$ orbitals are interchanged) so only the identity and $3C_2$ classes contain non-zero characters. The reducible representation obtained is

| E | $8C_3$ | $6C_4$ | $3C_2$ | $6C_2'$ |
|---|---|---|---|---|
| 12 | 0 | 0 | −4 | 0 |

which has components $2T_1 + 2T_2$ (under $O_h$ symmetry these become $T_{1g} + T_{1u} + T_{2g} + T_{2u}$).

**Problem 7.17** Show that the twelve fluorine $p_\pi$ orbitals of Figure 7.13 generate the above reducible representation and that it has $2T_1$ and $2T_2$ components in the point group O.

Appropriate linear combinations are obtained by the usual projection operator method but a difficulty arises because, in O symmetry, two quite independent sets of functions transform as $T_1$ and two other sets as $T_2$. The problem of distinguishing between them is readily solved by working, instead, in $O_h$ symmetry — where all sets are symmetry-distinguished — but this is a rather tedious task because this group has forty-eight symmetry operations, each of which has to be separately considered. In Appendix 4 we describe an alternative, shortcut, method of obtaining these linear combinations. This method, depending on an ascent-in-symmetry, is most useful for high symmetry systems in which a large number of basis functions is being considered.

Table 7.5 Symmetry-adapted combinations of fluorine $p_\pi$ orbitals, where $p_y(A)$ means the $p_y$ orbital on atom A as indicated in Figure 7.13

$T_{1g}$ *orbitals*

$$t_{1g}(1) = \tfrac{1}{2}[p_x(A) - p_x(C) + p_z(E) - p_z(F)]$$
$$t_{1g}(2) = \tfrac{1}{2}[p_x(A) - p_x(C) + p_z(B) + p_z(D)]$$
$$t_{1g}(3) = \tfrac{1}{2}[p_x(B) - p_x(D) + p_z(E) - p_z(F)]$$

$T_{1u}$ *orbitals*

$$t_{1u}(1) = \tfrac{1}{2}[p_z(B) + p_z(D) + p_z(E) + p_z(F)]$$
$$t_{1u}(2) = \tfrac{1}{2}[p_y(A) + p_y(C) + p_y(E) + p_y(F)]$$
$$t_{1u}(3) = \tfrac{1}{2}[p_x(A) + p_x(B) + p_x(C) + p_x(D)]$$

$T_{2g}$ *orbitals*

$$t_{2g}(1) = \tfrac{1}{2}[p_x(A) - p_x(C) - p_z(E) + p_z(F)]$$
$$t_{2g}(2) = \tfrac{1}{2}[p_y(A) - p_y(C) + p_z(B) - p_z(D)]$$
$$t_{2g}(3) = \tfrac{1}{2}[p_x(B) - p_x(D) - p_y(E) + p_y(F)]$$

$T_{2u}$ *orbitals*

$$t_{2u}(1) = \tfrac{1}{2}[p_z(B) + p_z(D) - p_z(E) - p_z(F)]$$
$$t_{2u}(2) = \tfrac{1}{2}[p_y(A) + p_y(C) - p_y(E) - p_y(F)]$$
$$t_{2u}(3) = \tfrac{1}{2}[p_x(A) + p_x(B) - p_x(C) - p_y(D)]$$

The appropriate linear combinations are given in Table 7.5, with one of each symmetry species being shown in Figures 7.14 to 7.17. Our interest in these combinations lies in the interactions between adjacent fluorine $p_\pi$ orbitals because we shall use these to predict relative energies for comparison with the results of photoelectron spectroscopy. Figures 7.14 to 7.17 show an interesting situation. The interactions between the sulphur $p_\pi$ orbitals are of two types. For

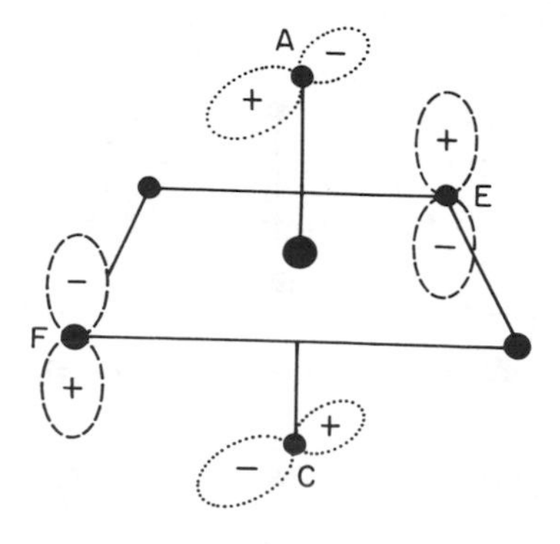


Figure 7.14 One of the $T_{1g}$ fluorine $p_\pi$ orbital combinations

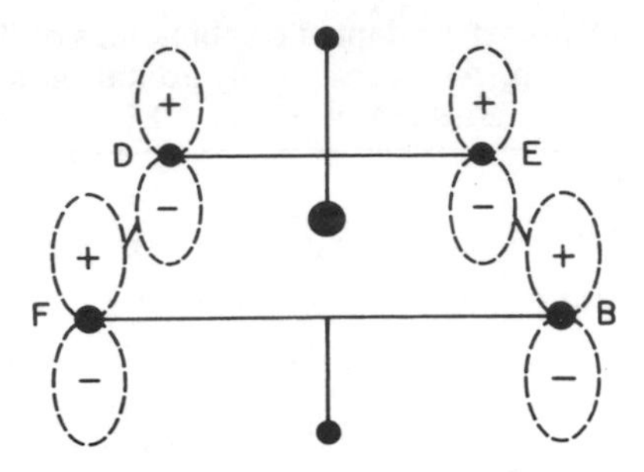


Figure 7.15 One of the $T_{1u}$ fluorine $p_\pi$ orbital combinations

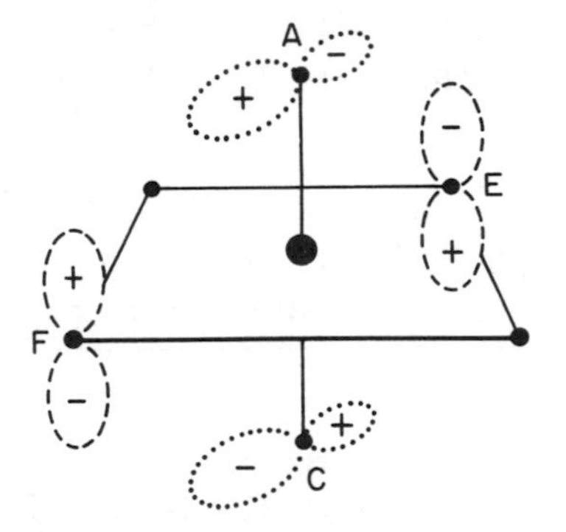


Figure 7.16 One of the $T_{2g}$ fluorine $p_\pi$ orbital combinations

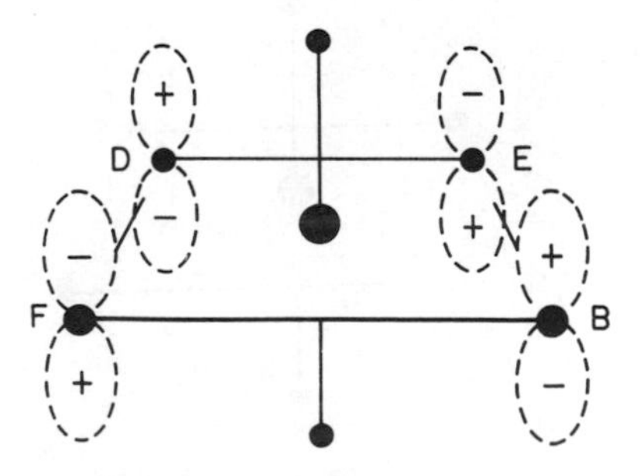


Figure 7.17 One of the $T_{2u}$ fluorine $p_\pi$ orbital combinations

the $T_{1u}$ (Figure 7.15) and $T_{2u}$ (Figure 7.17) sets the component $p_\pi$ orbitals are arranged parallel to each other; their interactions are therefore of $\pi$ type. For the $T_{1g}$ (Figure 7.14) and $T_{2g}$ (Figure 7.16) orbitals the axes of adjacent atomic $p_\pi$ orbitals are at right angles to each other so that their interaction is a mixture of $\sigma$ and $\pi$ types (as shown in Figure 7.18, the $p_\pi$ orbitals may be treated as vectors and the neighbouring interactions resolved into $\sigma$ and $\pi$ components). Qualitatively, $\sigma$ interactions are usually greater than $\pi$ and so we expect that the energy difference between the $T_{2u}$ and $T_{1g}$ orbitals would be greater than that between $T_{1u}$ and $T_{2g}$, provided that the interactions are comparable in other respects. The other important factor is relative nodality. As evident from Figures 7.14 to 7.17, the $T_{1u}$ and $T_{2g}$ orbitals are no-node combinations (apart from the nodes inherent in the $p_\pi$ orbitals themselves) — the positive lobe of a p orbital is next to the positive lobe of an adjacent p orbital — whereas the $T_{1g}$ and $T_{2u}$ orbitals each have two additional nodes — the positive lobe of one p orbital is adjacent to the negative lobe of its neighbour. We have:

$T_{1g}$ : 2 nodes, $\sigma$ and $\pi$ interactions
$T_{2u}$ : 2 nodes, $\pi$ interaction only
$T_{1u}$ : 0 nodes, $\pi$ interaction only
$T_{2g}$ : 0 nodes, $\sigma$ and $\pi$ interactions

Our discussion leads us to expect that the stability of these orbitals varies in the order

$$T_{2g} > T_{1u} > T_{2u} > T_{1g}$$

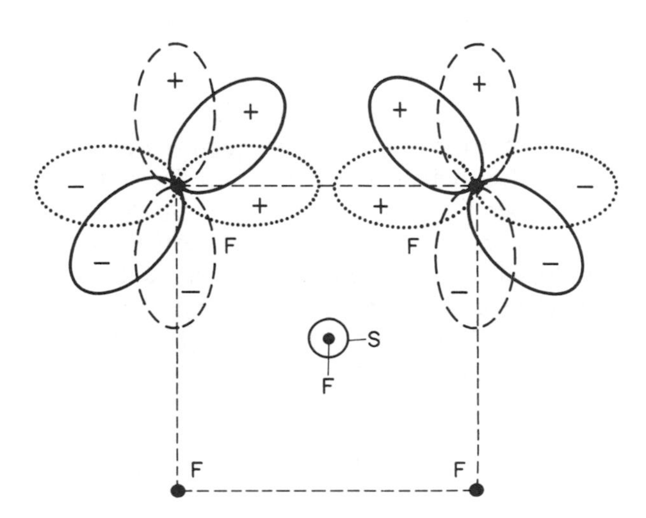


Figure 7.18 The overlap of 'coplanar' $p_\pi$ orbitals on adjacent fluorine atoms (shown solid) may be expressed as a sum of a $\pi$ overlap (between orbitals shown dashed) and a $\sigma$ overlap (between orbitals shown dotted). This diagram shows $SF_6$ viewed down a $C_4$ axis

In arriving at this order we have ignored any possible bonding of the fluorine $p_\pi$ orbitals with the sulphur atom. The only symmetry in common with the sulphur valence shell orbitals is $T_{1u}$. The latter are involved in S—F $\sigma$ bonding and so we shall consider interaction between the $p_\pi$-derived $T_{1u}$ set and the (more stable) S—F $\sigma$-bonding $T_{1u}$ set. Any interaction will lead to a further stabilization of the S—F $\sigma$-bonding set and the corresponding destabilization of the $p_\pi$-derived $T_{1u}$ orbitals. This destabilization could change our $p_\pi$ orbital energy sequence; although we remain confident of the order

$$T_{2g} > T_{2u} > T_{1g}$$

the $T_{1u}$ set could slot in between the $T_{2g}$ and $T_{2u}$ — as before — or between the $T_{2u}$ and $T_{1g}$ (assuming that the destabilization is not too great). In practice, the observed sequence for $SF_6$ seems to be[1]

$$T_{2g} > T_{2u} > T_{1u} > T_{1g}$$

or, including the orbitals we have associated with $\sigma$ bonding,

$$A_{1g} > T_{1u} > T_{2g} > E_g > T_{2u} > T_{1u} > T_{1g}$$

We see that this order is in good agreement with that given by our qualitative model, although we were not able to predict that the $\sigma$ interaction energy level sequence would overlap with the $\pi$ levels. (The order given is of calculated orbital energies; in the vertical ionization potentials, observed in the photoelectron spectrum, the $T_{2u}$ and $T_{1u}$ are identical.)

**Problem 7.18** Draw an orbital energy level diagram for $SF_6$ (cf., for instance, Figure 6.8).

Sulphur hexafluoride is the last, and most complicated, molecule of which we shall consider the electronic structure in detail. The molecules that we have selected were chosen because they enabled us to introduce particular aspects of group theory, not for their own intrinsic interest. However, the methods that we have developed are of general applicability and can be used to gain insight into the electronic structure of quite complicated molecules. Particularly useful here is to assume the highest reasonable symmetry for a molecule or molecular fragment and to consider the effects of a reduction in symmetry to the real-life level as a minor perturbation. Reductions in symmetry will be discussed in Chapter 8. First, however, we extend our discussion of octahedral molecules to transition metal complexes of this symmetry.

## 7.3 OCTAHEDRAL TRANSITION METAL COMPLEXES

It is probably true that a majority of transition metal complexes have octahedral symmetry, at least approximately. Entire books have been written on this subject but here we shall only describe the more important features.

At the simplest level an octahedral transition metal complex may be regarded as built up from a transition metal ion, $M^{n+}$, surrounded by six atoms or ions arranged at the corners of a regular octahedron. The six surrounding atoms may indeed be single atoms or they may be an atom through which a molecule is attached to the transition metal ion. In this picture the metal ion is bonded to the six surrounding *ligands* (a collective noun covering both bonded atoms and molecules) by pure electrostatic attraction. This simple model leads to *crystal field theory* and it is this which we shall now discuss in outline.

Transition metals are characterized by the fact that they exhibit variable valencies in their salts. The corresponding transition metal cations have different numbers of d electrons, the number of d electrons varying with the valence state of the cation. Loosely speaking, if a transition metal ion is oxidized then it loses a d electron; if it is reduced it gains one. We therefore focus attention on the d electrons and on the d orbitals.

In an octahedral $ML_6$ molecule a set of d orbitals on the central metal atom divides into two sets; one, consisting of the $d_{xy}$, $d_{yz}$ and $d_{zx}$ orbitals, has $T_{2g}$ symmetry, as indicated in Tables 7.1 and 7.2. In Figures 7.19 to 7.22 we illustrate the transformations of members of this set and detail their individual contributions to the characters.

**Problem 7.19** Check the transformations of the $d_{xy}$, $d_{yz}$ and $d_{zx}$ orbitals shown in Figures 7.19 to 7.22 and thus show that these orbitals transform as $T_2$ in O. Because they are centrosymmetric orbitals, it follows that they transform as $T_{2g}$ in $O_h$.

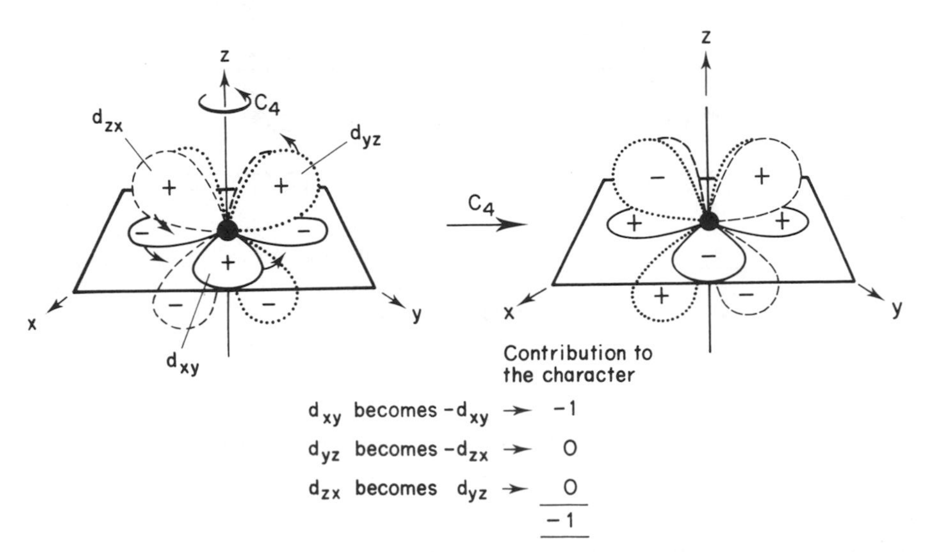


Figure 7.19 Transformation of the $T_{2g}$ set of d orbitals of a central metal atom under a $C_4$ rotation operation of the octahedron

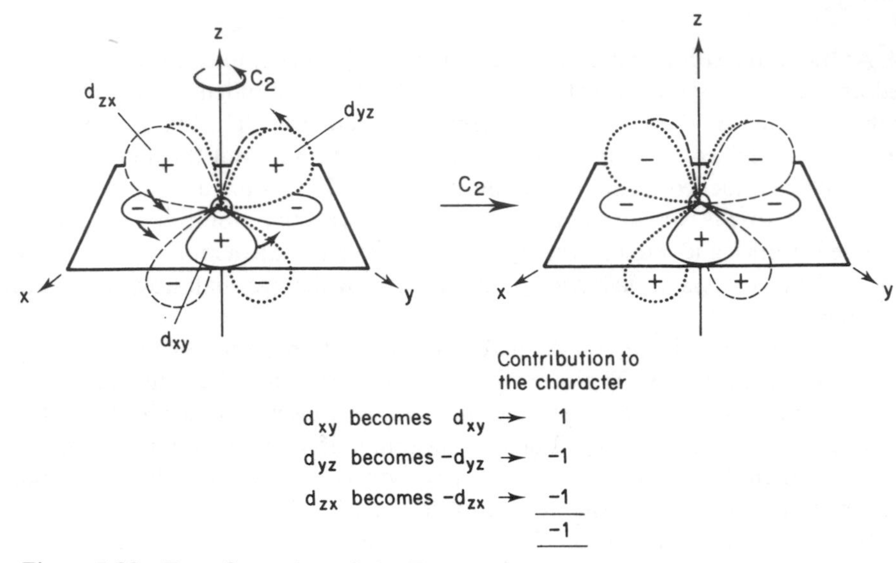


Figure 7.20 Transformation of the $T_{2g}$ set of d orbitals of a central metal atom under a $C_2$ rotation operation of the octahedron

The second set of d orbitals,* $d_{(x^2-y^2)}$ and $d_{(1/\sqrt{3})(2z^2-x^2-y^2)}$, transform together as the $E_g$ irreducible representation. Their transformations are illustrated in Figures 7.23 to 7.26 where their individual contributions to the characters are also detailed. The only point of difficulty arises in connection

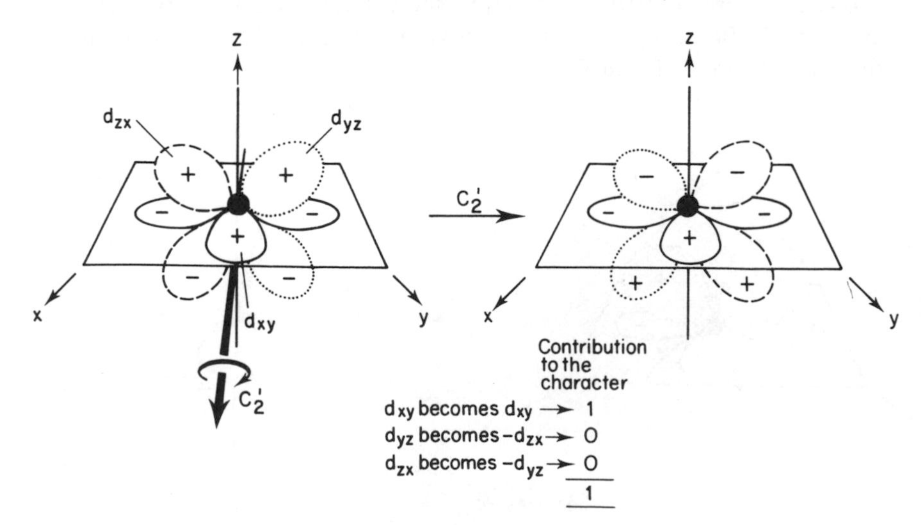


Figure 7.21 Transformation of the $T_{2g}$ set of d orbitals of a central metal atom under a $C_2'$ rotation operation of the octahedron

*The orbitals are usually called $d_{(x^2-y^2)}$ and $d_{z^2}$. In the present context we have to recognize that the label $z^2$ is a shorthand symbol for $(1/\sqrt{3})(2z^2-x^2-y^2)$.

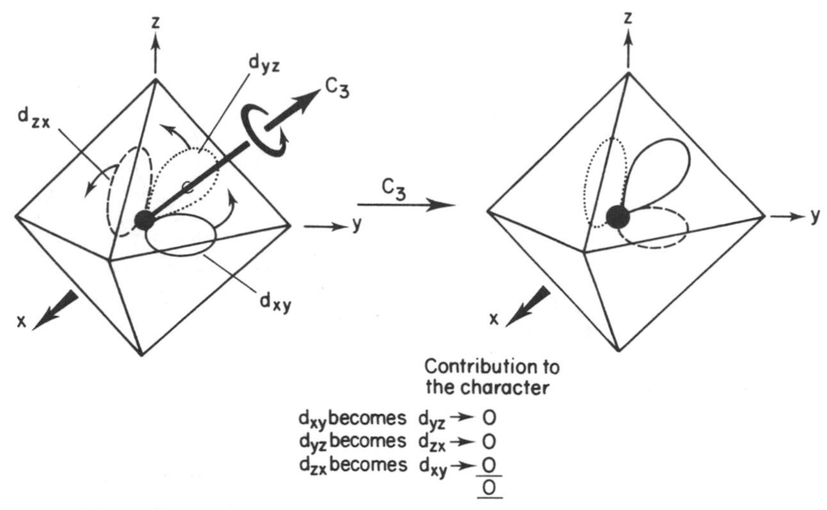


To simplify the diagram only one lobe of each of $d_{xy}$, $d_{yz}$ and $d_{zx}$ is shown

Figure 7.22 Transformation of the $T_{2g}$ set of d orbitals of a central metal atom under a $C_3$ rotation operation of the octahedron

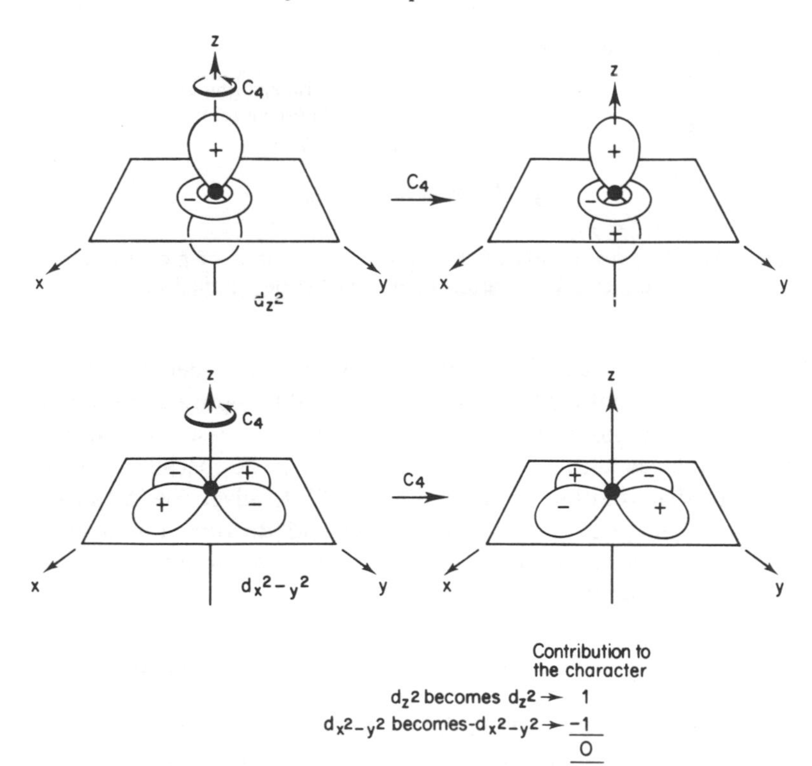


Figure 7.23 Transformation of the $E_g$ set of d orbitals of a central metal atom under a $C_4$ rotation operation of the octahedron

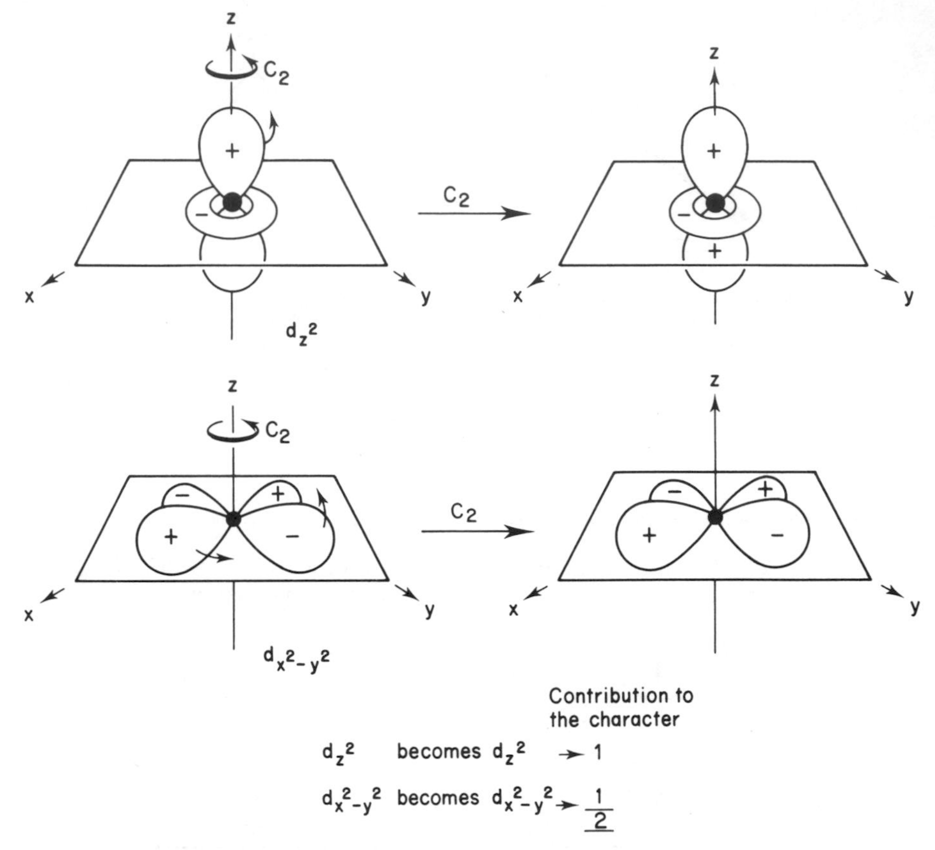


Figure 7.24 Transformation of the $E_g$ set of d orbitals of a central metal atom under a $C_2$ rotation operation of the octahedron

with the $C_3$ rotation and resembles that discussed in detail in Section 6.1, where we also met a doubly degenerate irreducible representation which gave a character of $-1$ under a $C_3$ rotation. In the present case it is helpful to write the $E_g$ orbitals as $d_{(x^2-y^2)}$ and $d_{(1/\sqrt{3})[(z^2-x^2)-(y^2-z^2)]}$ because this helps to demonstrate that rotation of the pair by $120^\circ$ to give, for instance, $d_{(y^2-z^2)}$ and $d_{(1/\sqrt{3})[(x^2-y^2)-(z^2-x^2)]}$ (i.e. $x \to y \to z \to x$) leads to functions which may be expressed in terms of the original. It is easy to show by expansion of the coefficients that, for instance,

$$d_{(y^2-z^2)} = -\frac{1}{2}\, d_{(x^2-y^2)} - \frac{\sqrt{3}}{2}\, d_{(1/\sqrt{3})[(z^2-x^2)-(y^2-z^2)]}$$

so that the coefficient with which $d_{(x^2-y^2)}$ appears in this expression ($-\frac{1}{2}$) is its contribution to the character under the $C_3$ rotation. The contribution of $d_{(1/\sqrt{3})(2z^2-x^2-y^2)}$ to $d_{(1/\sqrt{3})(2x^2-y^2-z^2)}$ is similarly shown to be $-\frac{1}{2}$ so that the aggregate character is $-1$.

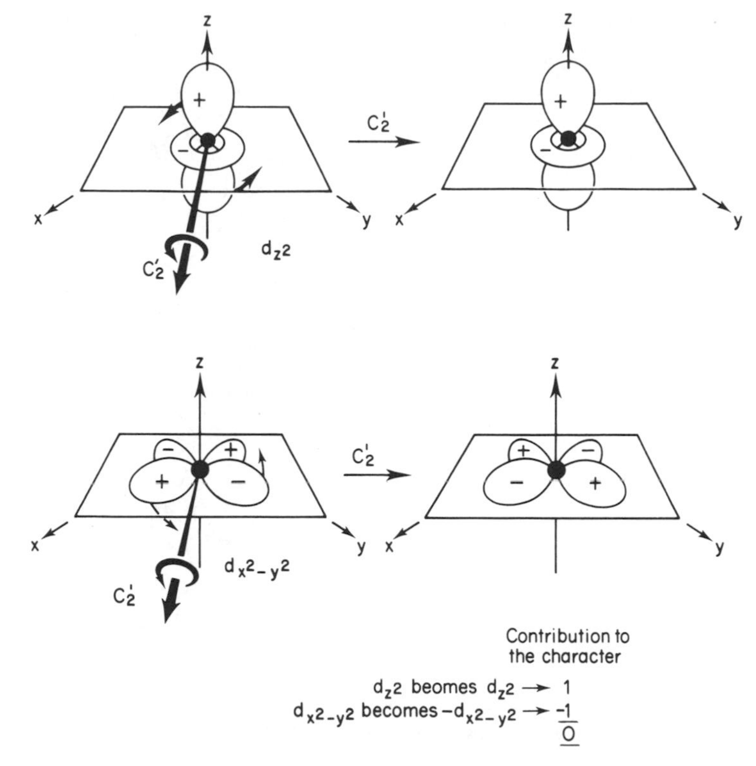


Figure 7.25 Transformation of the $E_g$ set of d orbitals of a central metal atom under a $C_2'$ rotation operation of the octahedron

**Problem 7.20** Check the transformation of the $d_{x^2-y^2}$ and $d_{z^2}$ orbitals given in Figures 7.23 to 7.26 and thus show that these orbitals transform as E in O. Because they are centrosymmetric orbitals it follows that they transform as $E_g$ in $O_h$.

Crystal field theory, being a purely electrostatic theory which does not admit of the existence of bonding and antibonding molecular orbitals, asserts that since the d electrons (like all other metal electrons) are non-bonding, they will occupy preferentially that arrangement in which electrostatic repulsion with the ligands (most simply represented as point negative charges) and with each other is a minimum. It is convenient to consider these two factors separately. Consider first the requirement of minimum electrostatic repulsion between the metal d electrons and the negatively charged ligands. In Figure 7.27 we show a representatative $E_g$ orbital (the $d_{x^2-y^2}$) and a representative $T_{2g}$ (the $d_{xy}$). Symmetry ensures that whatever we conclude about these holds also for the other member(s) of their respective sets. As Figure 7.27 suggests, it is in the

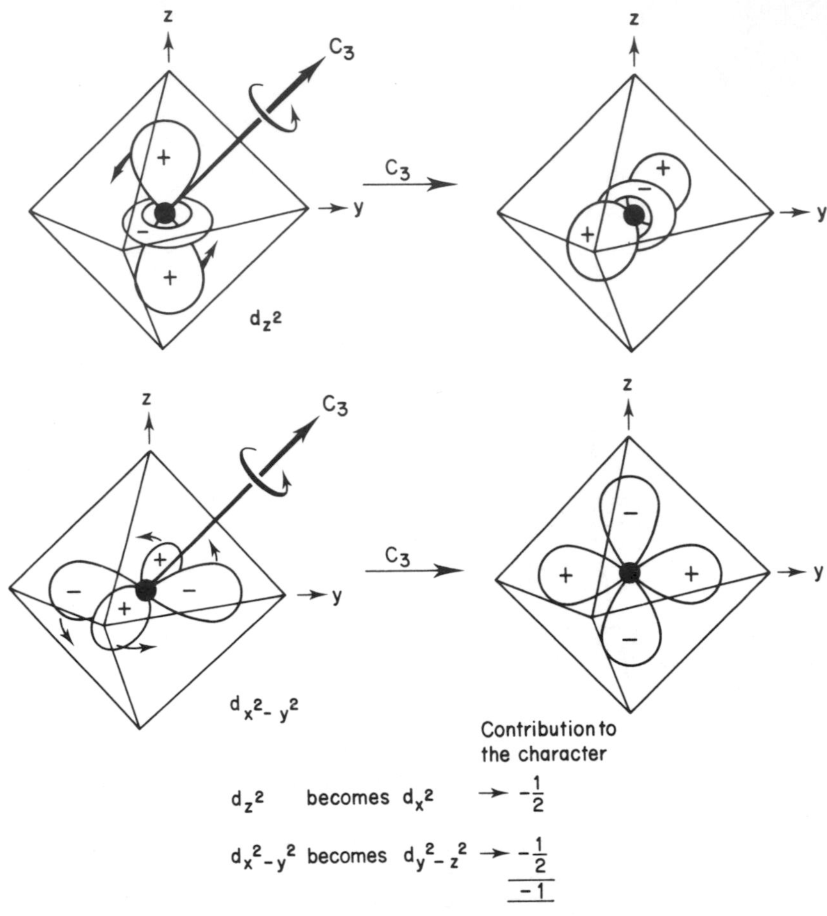


Figure 7.26 Transformation of the $E_g$ set of d orbitals of a central metal atom under a $C_3$ rotation operation of the octahedron. It will help to understand this diagram if it is recognized that the $C_3$ operation shown has the effect of converting $z \rightarrow x$, $x \rightarrow y$ and $y \rightarrow z$

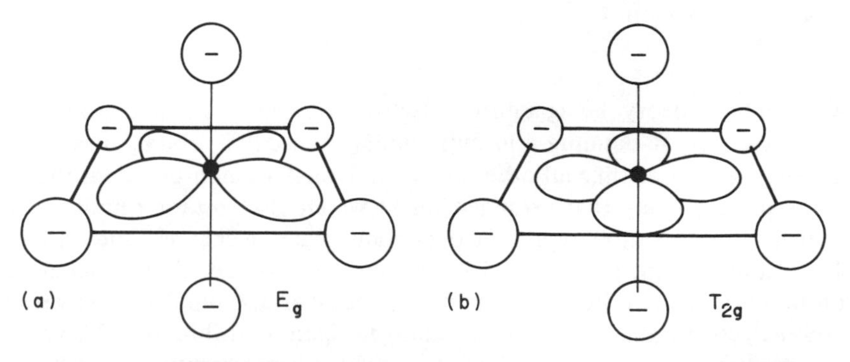


Figure 7.27 Representative (a) $E_g$ and (b) $T_{2g}$ orbitals of a central metal atom in an octahedral metal complex. In this figure ⊖ indicates negative electrical charge; the d orbitals are envisaged as also containing electron density so that electron–electron repulsion occurs

$E_g$ orbitals (Figure 7.27a) that an electron gets closest to the ligands and so experiences the greatest electrostatic repulsion. This conclusion, which is confirmed by detailed calculations, means that the $T_{2g}$ set (Figure 7.27b) of d orbitals has a lower energy than the $E_g$ set. The energy splitting between the two sets is usually denoted either $\Delta$ or 10Dq. If d electron–ligand repulsion were the only factor to be considered then the d electrons in octahedral transition metal complex ions would occupy the lower, $T_{2g}$, set of d orbitals until these were filled up. However, this preference is opposed by the effects of electron repulsion between the d electrons themselves. This electron repulsion is minimized if as far as possible, the d electrons occupy different d orbitals with parallel spin. That is, occupation of the $E_g$ orbitals will start as soon as the $T_{2g}$ set is half full. We have here a straight conflict between the two opposing forces. When the d electron–ligand repulsion wins we have the so-called 'strong field' case; when the d–d electron repulsion dominates we have the 'weak field' case. Alternative names are to talk of 'low spin' and 'high spin' complexes, the names originating from the fact that weak field complexes have more unpaired electrons — a higher resultant spin — than strong field complexes.

In summary, in crystal field theory, the relative magnitudes of $\Delta$(10Dq) and the d electron repulsion energies — the so-called 'pairing energy' — determine the way that the set of d orbitals are occupied. This is illustrated in Figure 7.28, where the clear difference between high spin and low spin electron arrangements for ions with between four and seven d electrons is evident. Differences also exist for ions with two, three and eight d electrons, but these

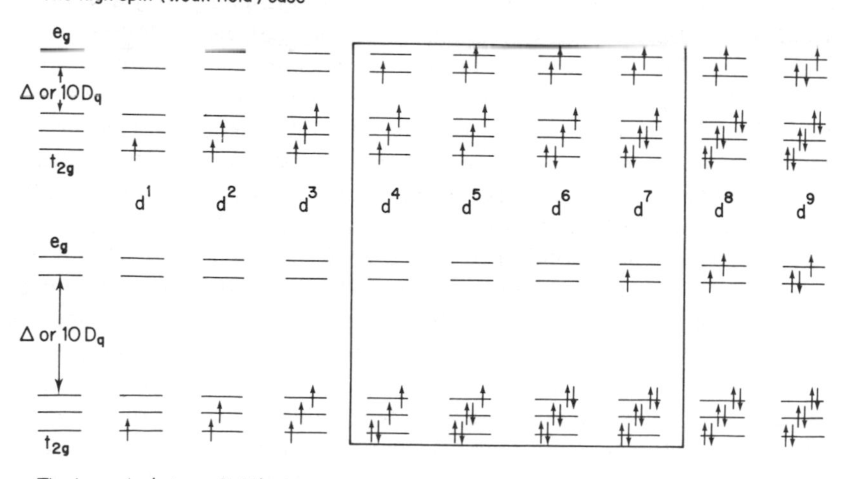


Figure 7.28 Differences in arrangement of electrons in the d orbitals of a metal atom in an octahedral complex occur for the $d^4$–$d^7$ configurations (those within the box)

differences appear in the detailed mathematics rather than in obvious orbital occupancies. Consequent upon these differences between high and low spin cases is a variety of associated spectral, magnetic, structural, kinetic and thermodynamic differences.

Inclusion of covalent bonding in the interaction between metal ion and ligands in a transition metal complex leads to *ligand field theory*. It differs from crystal field theory in that quantities which are well defined in crystal field theory become less well defined in ligand field theory (and are generally treated as parameters to be deduced from experiment). Qualitatively, Figure 7.28 remains appropriate except that the $E_g$ set of d orbitals is now identified as the antibonding counterpart of an $E_g$ set involved in metal–ligand $\sigma$ bonding.

In contrast to our discussion of $SF_6$, the valance shell of the central atom in transition metal complexes consists of s, p and d atomic orbitals. This means that the nine available metal orbitals span the $A_{1g}$, $T_{1u}$, $E_g$ and $T_{2g}$ irreducible representations. It will be recalled that the $\sigma$ orbitals of the surrounding six atoms — be they fluorine in $SF_6$ or ligands in a complex — span $A_{1g} + T_{1u} + E_g$. In a transition metal complex these ligand orbitals are full; this is evident if the ligand is a closed shell anion such as $F^-$, $Cl^-$, etc., and is equally true if it is a molecule such as $H_2O$ or $NH_3$, where the ligand $\sigma$ orbital is identified as a lone pair of electrons on the electronegative atom. This means that we can regard the interaction with the metal orbitals as stabilizing the ligand orbitals — lowering their energy. In this case the metal orbitals are to be regarded as being destabilized and the $E_g$ orbitals, which in crystal field theory are pure d orbitals, are to be regarded as antibonding combinations of ligand $\sigma$ and metal d orbitals.

Before concluding this section on transition metal ions it is of interest to note that in ligand field theory the d orbitals of $T_{2g}$ symmetry may also interact with ligand orbitals. It will be recalled that the fluorine $p_\pi$ orbitals in $SF_6$ transformed as $T_{1u} + T_{1g} + T_{2u} + T_{2g}$ so it is evident that in transition metal complexes any $T_{2g}$ interaction will be with ligand $\pi$ orbitals. If the relevant ligand $\pi$ orbitals are empty — and therefore of high energy — then the effect of ligand–metal $T_{2g}$ interactions will be to depress (stabilize) the lower $T_{2g}$ orbitals — those corresponding to the $T_{2g}$ d orbitals shown in Figure 7.28 — and raise the energy of the (empty) $T_{2g}$ ligand $\pi$ orbitals. The effect on the molecular orbitals corresponding to the d orbitals of Figure 7.28 will be to increase the splitting $\Delta$. If the ligand $\pi$ orbitals are filled — and therefore of relatively low energy — the effect will be to decrease the $E_g$–$T_{2g}$ splitting. These two cases are illustrated in Figure 7.29. The $\pi$ bonding that we have just described seems to be of importance because it is found that it is ligands with available but empty $\pi$ orbitals that give large values of $\Delta$ and thus low spin, high field complexes, and those with filled $\pi$ orbitals that give small values of $\Delta$ and thus high spin, low field, complexes. Examples of the former are the $CN^-$ and CO ligands (the empty $\pi$ orbitals being C—N or C—O $\pi$ antibonding) and of the latter are $Br^-$ and $Cl^-$ — here the filled $\pi$ orbitals are the atomic $p_\pi$ orbitals of the halogens.

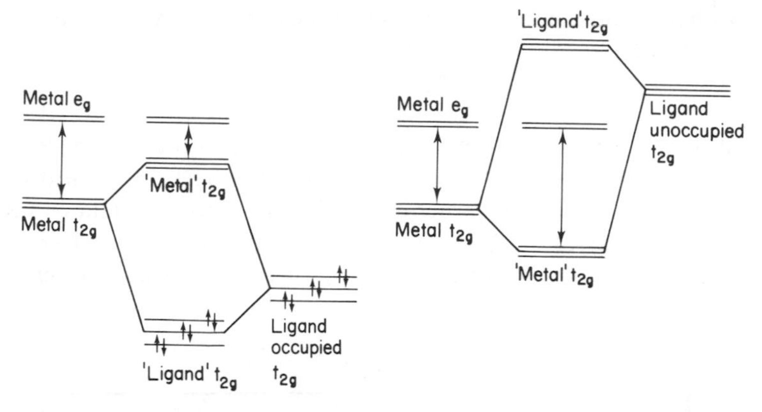


Figure 7.29 The effect of ligand $\pi$ orbitals on the $e_g$–$t_{2g}$ splitting (indicated by the double-headed arrows) depends on the relative energies of the metal and ligand $t_{2g}$ orbitals

## 7.4 THE BONDING IN TETRAHEDRAL MOLECULES

Early in the present chapter we mentioned that it is often possible to discuss octahedral and tetrahedral transition metal ions together because their geometries are both derived from a cube. Tetrahedral complexes, of general formula $ML_4$, are of widespread occurrence but are not as common as octahedral ones. Together, species with geometries which approximate to either octahedral or tetrahedral account for at least 80 per cent. of all coordination compounds.

The Cartisian coordinate axes which we used for an octahedron (Figure 7.3(b)) were those of the corresponding cube. Similarly, it is advantageous to use cube-derived axes for a tetrahedron, notwithstanding the fact that this choice is not in accord with taking the z axis as the axis of highest rotational symmetry (which would mean $C_3$ for the tetrahedron).

Because a tetrahedron is derived from a cube it is not surprising that the symmetry operations which turn a tetrahedron into itself are also symmetry operations of the cube (but the converse is not true). The corresponding symmetry operations are:

| Cube (and octahedron) | E | $8C_3$ | $6C_4$ | $3C_2$ | $6C_2'$ | i | $8S_6$ | $6S_4$ | $3\sigma_h$ | $6\sigma_d$ |
|---|---|---|---|---|---|---|---|---|---|---|
| Tetrahedron | E | $8C_3$ | | $3C_2$ | | | | $6S_4$ | | $6\sigma_d$ |

The group of operations of the tetrahedron is given the shorthand label $T_d$ (pronounced 'tee-dee').

**Problem 7.20** Draw diagrams to show all of the symmetry operations of a tetrahedron. (*Hint:* Figure 1.4 is helpful. The mid-point of each face of this figure corresponds to a corner of the octahedron shown in Figures 7.3 and 7.4.)

It will be noticed that, although there exists a group of the pure rotations of the tetrahedron (E, $8C_3$, $3C_2$, a group called T), the group $T_d$ is not a direct product group of T with any other group (if it were there would be three, not two, additional classes of $T_d$ compared with T and they would have 1, 8 and 3 operations in them). The character table of the $T_d$ group is given in Table 7.6.

Table 7.6

| $T_d$ | E | $8C_3$ | $3C_2$ | $6S_4$ | $6\sigma_d$ |
|---|---|---|---|---|---|
| $A_1$ | 1 | 1 | 1 | 1 | 1 |
| $A_2$ | 1 | 1 | 1 | −1 | −1 |
| E | 2 | −1 | 2 | 0 | 0 |
| $T_1$ | 3 | 0 | −1 | 1 | −1 |
| $T_2$ | 3 | 0 | −1 | −1 | 1 |

We shall not give a detailed discussion of the bonding in tetrahedral molecules but summarize below in Table 7.7 the ways in which the various orbitals transform.

Table 7.7

| Orbitals of an atom at the centre of the tetrahedron | Symmetry |
|---|---|
| s | $A_1$ |
| $(p_x, p_y, p_z)$ | $T_2$ |
| $(d_{x^2-y^2}, d_{(1\sqrt{3})(2z^2-x^2-y^2)})$ | E |
| $(d_{xy}, d_{yz}, d_{zx})$ | $T_2$ |
| **Orbitals of the four atoms at the corners of the tetrahedron** | **Symmetry** |
| $\sigma$ | $A_1 + T_2$ |
| $\pi$ | $E + T_1 + T_2$ |

**Problem 7.22** Check that the transformations given in Table 7.7 are correct.

**Problem 7.23** Use the projection operator method to derive explicit forms for the $\sigma$ orbitals of the atoms at the corners of a tetrahedron.

(*Note:* the derivation of the form of the $\pi$ orbitals is a much more difficult problem and is best tackled by the techniques described in Appendix 4.)

**Problem 7.24** Use the data in Table 7.7 to describe the bonding in methane, $CH_4$.

It will be noted that double and triple degeneracies exist in a tetrahedral environment and, rather importantly, that the p and three of the d orbitals of a central atom both transform as $T_2$. In this lies, ultimately, the explanation of the fact that tetrahedral transition metal complexes tend to be more highly coloured than do octahedral. Because d and p orbitals have the same symmetry they can mix; this mixing makes some electronic transitions more 'allowed' in a tetrahedron than they are in an octahedron.

Just as for an octahedron, in a tetrahedral environment the d orbitals of a transition metal split into two sets; $d_{x^2-y^2}$ and $d_{(1/\sqrt{3})(2z^2-x^2-y^2)}$ are of E symmetry and, as we have commented, $d_{xy}$, $d_{yz}$ and $d_{zx}$ are of $T_2$. If a diagram analogous to Figure 7.28 is drawn for a tetrahedron then it is concluded that in this situation the $T_2$ set is of higher energy than the E — the inverse of the splitting found for an octahedron.

**Problem 7.25** Draw diagrams analogous to Figure 7.2 for a tetrahedral transition metal complex and thus show that it is reasonable that the $d_{xy}$, $d_{yz}$ and $d_{zx}$ orbitals are of higher energy than $d_{x^2-y^2}$ and $d_{(1/\sqrt{3})(2z^2-x^2-y^2)}$. (*Hint:* the lobes of the $d_{x^2-y^2}$ orbital point towards the mid-points of the faces and of $d_{xy}$ towards the mid-points of the edges of the cube corresponding to the tetrahedron.)

The splitting of the d orbitals in a tetrahedron is only about one half of that for the corresponding ligands arranged octahedrally (more accurately, $\frac{4}{9}$). This reduction in separation means that low spin (high field) tetrahedral complexes are virtually unknown — all are high spin (low field).

It is the fact that in both a tetrahedral and octahedral environment the d orbitals of a transition metal split into a set of two (E in $T_d$, $E_g$ in $O_h$) and a set of three ($T_2$ in $T_d$, $T_{2g}$ in $O_h$) and that the d orbitals of E symmetry in $T_d$ are the same as those of $E_g$ in $O_h$ (and similarly for the $T_2$ and $T_{2g}$ orbitals) which enables a common discussion of the two symmetries in specialized texts. In this common discussion the orbital sets are referred to as E and $T_2$ (one can think of the discussion of octahedral molecules taking place in the group O for there these are the correct symmetry labels). The two geometries are then distinguished by the fact that the E–$T_2$ splittings are of opposite sign.

## 7.5 THE DETERMINATION OF THE POINT GROUP OF A MOLECULE

We have now met in our discussion all of the different types of point group operations. It is therefore a convenient point at which to briefly review the chemically important point groups and the allocation of a molecule to the correct one. That is, we shall tackle the question: 'How do I decide what the symmetry of a particular molecule is?' Unless we can guarantee to get the correct answer we could end up trying to use an incorrect character table.

When a molecule has a single rotational axis, $C_n$, this is usually quite evident (it seems to become easier as $n$ increases). It may be that this $C_n$ axis is the only symmetry element, in which case the point group is $C_n$. Frequently, however, there will be other symmetry elements. If the only ones are mirror planes which contain the $C_n$ axis (there must be $n$ such $\sigma_v$ planes because, if we had only a single one, the existence of the $C_n$ axis would create the other $n-1$), then the point group is $C_{nv}$. Were there just a single mirror plane perpendicular to the $C_n$ axis (a $\sigma_h$ plane) we would have a $C_{nh}$ point group. The simultaneous existence of $n$ $\sigma_v$ mirror planes and a $\sigma_h$ plane means that there must be further symmetry elements, in particular, $n$ $C_2$ axes perpendicular to the original $C_n$ axis equally spaced around it (and symmetrically related to the $n$ $\sigma_v$ mirror planes). Such point groups are $D_{nh}$ (D for *d*ihedral). Can these additional $n$ $C_2$ axes exist without the $n$ $\sigma_v$ and $\sigma_h$? The answer is that they can; the point group produced is $D_n$. What of the simultaneous existence of the $C_n$, the $n$ $\sigma_v$'s and the $n$ $C_2$'s? These combinations also exist and lead to the point group $D_{nd}$. (The $n$ $\sigma_v$ axes now symmetrically interleave the $n$ $C_2$ axes and so are, more correctly, referred to as $\sigma_d$ mirror planes; in similar fashion the '$\sigma_v$' mirror planes in the $D_{nh}$ point groups should be called $\sigma_d$, although most authors do not do this. In such groups with $n$ even, the vertical mirror plane reflection operations divide into two classes, one of which is usually arbitrarily denoted $(n/2)\sigma_v$ and the other $(n/2)\sigma_d$.) The combination of $C_n$, $\sigma_h$ and $n$ $C_2$ does not exist — the existence of these elements requires the coexistence of $n$ $\sigma_v$ and we are back to $D_{nh}$.

In the present chapter we met for the first time $S_n$ operations. A set of such operations, together with the identity, can comprise a group, provided that $n$ is even. Such groups are called $S_n$, although the case of $n = 2$, $S_2$, is usually called $C_i$ because the operation $S_2$ is precisely equivalent to inversion in a centre of symmetry (Figure 7.30).

If it is clear that a molecule has non-coincident $C_n$ and $C_{n'}$ axes, where $n$ and $n'$ are both greater than two ($n$ can be equal to $n'$), then the point group of the molecule is one of those for which the x, y and z axes are interchanged by some of the operations of the group. The Cartesian axes then transform together as a triply degenerate irreducible representation. If the molecule contains a $C_5$ axis (it would actually have several) then the point group would be an icosahedral one — I or $I_h$ (pronounced 'eye-aich'). (An icosahedron of $I_h$ symmetry is shown in Figure 7.31.) If it contains a $C_4$ axis (there would be

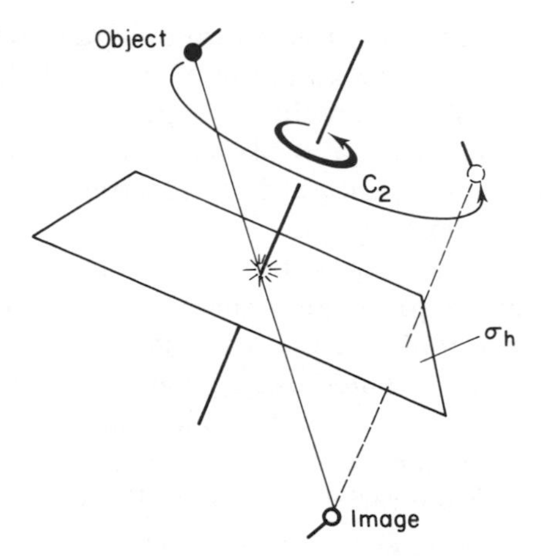


Figure 7.30 The operation i is equivalent to 'rotate about an arbitrary $C_2$ containing the i and reflect in a $\sigma_h$ containing the i'. That is, $i \equiv S_2$

three of them in all) then the point group would be cubic (or, equivalently, octahedral) — O or $O_h$. Finally, if its pure rotation axis of highest symmetry is a $C_3$ axis (there would be a total of four of these) then its symmetry would be that of a tetrahedral group — T, $T_d$ or $T_h$. The distinction between each of the two icosahedral, the two octahedral or three tetrahedral groups is quite

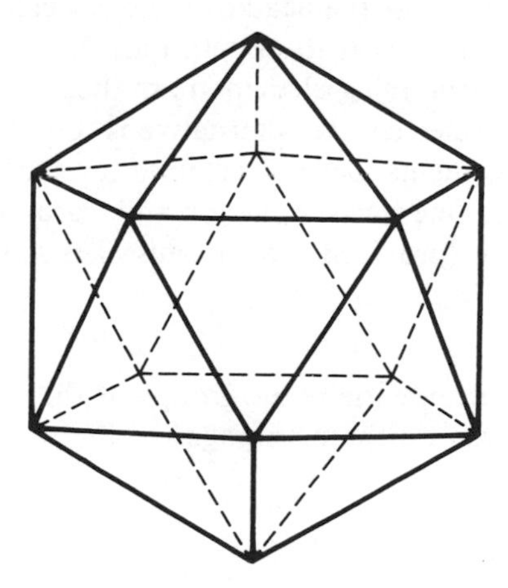

Figure 7.31 An icosahedron. A fivefold rotational axis passes through each pair of opposite corners and a threefold through the mid-points of each pair of opposite faces

simple. Groups with no suffix are groups with only pure rotation operations — they have no mirror planes and no centre of symmetry for example. They are rare. Groups with suffixes contain improper rotation operations. The distinction between $T_d$ and $T_h$ is that the latter contains a centre of symmetry whereas the former does not. The octahedral and tetrahedral groups, together, are often referred to as 'cubic' groups.

Linear molecules all have an axis — the molecular axis — about which rotation by any angle, no matter how large or small, is a symmetry operation. The 'fundamental' rotation operation, from which all others may be built up, is therefore a rotation by an infinitesimally small angle. It takes an infinite number of these rotations to return the molecule to its original position (rather than an equivalent, rotated, one). This axis is therefore a $C_\infty$ axis. Groups with such an axis are covered by our earlier discussion. If they have a centre of symmetry we obtain the $D_{\infty h}$ point group; if they have not, they are $C_{\infty v}$. Because they each have an order of infinity, the reduction of reducible representations in these groups has to be handled differently from the method we have developed. This is discussed in Appendix 5.

One point group remains. It is that which, in addition to the identity element, contains only the operation of reflection in a single mirror plane. It is denoted $C_s$ ('cee-ess').

The way that most experienced workers identify the point group of a molecule is to list as many symmetry operations as they can immediately see. This list is usually mental but the beginner may prefer to use a pencil and paper. Even if incomplete, this list may at once identify the point group; if not, it will certainly reduce the number of possibilities to two or three. A glance at the list of operations across the head of the character tables of the possible groups (Appendix 3) will reveal the operations in which they differ. These operations (or the corresponding elements) are then explicitly looked for and thus the correct group selected. An alternative is to mount a more systematic search for symmetry elements. Several schemes for such a search exist and we give one in Figure 7.32. One starts at the top and traces a path which ends with the correct point group (provided that no mistakes have been made!).

**Problem 7.26** Determine the point groups of the following molecules. The answers are given on the next page.

$C_2H_6$ (eclipsed)
$C_2H_6$ (staggered)
$C_6H_6$
$PF_5$ (trigonal bipyramid)
$Hg_2{}^{2+}$
$[PtCl_4]^{2-}$ (a planar molecule)
$TeCl_4$ (two pairs of equivalent chlorines)
$ClF_3$ (two equivalent fluorines)

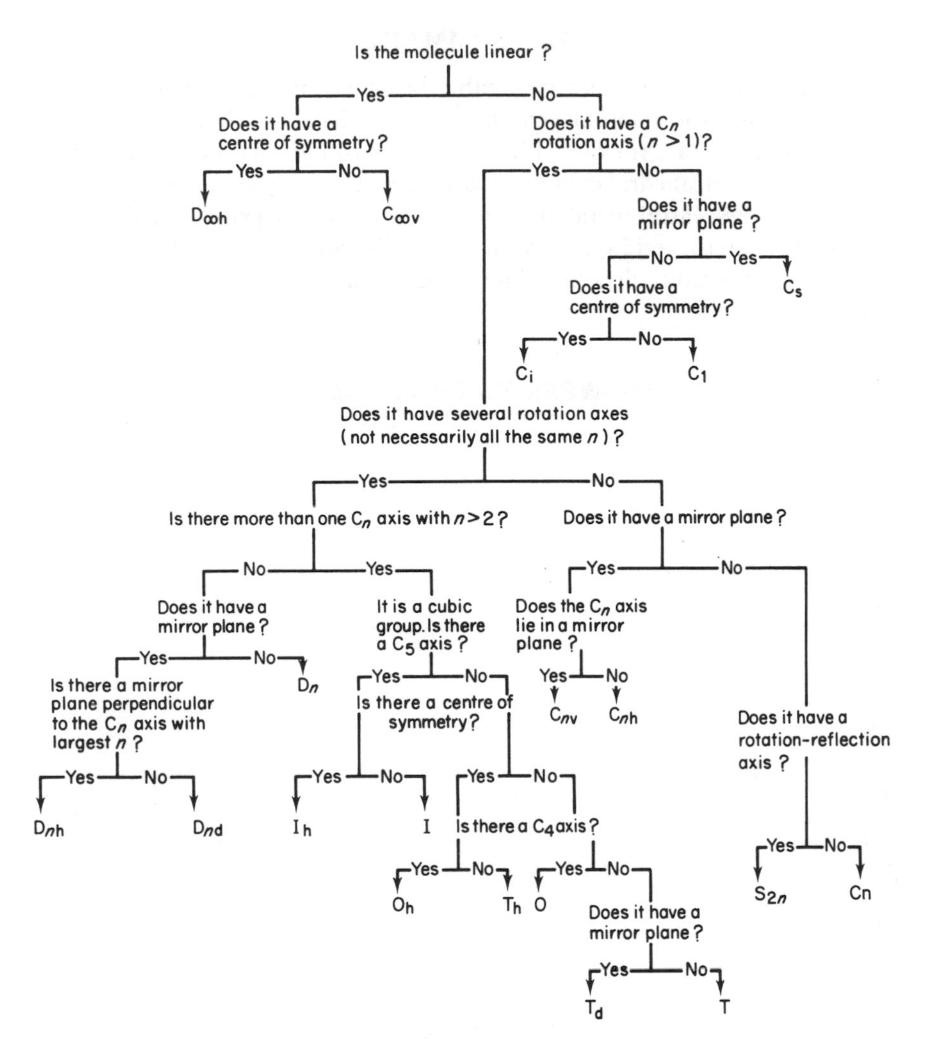


Figure 7.32 A 'yes–no' response table which enables a molecule to be assigned to the correct point group

The environment of the $Na^+$ ion in crystalline NaCl
$CO_2$
CO
$SO_4{}^{2-}$
$NO_3{}^-$ (planar)
$NO_2{}^-$
$C_3H_6$ (cyclopropane)
$Fe(C_5H_5)_2$ (ferrocene, eclipsed)
$Fe(C_5H_5)_2$ (ferrocene, staggered)

## 7.6 SUMMARY

In this chapter we have met two cubic point groups and their symmetry-enforced double and triple degeneracies (page 133). This high level of symmetry can compensate for quite high molecular complexity and discussions of octahedral and tetrahedral complexes are now invariably symmetry based, the d orbitals of the central metal atom being split into a degenerate pair (E) and a degenerate trio (page 153) ($T_2$). Finally, we discussed the identification of the point group of a molecule and a diagnostic scheme given in Figure 7.32 (page 167).

## ANSWERS TO PROBLEM 7.26

$D_{3h}$
$D_{3d}$
$D_{6h}$
$D_{3h}$
$D_{\infty h}$
$D_{4h}$
$C_{2v}$
$C_{2v}$
$O_h$
$D_{\infty h}$
$C_{\infty v}$
$T_d$
$D_{3h}$
$C_{2v}$
$D_{3h}$
$D_{5h}$
$D_{5d}$

## REFERENCE

1. W. von Niessen, W. P. Kraemer and G. H. F. Diercksend, *Chem. Phys. Letters*, **63**, 65, 1979.

# CHAPTER 8

# *Groups and subgroups*

## 8.1 INVARIANT AND NON—INVARIANT SUBGROUPS

So far in this book we have been concerned with particular molecules and, more important, particular symmetries. It is the purpose of the present chapter to adopt a less selective viewpoint and to examine, firstly, the relationships which exist between similar symmetries and, secondly, some of the consequences of low symmetry.

A molecule in which a central atom M is surrounded by six identical atoms or groups L, $ML_6$, is commonly, as we saw in Chapter 7, of octahedral symmetry, $O_h$. Suppose we now replace one of the L by a chemically similar, but different, atom or group X (for instance, both L and X could be halogen atoms). The molecule is then $ML_5X$ and has, at most, $C_{4v}$ symmetry (we here assume that the change from L to X does not lead to a gross structural change in the molecule — no $C_5$ axis is introduced, for example). Strictly, then, a discussion of the $ML_5X$ molecule should follow the general pattern developed in Chapter 5, where the elecronic structure of $BrF_5$ was considered, due allowance being made for the presence of X. However, the difference between L and X could be negligible (if they are isotopic variants of the same element, for instance). In such a case the difference in conclusions between a discussion based on $O_h$ symmetry and one based on $C_{4v}$ would be negligible. We conclude that there must be a continuity between the $C_{4v}$ and $O_h$ cases because, even if the difference between L and X is large, we could break this down into a series of small, hypothetical, steps. Similar arguments will apply whenever there is a similar relationship between two groups.

Just what is this relationship? It is clear that some of the symmetry elements (and, therefore, operations) of the $O_h$ point group are not present in $C_{4v}$ (thus we can think of the act of replacing L by X as destroying some symmetry elements). At a deeper level, the existence of operations common to the two groups will mean that there will be some relationship between their group multiplication tables and, very important, between their character tables. The group which has the smaller number of operations is referred to as a *subgroup* of the other; there are many fascinating relationships which exist between a group and its subgroups, some of which we shall meet in this chapter.

It is evident from group multiplication tables that the symmetry operations of a point group are not, in general, independent of one another — if we

Table 8.1

| $C_{2v}$ | E | $C_2$ | $\sigma_v$ | $\sigma_v'$ |
|---|---|---|---|---|
| E | E | $C_2$ | $\sigma_v$ | $\sigma_v'$ |
| $C_2$ | $C_2$ | E | $\sigma_v'$ | $\sigma_v$ |
| $\sigma_v$ | $\sigma_v$ | $\sigma_v'$ | E | $C_2$ |
| $\sigma_v'$ | $\sigma_v'$ | $\sigma_v$ | $C_2$ | E |

remove one symmetry operation we will usually as a consequence remove others also. Thus, for example, if in the case of the $C_{2v}$ point group we try to arbitrarily remove one mirror plane reflection operation we find that we simultaneously remove a second operation. This is shown in Table 8.1, where we see that a simple deletion of the $\sigma_v$ column and row still leaves a $\sigma_v$ entry (as a product of $C_2$ and $\sigma_v'$). We can only totally remove $\sigma_v$ entries from the table if we remove either $\sigma_v$ *and* $\sigma_v'$ or $\sigma_v$ *and* $C_2$. Deletion of the former pair leaves the point group $C_2$ (operations E, $C_2$) as a subgroup of $C_{2v}$ and deletion of the latter pair gives $C_s$ (operations E, $\sigma$). Because there is only one possible way of producing the subgroup $C_2$ from $C_{2v}$, $C_2$ is said to be an *invariant subgroup* of $C_{2v}$. More rigorously, an invariant subgroup contains only complete classes of the parent group. We thus understand why the point group $C_s$ is also an invariant subgroup of $C_{2v}$, despite the fact that it could be derived from either $\sigma_v$ or $\sigma_v'$; it is because these two mirror plane reflection operations are not in the same class in $C_{2v}$.

Not all subgroups are invariant. The $C_{3v}$ point group, which we met in Chapter 6, provides an example. We give the multiplication table for this group in Table 8.2; the operations are those indicated in Figure 6.1. It differs from all the other multiplication tables that we have given in that it is not symmetric about the leading diagonal (top left to bottom right). Put another way, for some combinations of operations the result depends on the order in which the operations are applied. Thus,

$$C_3^+ \ \sigma_v(1) = \sigma_v(2)$$

but

$$\sigma_v(1) \ C_3^+ = \sigma_v(3)$$

Table 8.2

| | | First operation | | | | | |
|---|---|---|---|---|---|---|---|
| | $C_{3v}$ | E | $C_3^+$ | $C_3^-$ | $\sigma_v(1)$ | $\sigma_v(2)$ | $\sigma_v(3)$ |
| Second operation | E | E | $C_3^+$ | $C_3^-$ | $\sigma_v(1)$ | $\sigma_v(2)$ | $\sigma_v(3)$ |
| | $C_3^+$ | $C_3^+$ | $C_3^-$ | E | $\sigma_v(2)$ | $\sigma_v(3)$ | $\sigma_v(1)$ |
| | $C_3^-$ | $C_3^-$ | E | $C_3^+$ | $\sigma_v(3)$ | $\sigma_v(1)$ | $\sigma_v(2)$ |
| | $\sigma_v(1)$ | $\sigma_v(1)$ | $\sigma_v(3)$ | $\sigma_v(2)$ | E | $C_3^-$ | $C_3^+$ |
| | $\sigma_v(2)$ | $\sigma_v(2)$ | $\sigma_v(1)$ | $\sigma_v(3)$ | $C_3^+$ | E | $C_3^-$ |
| | $\sigma_v(3)$ | $\sigma_v(3)$ | $\sigma_v(2)$ | $\sigma_v(1)$ | $C_3^-$ | $C_3^+$ | E |

We therefore have to take care to specify that the operations at the head of the columns are on the right in expressions such as those above. Equivalently, they are the *first* operation. This may seem strange but, if so, it is only because we are accustomed to reading from left to right so that in the first example above we *read* $C_3^+$ before $\sigma_v(1)$. However, if we let our operations operate on some function, $\psi$ say, then we have

$$C_3^+ \ \sigma_v(1)\psi$$

and, clearly, here $\sigma_v(1)$ must operate *before* $C_3^+$.

**Problem 8.1** Using Figure 6.1 check that the $C_{3v}$ group multiplication table given in Table 8.2 is correct.

In the multiplication table, Table 8.2, deletion of a single $\sigma_v$ operation requires that the other two are also deleted, to give the group $C_3$ as an invariant subgroup. However, deletion of the two $C_3$ operations causes the multiplication table to break up into three disconnected multiplication tables. This is because we can only remove the $C_3^+$ and $C_3^-$ entries from Table 8.2 by both deleting the $C_3^+$ and $C_3^-$ columns and rows and then deleting one of the pairs [$\sigma_v(1)$ and $\sigma_v(2)$] or [$\sigma_v(2)$ and $\sigma_v(3)$] or [$\sigma_v(3)$ and $\sigma_v(1)$]. For each of these three choices we arrive at $C_s$ as a subgroup. That is, there are three different but equivalent ways that $C_s$ can be obtained as a subgroup, so it is *not* an invariant subgroup of $C_{3v}$ — i.e. one that can be obtained in only one way (although, as we have seen, it is an invariant subgroup of $C_{2v}$).

**Problem 8.2** Check the above assertions by deleting from Table 8.2:
(a) All $C_3$ operations. Is it possible to just delete $C_3^+$ but leave $C_3^-$?
(b) One $\sigma_v$ operation.
(c) One $\sigma_v$ operation and the $C_3$ operations.

The distinction between invariant and non-invariant subgroups may seem rather academic. In fact, it has quite a variety of consequences, of which we will give just two examples. The first example concerns molecular dynamics. Suppose that a molecule of $C_{3v}$ symmetry is momentarily distorted — by a molecular vibration, for instance — to give a molecule of $C_s$ symmetry. Thus, in the ammonia molecule, one N—H bond might be momentarily longer (or shorter) than the other two. Because the symmetry of the molecule has been reduced to that of a non-invariant subgroup there exist other different but equivalent distortions (in the case of our ammonia molecule there are two such equivalent distortions, corresponding to distortion of one of the two other N—H bonds to give a different but equivalent arrangement of $C_s$ symmetry).

That is, because there are three different but equivalent $C_s$ subgroups of $C_{3v}$ there will be three equivalent distortions; the molecule would be of the same energy in each of the three equivalent configurations. In this situation, the distortion can 'rotate' from one bond to the next with no nett cost in energy. That is, the presence of non-invariant subgroups means that a molecule may indulge in some unexpected gymnastics. A full discussion of this is well beyond the scope of the present book although we shall go into more detail in Chapter 9, but it is clear that special care has to be taken in a detailed analysis of the vibrational and rotational properties of molecules with symmetries which have non-invariant subgroups.

Our second example is concerned with the character tables of invariant subgroups. When the operations of a point group can be written as a product of the operations of two of its invariant subgroups, then its character table can also be derived from those of these subgroups. Consider the $C_{2v}$ point group. We have seen that it has two invariant subgroups, $C_2$ and $C_s$. It follows that all of the operations of $C_{2v}$ can be derived from those of these two subgroups. We take each of the operations of one invariant subgroup and combine it, in turn, with each of the operations of the other invariant subgroup. Thus, in our case, we have the steps shown in Table 8.3. That is, the operations of $C_{2v}$ are products of the operations of $C_2$ and $C_s$. Using the language of Section 4.3, the group $C_{2v}$ is said to be the direct product of the groups $C_2$ and $C_s$, a relationship usually written as $C_{2v} = C_2 \times C_s$. In Section 4.3 we also saw that a similar property holds for the corresponding character tables. Thus, in the present case we take the character table for $C_2$ and multiply the whole of it by the characters of the $C_s$ table to give a table four times the size of that of $C_2$. We show this in Table 8.4 where, for simplicity, we have written out on the left the $C_2$ character table four times. Each one is then multiplied by the corresponding $C_s$ character to give the $C_{2v}$ table.

The $C_{2v}$ character table given in Table 8.4 is the same as that met in Chapter

Table 8.3

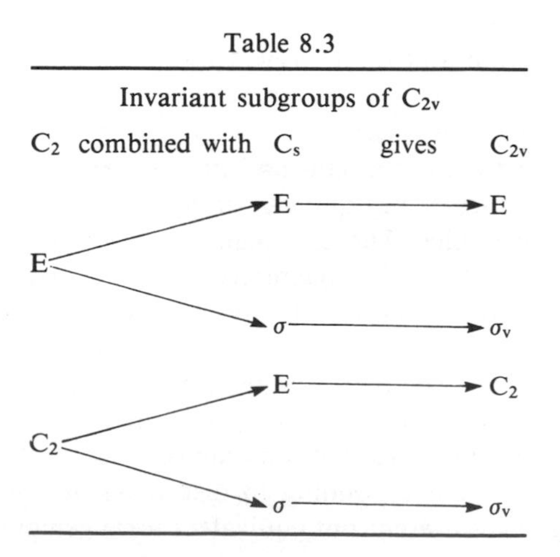

Table 8.4

| $C_2$ | E | $C_2$ | E | $C_2$ |
|---|---|---|---|---|
| A | 1 | 1 | 1 | 1 |
| B | 1 | −1 | 1 | −1 |
| A | 1 | 1 | 1 | 1 |
| B | 1 | −1 | 1 | −1 |

×

| $C_s$ | E | $\sigma$ |
|---|---|---|
| A′ | 1 | 1 |
| A″ | 1 | −1 |

=

| $C_{2v}$ | E | $C_2$ | $\sigma_v$ | $\sigma_v'$ |
|---|---|---|---|---|
| $A_1$ | 1 | 1 | 1 | 1 |
| $B_1$ | 1 | −1 | 1 | −1 |
| $A_2$ | 1 | 1 | −1 | −1 |
| $B_2$ | 1 | −1 | −1 | 1 |

2 (Table 2.4) with the $A_2$ and $B_1$ irreducible representations interchanged in position.

**Problem 8.3** Check through the individual steps in Tables 8.3 and 8.4.

We have, of course, already met examples of this relationship between character tables. In Chapter 4 we exploited the fact that $D_2$ and $C_i$ are both invariant subgroups of $D_{2h}$ (Tables 4.3 and 4.4); in Chapter 7 we used the fact that $O_h$ has invariant subgroups O and $C_i$ (Table 7.2 and the preceding discussion).

**Problem 8.4** Show that the operations of the group $C_{3v}$ are the product of operations of the groups $C_3$ (E, $C_3^+$, $C_3^-$) and $C_s$ (E, $\sigma$). However, because $C_s$ is not an invariant subgroup of $C_{3v}$, the character table of $C_{3v}$ is *not* the direct product of the character tables of $C_3$ and $C_s$. This is immediately seen when the character tables of $C_{3v}$ and $C_3$ are compared (Appendix 3).

At the beginning of this chapter we recognized that $C_{4v}$ is a subgroup of $O_h$. It is not an invariant subgroup because there are eight $C_4$ operations in $O_h$ but only two in $C_{4v}$. It is also evident that the character table of $O_h$ is not a direct product of that of $C_{4v}$ with any other group because that of $O_h$ contains triply degenerate irreducible representations whereas $C_{4v}$ does not. This is another illustration of the rule that the character table of a group is never the direct product of the character table of a non-invariant subgroup with that of another group.

## 8.2 CORRELATION TABLES

Having discussed how the character table of a group may be related to that of its subgroups we now consider the opposite problem: how is the character

Table 8.5

| $C_{3v}$ | E | $2C_3$ | $3\sigma_v$ |
|---|---|---|---|
| $A_1$ | 1 | 1 | 1 |
| $A_2$ | 1 | 1 | −1 |
| E | 2 | −1 | 0 |

| $C_s$ | E | $\sigma$ |
|---|---|---|
| A′ | 1 | 1 |
| A″ | 1 | −1 |

table of a subgroup related to that of the parent group? Again, the general form of the relationship is best seen by considering an example. The example which we choose corresponds to the physical situation described earlier in this chapter — that in which a molecule of $C_{3v}$ symmetry is distorted to give a structure with $C_s$ symmetry (i.e. a distortion leading to the loss of the threefold axis). The character tables of the $C_s$ and $C_{3v}$ point groups are given in Table 8.5. (Note that in the $C_s$ character table a single prime as a superscript indicates something which is symmetric with respect to a mirror plane reflection and a double prime indicates antisymmetry. This use of primes reappears in other point groups — see Appendix 3.) In the $C_{3v}$ character table in Table 8.5 we have indicated the loss of the $C_3$ axis by deleting the column associated with the corresponding operations. Since loss of this axis also leads to the loss of two $\sigma_v$ mirror planes we have also deleted the figure 3 from the $3\sigma_v$ entry. It is clear from Table 8.5 that the remaining characters of the $A_1$ irreducible representation of the $C_{3v}$ point group are those of the A′ irreducible representation of the $C_s$ point group. We say that the '$A_1$ irreducible representation of $C_{3v}$ *correlates* with the A′ irreducible representation of $C_s$'. This means that any function or object which transforms as $A_1$ in $C_{3v}$ *must* transform as A′ in $C_s$ when the molecular symmetry changes. Similarly, Table 8.5 shows that the $A_2$ irreducible representation of $C_{3v}$ correlates with A″ of $C_s$.

The E irreducible representation of $C_{3v}$ is both interesting and important for it does not correlate uniquely with a single irreducible representation of $C_s$. Rather, it gives rise to a reducible representation, one which is readily seen to have A′ + A″ components. In summary, then, we have the correlations shown in Table 8.6.

This example illustrates the general theorem that each irreducible representation of a group gives rise to a representation, which may be either reducible or irreducible, of each of its subgroups. Tables showing these correlations — so-called *correlation tables* — are available, but it is very easy to work them out using the example given above as a model. Working out sometimes has an

Table 8.6

| $C_{3v}$ | | $C_s$ |
|---|---|---|
| $A_1$ | ⟶ | A′ |
| $A_2$ | ⟶ | A″ |
| E | ⟶ | A′ + A″ |

advantage over the use of tables. The $D_{2h}$ group was described in Chapter 4 (its character table is given in Table 4.1). This group has $C_{2v}$ as a subgroup and we can correlate the irreducible representations of the two groups very easily.

**Problem 8.5** Use either Tables 4.1 and 2.4 or Appendix 3 to correlate the irreducible representations of the $D_{2h}$ and $C_{2v}$ groups. (*Hint:* if you find this problem more difficult than expected, read the next part of this section.)

As you may have discovered when tackling Problem 8.5, there is a catch in correlating from $D_{2h}$ to $C_{2v}$. the $D_{2h}$ group has three different $C_2$ axes. The precise correlation between the two groups depends on which of the three twofold axes we retain in going from $D_{2h}$ to $C_{2v}$. This does not indicate any fundamental problem, rather that we may have to relable coordinate axes (and associated basis functions) in moving between the two groups. In compilations of correlation tables it is usual to indicate all three possible $D_{2h} \rightarrow C_{2v}$ correlations, but one still has to decide which correlation is appropriate before using the tables. In such cases even experienced workers may find that they are less likely to make a mistake by working out the correlations for themselves rather than by using the tables!

There is another way of showing correlations, and that is by use of a diagram. That for the correlation between irreducible representations of the $C_{3v}$ and $C_s$ point groups is given in Table 8.7. Such diagrams emphasize another aspect of the consequences of a decrease in symmetry. Table 8.7 shows, for example, that a function transforming as $A_1$ in $C_{3v}$ and one of the two functions transforming as E have a common symmetry in $C_s$ — that described by the A′ irreducible representation. This means that in $C_s$ symmetry these two functions can interact with each other, an interaction which is symmetry-forbidden in $C_{3v}$ symmetry.

Another aspect of a reduction in symmetry, equally evident from either Table 8.6 or Table 8.7, is that a decrease in symmetry may lead to a decrease in degeneracy. In our example, the degeneracy of functions transforming as E in $C_{3v}$ is lost in $C_s$. A particularly important case is that of octahedral transition metal coordination compounds, discussed in Chapter 7. Although much

Table 8.7

| $C_{3v}$ | $C_s$ |
|---|---|
| $A_1$ | A′ |
| $A_2$ | |
| E | A″ |

Table 8.8

| $D_{3d} \xleftarrow[\text{(along } C_3)]{\text{trigonal distortion}}$ | $O_h$ | $\xrightarrow[\text{(along } C_4)]{\text{tetragonal distortion}} D_{4h}$ |
|---|---|---|
| $A_{1g}$ | $A_{1g}$ | $A_{1g}$ |
| $A_{2g}$ | $A_{2g}$ | $B_{1g}$ |
| $E_g$ | $E_g$ | $A_{1g} + B_{1g}$ |
| $A_{2g} + E_g$ | $T_{1g}$ | $A_{2g} + E_g$ |
| $A_{1g} + E_g$ | $T_{2g}$ | $B_{2g} + E_g$ |
| $A_{1u}$ | $A_{1u}$ | $A_{1u}$ |
| $A_{2u}$ | $A_{2u}$ | $B_{1u}$ |
| $E_u$ | $E_u$ | $A_{1u} + B_{1u}$ |
| $A_{2u} + E_u$ | $T_{1u}$ | $A_{2u} + E_u$ |
| $A_{1u} + E_u$ | $T_{2u}$ | $B_{2u} + E_u$ |

of the basic theory of such compounds is conveniently developed assuming full octahedral symmetry ($O_h$), real-life examples usually show some minor distortion. The most important cases are those in which such a distortion is either along a fourfold or a threefold axis (either distortion therefore destroying all other fourfold and threefold rotation axes), as shown in Figure 8.1. The appropriate correlation table is given in Table 8.8.

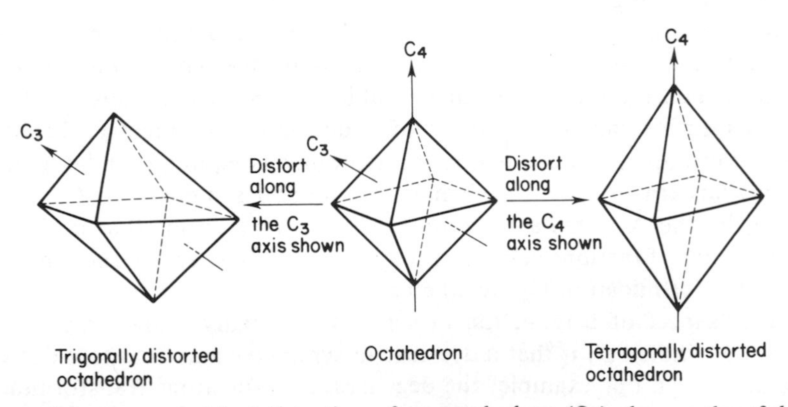


Figure 8.1 A symmetrical distortion of an octahedron ($O_h$) along a threefold axis gives a figure with $D_{3d}$ symmetry whilst a symmetrial distortion along a fourfold axis gives a $D_{4h}$ figure.

**Problem 8.6** Use the character tables of the $O_h$, $D_{4h}$ and $D_{3d}$ point groups in Appendix 3 to check the correlations given in Table 8.8.

We see from Table 8.8 that in $D_{4h}$ symmetry all degeneracies present in $O_h$ symmetry are at least partially removed. One important consequence of this is that when a single spectral band is predicted in the electronic absorption spectrum of an octahedral transition metal complex this band would be expected to show a splitting if the real symmetry is $D_{4h}$ and a degeneracy is involved in the transition (for instance, the excited state might be triply degenerate). Such a splitting could take the form of the observation of a separate peak, a shoulder or an asymmetry on the band. The $D_{3d}$ case shows, however, that it is not always true that a reduction in symmetry causes all degeneracies to be relieved (i.e. a splitting to occur); thus the $E_g$ and $E_u$ irreducible representations of $O_h$ persist in $D_{3d}$. There is, however, a trap for the unwary. In the point group $O_h$ we chose a $C_4$ axis as the z axis; we would do the same in $D_{4h}$. In $D_{3d}$, the axis of highest symmetry is a $C_3$ axis and this is the z axis. It follows that, although $E_g$ of $O_h$ becomes $E_g$ of $D_{3d}$ it is *not* true that the basis functions for $E_g$ in $O_h$, $x^2 - y^2$ and $(1/\sqrt{3})(2z^2 - x^2 - y^2)$, are basis functions for $E_g$ in $D_{3d}$. A detailed analysis, using the methodology of Appendix 2, is needed to describe the correlations between basis functions in $O_h$ and $D_{3d}$.

In practice, the correlations which exist between groups are quite important for two reasons. Firstly, as indicated above, they enable the properties of low symmetry molecules to be related to those of high symmetry species. Another aspect of this occurs when a molecule is of high symmetry but is trapped in a low symmetry environment — an octahedral molecule on a low symmetry lattice site in a crystal, for example. Secondly, some of the problems of degenerate representations — and we met some in the last chapter — can often be neatly side-stepped by pretending that a molecule has a lower symmetry than is in fact the case — so that the degeneracy is split (or 'relieved') — and, after working in the low symmetry group, using a correlation relationship to apply the result to the high symmetry case.

Another interesting aspect of the relationship of a group to its subgroup is that the number of symmetry operations in a group (the order of the group) is a simple multiple of the number of symmetry operations of any of its subgroups. The multiplicative factor — which is always an integer — is called the *index* of the subgroup (relative to the particular parent group). Thus, the $C_3$ group (of order 3) is a subgroup of index 2 of the point group $C_{3v}$ (of order 6). However, the same $C_3$ group (of order 3) is a subgroup of index 40 of the point group $I_h$ (of order 120).

**Problem 8.7** Use Appendix 3 to determine the index of each of the following subgroups of $O_h$.

$$D_{4h},\ C_{4v},\ D_{3d},\ C_{3v},\ D_{2h},\ C_{2v}$$

An important application of the concept of index concerns rotational subgroups. A point group may contain operations which are just proper rotation operations (such as $C_2$, $C_3$, and so on) or it may consist of some operations which are pure rotations and others which are improper rotations (such as $\sigma_v$, i, $S_4$). By deleting all of the improper rotations one can always obtain a subgroup of a group which contains both proper and improper rotations. What remains is the *pure rotational subgroup* of the parent group. This subgroup is always of index 2. The importance of rotational subgroups is their relationship to the (infinite) group consisting of all the pure rotation operations associated with a sphere. A rotation of any angle about any radial axis is a symmetry operation of a sphere. In particular, infinitesimally small rotations are symmetry operations. This property is closely associated with the importance of angular momentum in the theory of atomic structure. When there is an atom at the centre of mass of a molecule then all the pure rotational symmetry elements associated with the molecular point group pass through it. The corresponding rotation operations are all that remain of the infinity of rotation operations which would have turned this atom into itself if the rest of the molecule were not present. What remains of the consequences of the angular momentum in the free atom are therefore manifest in the molecule in its pure rotational subgroup. In particular, this group provides a method of determining how the degeneracies which may be associated with the free atom (and these degeneracies may be quite large) are split up when the atom is placed in the molecule. For the metal atom at the centre of a transition metal complex, in particular, this is quite invaluable information. As we saw in Chapter 7, these compounds often contain unpaired electrons. Besides behaving like tiny bar magnets themselves, the motion of these electrons can have a magnetic effect, like an electrical current in a solenoid. This additional, orbital, magnetism is closely connected with any angular momentum possessed by the unpaired electrons.

**Problem 8.8** Use the symmetry operations listed at the top of the character tables in Appendix 3 to show that deletion of improper rotation operations in the following point group leads to a pure rotational subgroup of index 2:

$$I_h, T_h, D_{5h}, C_{2h}, D_{3d}$$

(*Note:* in several of these examples it is possible to obtain subgroups by deletion of all improper and some proper rotations. Such subgroups are not of index 2. The statements made in the text refer to the largest pure rotational subgroup of a given group.)

## 8.3 SUMMARY

There are relationships between a group and its subgroups. The operations of a group can be immediately obtained from the operations of its subgroups (page 172) as can its character table (page 172), provided that the subgroups are invariant (page 170). Correlations exist between the irreducible representations of a group with those of its subgroups and are useful in discussions associated with molecules which approximate to high symmetry species (page 173).

Groups containing improper rotation operations always have a pure rotational subgroup of index 2 (page 177).

# CHAPTER 9
# *Molecular vibrations*

So far in this book we have been largely concerned with the techniques of group theory and their application to the problem of the electronic structure of molecules. There are a few additional techniques which we have yet to meet, but we are already in a position to discuss one important application of group theory in quite a different area — the analysis of molecular vibrations. Although a chemist may be interested in vibrational spectra in a qualitative way — some molecular fragments reveal their presence by characteristic 'fingerprint' peak patterns — our concern is with a more detailed analysis of the relationship between spectra and structure. Vibrational spectra provide a way of determining the geometrical arrangement of groups attached to an atom. Because such geometrical arrangements are often a problem in inorganic chemistry it is in this area that the methods of this chapter find current research application. In this chapter, then, our concern is with a prediction of the number of infrared and Raman peaks (i.e. spectroscopic features resulting from the excitation of molecular vibrations) that is expected for a particular molecular geometry; the predictions will vary with geometry and so provide a method of distinguishing between alternative possible geometries. However, the spectral activity of vibrations is a subject which we must defer to Chapter 10 because this discussion requires a further development of our basic ideas. In the present chapter we shall be concerned solely with determining the symmetries of the normal modes of vibration of a molecule. Having determined these symmetries we shall find in Chapter 10 that the selection rules follow immediately.

## 9.1 NORMAL MODES

A normal mode is a 'natural' vibration of a molecule; just as a tuning fork has a 'natural' frequency and motion (or mode), so too has a molecule, a difference being that all molecules (except diatomic ones) have more than one 'natural' frequency. More precisely, a normal mode is one which has the property that if each atom in a molecule is displaced from its equilibrium position by a displacement which corresponds to its maximum amplitude in a normal mode, then when the atoms are simultaneously 'let go' the atoms will all undergo a motion at the same frequency. Further, once having been 'let go' they will all simultaneously pass through the equilibrium configuration and, later, simultaneously again reach the positions of maximum amplitude. It follows that the motions of symmetry-related atoms in the molecule will be simply related to each other, so that it is possible to place a symmetry label on each normal mode. Our concern, then, is with the determination of these symmetry labels.

The description which we have just given of the vibrations of molecules differs from that which seems intuitively more realistic. In reality, we would expect the motion of individual atoms in a molecule to be much more complicated than this; we envisage atoms moving in apparently random directions, amplitudes and phases. It turns out that, provided the amplitudes are not too great, such a complicated motion can be regarded as a sum of normal motions occurring with different phases, just as the sound — and complicated sound wave motion — from a violin can be regarded as a sum of different harmonic frequencies. Clearly, normal modes of vibration are of key importance in the description of molecular vibrations. These normal modes are quantized — just like the harmonics of a vibrating stretched string — and it is possible to add a further quantum of vibrational energy to any mode. For some of the modes this quantum can be added by infrared radiation and for some it can be added by a Raman mechanism. These processes are the basis of infrared and Raman spectroscopies. In adding an additional quantum we can, to a good approximation, ignore the other vibrations already occurring within a molecule. That is, we can pretend that the molecule is not vibrating at all! The formal justification for this pretence is not usually given but it is not very difficult — and is quite good fun — and so we include it as an optional section at the end of the present chapter.

There are two distinct methods which may be adopted in tackling the problem of determining the symmetries of the normal modes of vibration of a molecule. Firstly, one may break the molecule up into fragments and consider the vibrations of each fragment in turn. We cannot generally expect to obtain the normal vibrations in this way because the normal vibrations will probably involve the movement of some atoms which are not in the fragment under consideration. As we shall see, an additional step is needed to obtain the normal vibrations.

The approach to molecular vibrations through the vibrations of molecular fragments requires that some care is taken in the choice of fragments. Thus, for methyl chloride $CH_3Cl$, one would consider separately the C—H bond stretching vibrations, the C—Cl bond stretch, the HCH angle change and the HCCl angle change vibrations. Such a breakdown is the most useful starting point for the study of the vibrations of a particular molecule since each type of motion is often associated with a characteristic spectral region (which, correctly, suggests that, despite what we said above, the vibrations of fragments are often not very different from normal vibrations). The use of this method depends on one's ability to spot all of the different vibrators (or 'internal coordinates' as they are usually called), and it is all too easy to miss some. It is here that the second method is useful since it provides a check on the first. In it one determines all of the symmetries of the normal modes together; it has no place for chemical experience or intutition about the motion of the atoms involved in each normal vibration.

We shall consider both methods. That which divides the vibration of the molecule into several smaller problems, that of the vibrations of fragments, is the simpler and we consider it first.

## 9.2 SYMMETRY COORDINATES

It is easiest to work with a particular example and so we shall consider the C—H stretching vibrations of the methyl chloride molecule, $CH_3Cl$, using the $C_{3v}$ character table given in Table 9.1. Although the frequencies of the A—B stretching vibrations in an $AB_3$ unit will be dependent on the masses of the atoms involved, the symmetries of the vibrations will not. Our discussion is therefore equally appropriate to other molecules of $C_{3v}$ symmetry, such as ammonia. The problem is easier to visualize if we exploit this generality and, instead of worrying about atoms, think instead of displacements (e.g. a bond stretch). These may be symbolically represented by arrows as shown in Figure 9.1. There are three C—H bond stretches to consider and so there are three arrows in Figure 9.1.

It is easy to show that the three arrows of Figure 9.1 form the basis for a reducible representation, the components of which are the $A_1 + E$ irreducible representation of the $C_{3v}$ point group. We must consider carefully what this

Table 9.1 The character table of the $C_{3v}$ point group

| $C_{3v}$ | E | $2C_3$ | $3\sigma_v$ | |
|---|---|---|---|---|
| $A_1$ | 1 | 1 | 1 | z |
| $A_2$ | 1 | 1 | −1 | $R_z$ |
| E | 2 | −1 | 0 | (x, y), ($R_y$, $R_x$) |

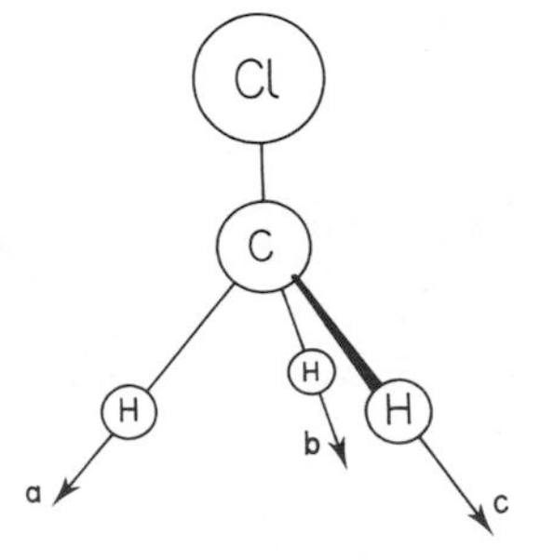


Figure 9.1 Three independent C—H stretch internal coordinates in $CH_3Cl$

means. As we have seen, vibrational modes may be labelled by that irreducible representation of the point group which describes the phase relationship between the motions of symmetry-related internal coordinates (bond stretches in the present context). A diagram such as Figure 9.1 merely indicates the existence of some independent vibrators (of which there are three in the present case). We could re-draw the diagram with the direction of some of the arrows reversed if we chose and this would not change the fundamental problem. When we obtain, by the usual projection operator technique, a function (or a corresponding diagram) which shows a symmetry-adapted combination of internal coordinates (a so-called 'symmetry coordinate') we can no longer change one phase arbitrarily; all must change simultaneously. There is nothing new in this — we have met analogous situations in previous chapters — but a particular confusion arises here because, as we shall immediately see in Figure 9.2(a), a diagram such as Figure 9.1 may be indistinguishable from one used to depict a symmetry coordinate. The moral is — read text and captions carefully! We shall not detail the application of the projection operator tech-

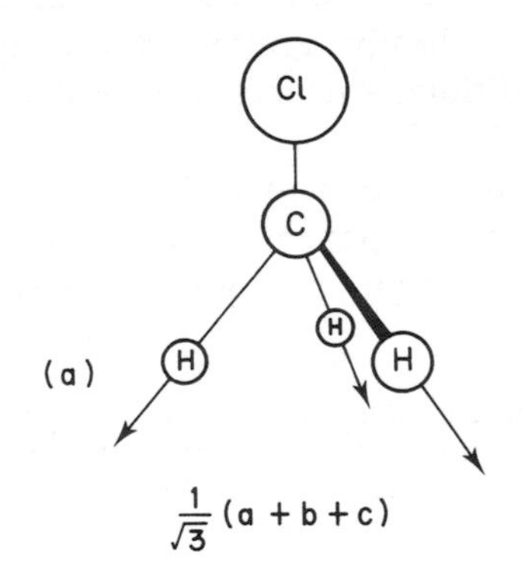


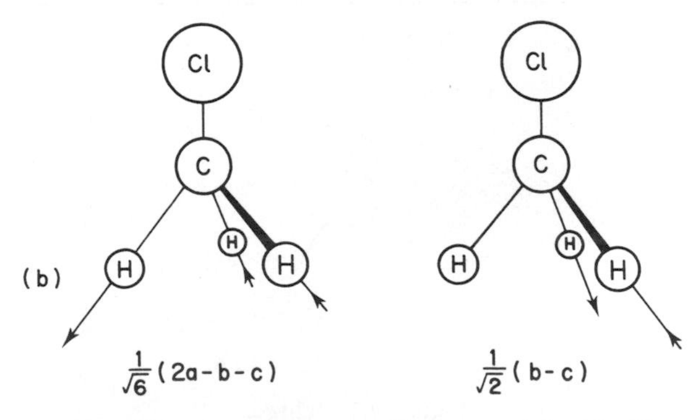


Figure 9.2 (a) The $A_1$ combination of the C—H stretching internal coordinates in $CH_3Cl$ and (b) the E combinations

nique to the present problem because this application is indistinguishable from the one that we described in Chapter 6. In Figure 6.5 we labelled the 1s orbitals of the hydrogen atoms in ammonia as a, b and c, and proceeded in the text to derive the $A_1$ and E combinations. They were found to be (Table 6.3)

$$\psi(A_1) = \frac{1}{\sqrt{3}}(a + b + c)$$

$$\psi_1(E) = \frac{1}{\sqrt{6}}(2a - b - c)$$

$$\psi_2(E) = \frac{1}{\sqrt{2}}(b - c)$$

In this problem we assumed that a, b and c all had the same phase. The present, vibrational, problem is, mathematically, just the same provided that we now take a, b and c to represent the bond extensions shown in Figure 9.1. The analogy that we have just drawn between the group theoretical analysis of molecular bonding and molecular vibrations is quite general. When the members of two different basis sets — for instance, the atomic orbitals in a bonding problem and the internal coordinates in a vibrational problem — transform isomorphously (i.e. there is a one-to-one correspondence between members of the two sets) then their transformations lead to symmetry-adapted combinations of identical symmetries and mathematical form. Corresponding to the symmetry-adapted group orbitals of bonding theory we have the symmetry coordinates of vibrational theory. Similarly, corresponding to the molecular orbitals of bonding theory (which are combinations of symmetry-adapted group orbitals of the same symmetry species) we have normal coordinates in vibrational theory. These normal coordinates are linear combinations of symmetry coordinates of the same symmetry species. However, because neither is symmetry-determined, the coefficients with which symmetry coordinates appear in a normal coordinate have no relationship to the coefficients with which symmetry-adapted orbitals appear in molecular orbitals.

**Problem 9.1** Use the results of the sections indicated to write down the symmetry species and symmetry coordinates of:

(a) The C—H stretching vibrations of ethylene (Section 4.5).
(b) The Br—F (equatorial) stretching vibrations of $BrF_5$ (Section 5.5).
(c) The S—F stretching vibrations of $SF_6$ (Section 7.2).

(*Note:* angle change vibrations can be handled similarly. However, there is a hidden problem associated with them which is discussed towards the end of the present section.

In Chapter 6, the molecular orbitals of the ammonia molecule were expressed as combinations of orbitals of the same symmetry species. (Thus, the $A_1$

N—H bonding molecular orbital of ammonia was taken as an in-phase sum of the $A_1$ symmetry-adapted combination of hydrogen 1s orbitals with a nitrogen 2s–2p combination of $A_1$ symmetry.) In just the same way a normal coordinate (of a particular symmetry species) is taken as a sum of contributions from symmetry coordinates of the same symmetry species. Thus, if there is only one symmetry coordinate of a particular symmetry type, then this symmetry coordinate *is* the normal coordinate.

The $A_1$ and E symmetry coordinates derived from the C—H bond stretching coordinates shown in Figure 9.1 are drawn in Figure 9.2. Evidently, a crucial difference between $A_1$ and E symmetry coordinates is that in the latter a C—H bond may contract as its neighbour stretches, whereas for the former all of the C—H vibrators stretch and contract together. Provided, then, that a C—H bond is sensitive to whether its neighbour is contracting or stretching, the spectral bands associated with the $A_1$ and E modes will have different energies and will therefore appear at different frequencies. The greater the sensitivity of (or, as it is usually put, the greater the coupling between) the C—H vibrators the greater the separation between the spectral peaks associated with the different vibrations. With strong coupling we would expect to see two peaks in the C—H stretching region of the spectrum, provided that the normal modes closely resemble the symmetry coordinates we have derived and that both modes are spectrally active. In the present case we shall show in Chapter 10 that the $A_1$ and E modes are both infrared and Raman active. Further, it turns out that the normal modes are closely approximated by our symmetry coordinates. It is the direct connection between the analysis and prediction for a particular spectral region which makes the molecular fragment approach so useful. However, it is not without its problems. Suppose we are considering the HCH bond angle change vibrations of methyl chloride. This basis set is pictured in Figure 9.3, where we have represented each bond angle change by a double-headed arrow. It is simple to show that this basis gives rise to a representation with $A_1 + E$ components. What is the form of the $A_1$ vibration?

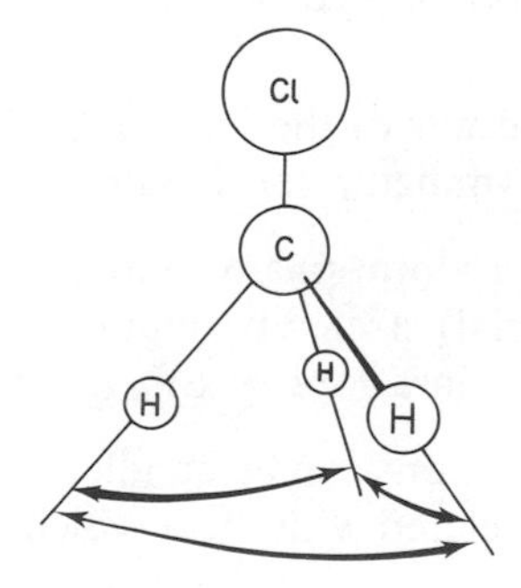


Figure 9.3 Three independent HCH bond angle change internal coordinates in $CH_3Cl$

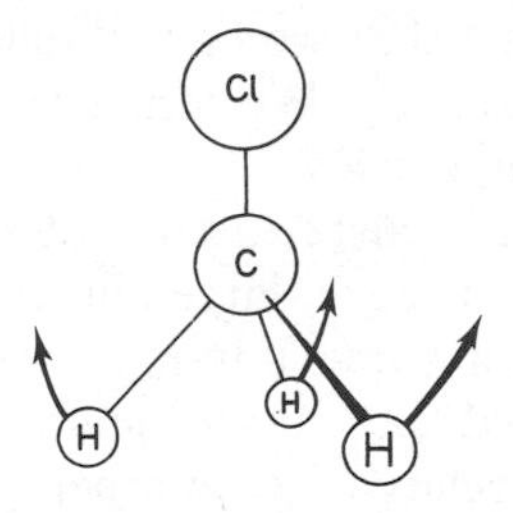


Figure 9.4 The $A_1$ combination of HCH (or HCCl) bond angle change internal coordinates in $CH_3Cl$

In this vibration, all HCH bond angles must increase and decrease in phase. It is easy to see that the only way that they can do this is as shown in Figure 9.4. However, as this figure shows, this vibration seems to be more evidently associated with changes in the HCCl bond angles! Not surprisingly, the HCCl bond angle change set of internal coordinates also gives rise to an $A_1$ vibration, which is also that shown in Figure 9.4! This figure makes it clear that when we fragment a molecule we may find ourselves inadvertently duplicating vibrations. This problem is not confined to angle change vibrations. In a cyclic system, for instance, vibrations in which some bond lengths increase whilst others decrease usually have a simultaneous change in at least one bond angle. In this situation there is not a 'right' and a 'wrong' vibration — we have duplicated a vibration and this duplication is inseparable from the method we are using.

The above discussion of the vibrations of $CH_3Cl$ is summarized and completed in Table 9.2. In this table we have followed convention by denoting bond-stretching internal coordinates by the symbol $\nu$ and bond angle change (deformation) coordinates by the symbol $\delta$. The total number of symmetry coordinates given in this table is ten (the E's are doubly degenerate), which is to be compared to the number of normal vibrations predicted by the $3N-6$ rule. In $CH_3Cl$, $N$ (the number of atoms) is five and so the $3N-6$ rule requires that this molecule has just nine normal vibrations. As we have seen, the

Table 9.2

| Internal coordinate | Symmetry coordinate symmetries |
|---|---|
| $\nu(C-H)$ | $A_1 + E$ |
| $\nu(C-Cl)$ | $A_1$ |
| $\delta(H-C-H)$ | $A_1 + E$ |
| $\delta(H-C-Cl)$ | $A_1 + E$ |

disparity arises from the fact that we have duplicated an $A_1$ symmetry coordinate in the two deformation sets. We conclude that the normal vibrations of $CH_3Cl$ are of $3A_1 + 3E$ symmetries.

Although this conclusion is correct, it is not entirely justified. That this is so becomes evident if we consider a large and complicated molecule. In such cases it is easy to overlook some internal coordinates or to be uncertain whether or not they have already been included — as when one part of a molecule rocks or twists relative to another part, for instance. Equally, duplication of symmetry species such as that which occurred in $CH_3Cl$ could easily go undetected. Clearly, a more systematic and reliable method is needed and it is this which we give in the next section.

From what has just been said, the value of the fragment model would appear dubious. This is not so; in practice, the model is not used to predict the symmetries of the normal modes of a molecule. Rather, it is used to predict the number of bands expected in the spectral regions associated with individual internal coordinates such as those listed on the left-hand side of Table 9.2. Even so, care has to be taken to spot duplication and here qualitative diagrams of symmetry coordinates such as Figures 9.2 to 9.4 are of great help.

**Problem 9.2** Associated with the equilateral triangular arrangement of three carbon atoms in cyclopropane, $C_3H_6$, are three $\nu$ (C—C) stretching vibrations and three $\delta$ (C—C—C) angle change vibrations. Derive the symmetry coordinates associated with these internal coordinates and decide how they may be brought into conformity with the requirements of the $3N - 6$ rule. (*Note:* although the symmetry of cyclopropane is $D_{3h}$ it is simplest to work in the point group $C_{3v}$ — identical results are obtained).

## 9.3 THE WHOLE-MOLECULE METHOD

The alternative technique, that of considering the entire molecule and generating all of the vibrations, is particularly useful as a check on the results obtained from a fragment analysis such as that described above. Not only does it correctly give the symmetry species of the molecular vibrations but vibrations duplicated (or others inadvertently omitted!) in a fragment analysis can usually be detected. In the entire-molecule method one starts by considering the total motional freedom within the molecule. That is, each atom is allowed to move in any direction. Of course, with this amount of freedom we are, implicitly, allowing the molecule to translate and rotate as well as vibrate, but it is a simple matter to select the vibrations from the totality of allowed molecular motions.

The method is illustrated for the methyl chloride case in Figure 9.5. Each atom is allowed to move in any direction by having a basis set of translations to which each atom contributes three translational displacements in perpen-

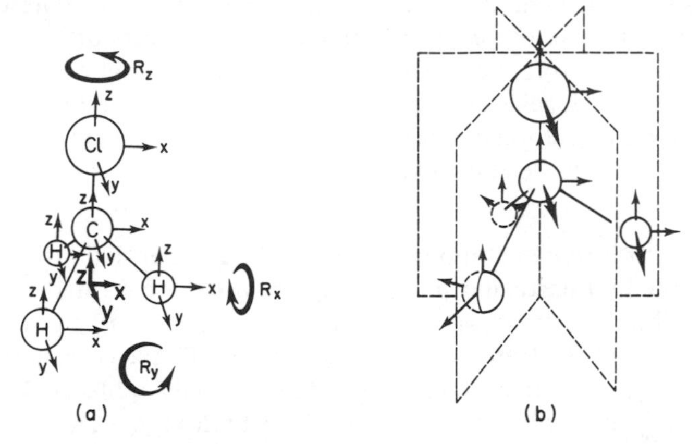


Figure 9.5 (a) The three independent translational displacements of each atom in $CH_3Cl$. In this diagram are also indicated the three rotational motions of the entire molecule (these rotational motions must be some combination of the atomic translational displacements). (b) A more educated choice of atomic displacements than that shown in Figure 9.5(a). It proves best to place the three independent translations (arrows) of each atom as far as possible along rotational axes or in, or perpendicular to, mirror planes

dicular directions. The most obvious — and sometimes used — arrangement of displacement vectors (arrows) is that shown in Figure 9.5(a). In it corresponding displacements of all atoms are parallel. However, the three displacements on each atom can be in any direction in space without in any way changing the final answer. A considerable simplification results if the arrows are chosen to point in symmetry directions (along rotational axes or in, or perpendicular to, mirror planes). The directions of the arrows shown in Figure 9.5(b) have been chosen with this in mind. Remembering that only arrows on atoms which are unshifted by a symmetry operation can contribute to the character, and bearing in mind the discussion of x and y axes in Section 6.1 (a discussion which applies equally to displacements along x and y), it is a simple matter to show that this set of arrows gives rise to the reducible representation

| E | $2C_3$ | $3\sigma_v$ |
|---|---|---|
| 15 | 0 | 3 |

and that this representation has $4A_1 + A_2 + 5E$ components.

**Problem 9.3** Check the reducible representation given above using the choice of displacement vectors shown in Figure 9.5(b). The advantage of using this set of vectors over a set such as that shown in Figure 9.5(a) can be seen if an attempt is made to generate the reducible representation using this latter set of vectors. Note that if the vectors of Figure 9.5(a) are modified by rotating the x and y displacements on each hydrogen atom about the local z axis so that the local $\sigma_v$ mirror plane bisects the angle between local x and y displacements, then the reducible representation may again be readily generated.

The irreducible representations spanned by bodily translations and rotations of a molecule in a particular point group are usually indicated alongside the character table, the symbols $T_x$, $T_y$ and $T_z$ (or just x, y, z) indicating the transformation of the translations and $R_x$, $R_y$ and $R_z$ the corresponding rotations. We have done this in Table 9.1 (the character table of the $C_{3v}$ point group); in the character tables in Appendix 3 both x, y, z and $T_x$, $T_y$, $T_z$ are indicated.

The symmetry species spanned by $T_z$, $T_x$ and $T_y$ can be obtained by the usual method of investigating their transformational properties; this is conveniently done by representing them by the solid arrows at the centre of the three hydrogens in Figure 9.5(a). Rotations may be similarly treated, representing them by the curved arrows in Figure 9.5(a). Usually this is found to be more difficult but, fortunately, there is an alternative — and simpler — way. There is a general rule that:

> A rotation about an axis $R_\alpha$ ($\alpha = x$, y or z) has the same character as the corresponding translation $T_\alpha$ for all proper rotation operations. To obtain the character for $R_\alpha$ under an improper rotation operation, however, one simply has to change the sign of the character of $T_\alpha$ — i.e. multiply it by $-1$.

Thus, it is easy to see from Figure 9.5(a) that $T_z$ has $A_1$ symmetry in $C_{3v}$. That is, it has the characters

| | E | $2C_3$ | $3\sigma_v$ |
|---|---|---|---|
| $T_z$: | 1 | 1 | 1 |

Application of the above rule shows that $R_z$ has the characters

| | E | $2C_3$ | $3\sigma_v$ |
|---|---|---|---|
| $R_z$: | 1 | 1 | $-1$ |

That is, it is of $A_2$ symmetry. For the $C_{3v}$ group we thus find (Table 9.1) that the three translations transform as $A_1 + E$ and the three rotations as $A_2 + E$.

**Problem 9.4** Check that the three rotations $R_x$, $R_y$ and $R_z$ transform as $A_2 + E$ in $C_{3v}$ by considering the behaviour of the three curved arrows in Figure 9.5(a) under the group operations.

**Problem 9.5** By inspecting the character tables in Appendix 3 check that the above rule is invariably true. Having done this, return to Figure 9.5(a) and attempt to understand, pictorially, its origin.

As is well known, and we have already used, the number of normal vibrations of a non-linear molecule is $3N - 6$, where $N$ is the number of atoms in the molecule. The $3N$ is the total motional freedom of the molecule and the 6 arises because the bulk translations and rotations of the molecule are included in the $3N$ degrees of freedom. We have just established that these translations and rotations transform as $A_1 + A_2 + 2E$ for $CH_3Cl$ so that we can obtain the symmetries of the normal vibrations by subtracting these from the symmetries generated by the $3N$ degrees of freedom, $4A_1 + A_2 + 5E$. We thus find that the vibrations transform as $3A_1 + 3E$, a result which agrees with the conclusions reached by the fragment analysis of the previous section.

## 9.4 THE VIBRATIONS OF ALREADY VIBRATING MOLECULES (this section may be omitted at a first reading)

• There is one final question to which we address ourselves. It arises when we recognize, as we did at the beginning of this chapter, that at room temperature any molecule with low frequency vibrations is likely to have some of these vibrations thermally excited. If at least one such vibration is not totally symmetric then it must reduce the molecular symmetry (only totally symmetric vibrations maintain, or, in some situations, increase, molecular symmetry).

**Problem 9.6** For each of the molecules of Problem 9.1 a totally symmetric vibration will have been obtained as part of the answer to that problem. Use the symmetry coordinates obtained to sketch out the form of the vibration and thus show that the vibration does not lead to a change in molecular symmetry. Similarly, show that all other vibrations obtained in answer to Problem 9.1 lead to a reduction in molecular symmetry. (*Hint:* exaggerate the vibrational distortion and determine the symmetry operations of the distorted molecule, using Figure 7.32 if necessary.)

If there is high probability that a molecule has a symmetry which is lower than that which we have assumed, how valid is an analysis based solely on the high symmetry case? The answer is a rather unexpected one. The symmetry group which we have been using is not the one that we think that we have been using! Consider the case of the methyl chloride molecule but now subject the molecule to an arbitrary distortion such as that shown in Figure 9.6. To this distorted molecule we now apply the operations of the $C_{3v}$ point group. The

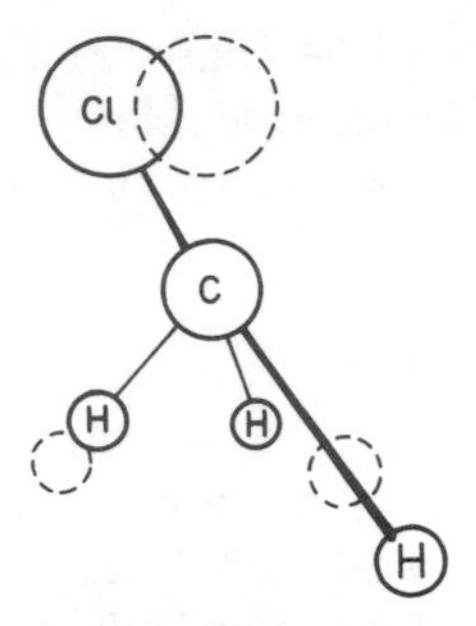


Figure 9.6 The distorted $CH_3Cl$ molecule to be used in Figure 9.7. Dotted circles show original 'atomic' positions. It is assumed that the distortion gives the molecule neither linear nor angular momentum

result of this is shown in the top half of Figure 9.7. Finally, we apply a permutation operation to the labels on the hydrogen atoms, the permutation chosen being that which brings the hydrogens back to their original positions. We describe a permutation by a symbol such as (123), a symbol which means 'replace the label 1 by label 2, replace the label 2 by the label 3 and replace the label 3 by the label 1'. Application of these permutation operations leads to the bottom half of Figure 9.7. The combined effect of point group and permutation operations has been that of giving us back the original molecule, but with the distortion differently related to the labels 1, 2 and 3. If the hydrogen atoms 1, 2 and 3 had been quite different atoms (e.g. if 1 were H, 2 were F and 3 were Br) then the six distorted molecules in the lower group of six would have been quite different. The fact that all three *are* hydrogen atoms, however, means that all these six distorted molecules have exactly the same energy. That is, they tell us something about the potential energy surfaces of the $CH_3Cl$ molecule. For any one distorted arrangement there are always five others with precisely the same energy. Molecular vibrations explore potential energy surfaces, so this sixfold repetition is information which is clearly relevant to a vibrational analysis. It is a straightforward, if somewhat tedious, task to show that the six operations

$$E\ (1)(2)(3)$$
$$C_3^+\ (132)$$
$$C_3^-\ (123)$$
$$\sigma_v\ (1)(23)$$

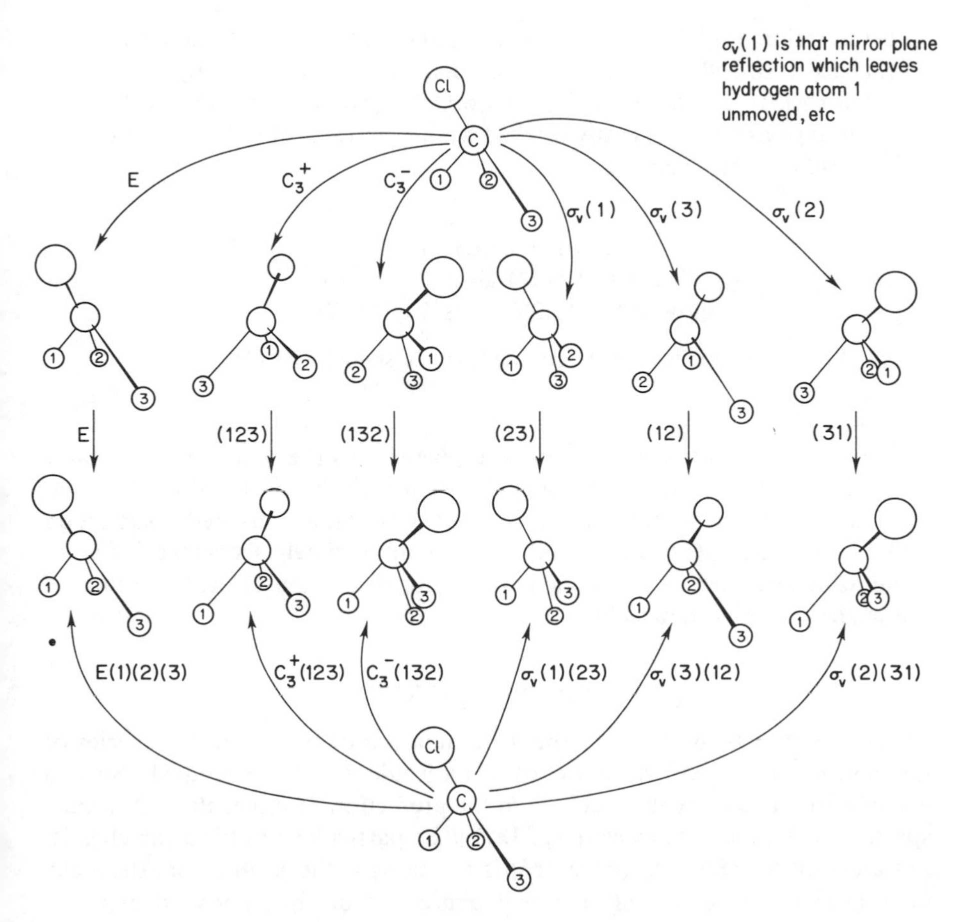


Figure 9.7 Rotation/permutation operations appropriate to distorted $CH_3Cl$

$$\sigma_v(2)(31)$$
$$\sigma_v(3)(12)$$

multiply to form a group; further, these new operations are in a one-to-one correspondence with the operations of the $C_{3v}$ point group and also multiply in the same way as their $C_{3v}$ counterparts.

**Problem 9.7** Show by deriving the group multiplication table that the set of operations given above form a group and that this table is isomorphic to the $C_{3v}$ multiplication table (Table 8.2). (*Hint:* although diagrams such as Figure 9.7 or a model may be of help in tackling this problem it is perhaps easiest to treat it as algebraic, using Table 8.2 for the operations and treating the permutations separately.) (*Note:* the permutation (1)(23) means '2 and 3 interchange while 1 remains unchanged'. It follows that, for example, (1)(23) followed by (123) is equal to (3)(12) whereas (123) followed by (1)(23) gives (2)(31). Such permutations are most easily worked out as follows, using (123) followed by (1)(23) as an example:

| | | | |
|---|---|---|---|
| Start (identity) | 1 | 2 | 3 |
| Application of (123) gives | 2 | 3 | 1 |
| Followed by (1)(23) gives | 3 | 2 | 1 |

which, on comparison with the identity, is seen to be (2)(31)

The two groups are isomorphous and give rise to the same character table (or, more strictly, they have isomorphic character tables). It is this close connection between the 'correct' group — that containing combined point group and permutation operations — and the more immediately accessible $C_{3v}$ point group which enabled us to use the latter in our discussion of the vibrations of the methyl chloride molecule.

## 9.5 SUMMARY

In this chapter the problem of the determination of the symmetry species of the normal modes of vibration of a molecule has been studied. Such a classification is an essential prelude to the prediction of molecular vibrational spectra. A fragment analysis (page 181) is of particular utility to the chemist but a complete treatment (page 186) is useful as a check on errors that may have been introduced in a fragment analysis. For this, knowledge of the transformational properties of the bulk translations and rotations of the molecule is essential (page 188). Although point groups are invariably used in vibrational analyses, a more detailed study shows that they are actually used because they are isomorphic to the correct vibrational symmetry groups (page 189).

CHAPTER 10

# *Direct products*

## 10.1 THE SYMMETRY OF PRODUCT FUNCTIONS

All of the previous chapters have been concerned with a discussion of the symmetries of individual objects, such as orbitals. Very commonly, however, the chemist is interested in products of such quantities. Thus, in a many-electron atom or molecule the electronic wavefunction will be a product wavefunction which, at the simplest level, takes the form

$$\psi = \phi_1\phi_2\phi_3 \ . \ . \ . \ \phi_n$$

where $\phi_1 \ . \ . \ . \ \phi_n$ are individual one-electron orbitals and $\psi$ is the single wavefunction which describes all $n$ electrons. Even in a one-electron problem we have to discuss products of one-electron wavefunctions — products of orbitals — as soon as we become interested in the overlap between two orbitals — the overlap integral between two orbitals $\phi_1$ and $\phi_2$, $S_{12}$, is given by

$$S_{12} = \int \phi_1\phi_2 \, dv$$

where the integral is over all space.

We know how to place symmetry labels on $\phi_1$, $\phi_2$, . . . , etc., but how do we place a symmetry label on a product function such as $\phi_1\phi_2$? The present chapter is concerned with the answer to this question and with some important consequences which stem from it.

It is easiest to progress by considering a specific example; the one that we choose is taken from the $C_{2v}$ point group, the character table which is given in Table 10.1.

Consider a product function $\phi_1\phi_2$ and suppose that $\phi_1$ has symmetry $A_2$; we will write this as $\phi_1(A_2)$. Similarly, we take $\phi_2$ to be of $B_2$ symmetry and write it $\phi_2(B_2)$. We have to determine the symmetry of their product $\psi$; i.e. we have

Table 10.1

| $C_{2v}$ | E | $C_2$ | $\sigma_v$ | $\sigma_v'$ |
|---|---|---|---|---|
| $A_1$ | 1 | 1 | 1 | 1 |
| $A_2$ | 1 | 1 | −1 | −1 |
| $B_1$ | 1 | −1 | 1 | −1 |
| $B_2$ | 1 | −1 | −1 | 1 |

to fill in the empty bracket in

$$\psi(\quad) = \phi_1(A_2)\phi_2(B_2)$$

In principle, the method is simple. We subject $\psi$ to all of the operations of the group and obtain a set of characters by relating the transformed function to the original one. If we apply the $C_2$ operation, for example, to $\psi$, this must mean that we are really applying it to $\phi_1(A_2)$ and to $\phi_2(B_2)$ simultaneously. Now we know, from Table 10.1, that under this operation $\phi_1(A_2) \rightarrow \phi_1(A_2)$ because the $A_2$ irreducible representation has a character of 1 for this operation. Similarly, $\phi_2(B_2) \rightarrow -\phi_2(B_2)$. Putting these two results together, we have that under the $C_2$ rotation operation

$$\phi_1(A_2)\phi_2(B_2) \rightarrow -\phi_1(A_2)\phi_2(B_2)$$

That is, $\psi(\quad) \rightarrow -\psi(\quad)$, so the character generated by the transformation of $\psi(\quad)$ under this operation is $-1$. It is clear that this $-1$ really occurs because the products of the characters of the $A_2$ and $B_2$ irreducible representations under the $C_2$ operation is $-1$. Similarly, because the $A_2$ and $B_2$ characters under the $\sigma_v$ operation are $-1$ and $-1$ respectively, their product, 1, is the character of $\psi(\quad)$ under this operation. Summarizing this and extending it to include the other operations, we have:

| | E | $C_2$ | $\sigma_v$ | $\sigma_v'$ |
|---|---|---|---|---|
| Characters generated by the transformation of $\phi_1(A_2)$ (i.e. the $A_2$ irreducible representation) | 1 | 1 | $-1$ | $-1$ |
| Characters generated by the transformation of $\phi_2(B_2)$ (i.e. the $B_2$ irreducible representation) | 1 | $-1$ | $-1$ | 1 |
| Characters of the transformation of $\psi(\quad)$ (i.e. the products of the two rows of characters above) | 1 | $-1$ | 1 | $-1$ |

We see that the representation generated is the $B_1$ irreducible representation; i.e. $\psi(\quad)$ can now be identified as $\psi(B_1)$.

**Problem 10.1** Using the procedure described above, fill in the empty brackets in the product functions

$$\psi(\quad) = \phi_1(B_1)\phi_2(B_2)$$
$$\psi(\quad) = \phi_1(B_1)\phi_2(A_2)$$
$$\psi(\quad) = \phi_1(A_2)\phi_2(A_1)$$

Your answers can be checked by reference to Table 10.2.

In this example we have found the general method of determining the symmetries of product functions; we simply multiply together the characters of the irreducible representations which describe the transformation of the individual

Table 10.2

| $C_{2v}$ | $A_1$ | $A_2$ | $B_1$ | $B_2$ |
|---|---|---|---|---|
| $A_1$ | $A_1$ | $A_2$ | $B_1$ | $B_2$ |
| $A_2$ | $A_2$ | $A_1$ | $B_2$ | $B_1$ |
| $B_1$ | $B_1$ | $B_2$ | $A_1$ | $A_2$ |
| $B_2$ | $B_2$ | $B_1$ | $A_2$ | $A_1$ |

functions. The act of multiplying two irreducible representations in this way is said to give rise to the *direct product* of the two individual representations; if we multiply three irreducible representations we form a triple direct product, and so on. The name 'direct product' is not new — we first met it in Section 4.3 where the operations of the group $D_{2h}$ were described as the direct product of the operations of the groups $D_2$ and $C_i$. These two usages of 'direct product' are related; the connection may be seen in the discussion of Section 2.3 where we saw the close relationship between the way that operations of a group multiply and the way that the corresponding irreducible representations multiply. It is, then, not surprising that the same phrase, direct product, should be applicable to each type of multiplication. The connection between the two multiplications is described more fully in Appendix 2.

As we shall see in the remainder of this chapter, direct products are very important in the application of symmetry to chemistry. For these applications, all that is needed is a list — a table — of two function direct products. Triple and higher direct products can readily be deduced from such a table. The direct product table for the irreducible representations of the $C_{2v}$ point group is given in Table 10.2. In this table we have used an obvious and conventional symbolism. The entry at a particular point in the table is the symmetry of the direct product of the species which label the column and row in which the entry falls.

**Problem 10.2** Check that Table 10.2 is correct. This will provide useful additional practice in the formation of direct products.

**Problem 10.3** Use Table 10.2 to obtain symmetry labels for the following product functions:

$$\psi(\ \ ) = \phi_1(A_1)\phi_2(B_1)\phi_3(B_2)$$
$$\psi(\ \ ) = \phi_1(A_2)\phi_2(B_1)\phi_3(B_2)$$
$$\psi(\ \ ) = \phi_1(A_2)\phi_2(B_2)\phi_3(B_1)$$
$$\psi(\ \ ) = \phi_1(A_2)\phi_2(B_1)\phi_3(A_1)\phi_4(B_1)$$
$$\psi(\ \ ) = \phi_1(A_2)\phi_2(B_1)\phi_3(A_2)\phi_4(B_1)$$
$$\psi(\ \ ) = \phi_1(A_2)\phi_2(A_2)\phi_3(B_1)\phi_4(B_1)$$

What do your results tell you about the importance of the order in which functions are listed on the right-hand side of these expressions?

Note that Table 10.2 is symmetric about the leading diagonal (top left to bottom right). Thus, the result obtained for the example considered earlier in this chapter is

$$A_2 \times B_2 = B_1$$

It is equally true that

$$B_2 \times A_2 = B_1$$

This follows because what we are really doing is to multiply sets of numbers together and the result obtained is independent of the order in which we multiply them — the origin of the diagonal symmetry of Table 10.2 is at once evident.

The method which we have used to obtain direct products is entirely general. However, another result that we find in Table 10.2 — that the product of two irreducible representations is always another irreducible representation — is not. Direct products involving two or more degenerate irreducible representations invariably give rise to a reducible representation as a product. Let us look at the $C_{4v}$ point group for an example. The $C_{4v}$ character table is given in Table 10.3 (we first met it in Table 5.6).

It is evident that the direct product $E \times E$ must be reducible because the number that will appear in the identity column (4) is larger than any character in the table.

The direct product $E \times E$ is

| | E | $2C_4$ | $C_2$ | $2\sigma_v$ | $2\sigma_v'$ |
|---|---|---|---|---|---|
| $E \times E$: | 4 | 0 | 4 | 0 | 0 |

which is readily seen to be a representation with components $A_1 + A_2 + B_1 + B_2$.

Table 10.3

| $C_{4v}$ | E | $2C_4$ | $C_2$ | $2\sigma_v$ | $2\sigma_v'$ |
|---|---|---|---|---|---|
| $A_1$ | 1 | 1 | 1 | 1 | 1 |
| $A_2$ | 1 | 1 | 1 | $-1$ | $-1$ |
| $B_1$ | 1 | $-1$ | 1 | 1 | $-1$ |
| $B_2$ | 1 | $-1$ | 1 | $-1$ | 1 |
| E | 2 | 0 | $-2$ | 0 | 0 |

**Problem 10.4** Show that the direct product table for the $C_{4v}$ group is

| $C_{4v}$ | $A_1$ | $A_2$ | $B_1$ | $B_2$ | E |
|---|---|---|---|---|---|
| $A_1$ | $A_1$ | $A_2$ | $B_1$ | $B_2$ | E |
| $A_2$ | $A_2$ | $A_1$ | $B_2$ | $B_1$ | E |
| $B_1$ | $B_1$ | $B_2$ | $A_1$ | $A_2$ | E |
| $B_2$ | $B_2$ | $B_1$ | $A_2$ | $A_1$ | E |
| E | E | E | E | E | $(A_1 + A_2 + B_1 + B_2)$ |

## 10.2 CONFIGURATIONS AND STATES

Let us look at the meaning of this E × E direct product in more detail. Suppose we have two electrons, one of which is to be placed in the degenerate pair of orbitals of E symmetry denoted individually $e_1$ and $e_2$. The second electron is to be placed in a different degenerate pair of orbitals of E symmetry which we individually denote by $E_1$ and $E_2$. The possible two-electron functions are

$$e_1E_1$$
$$e_1E_2$$
$$e_2E_1$$
$$e_2E_2$$

That is, they are four in number (in agreement with the number 4 which appeared in the identity column when we formed the E × E direct product). The group theory tells us that it is possible to take linear combinations of these four functions such that one combination has $A_1$ symmetry, one has $A_2$ symmetry, one has $B_1$ symmetry and one has $B_2$. These symmetry-adapted functions may be obtained by the projection operator method described in Chapters 4 and, more particularly, Chapter 5, because it deals with a non-Abelian group. We first simply choose one function — $e_1E_1$ for instance — and work out how it transforms under the operations of the group. For this, we need to know how the individual functions $e_1$ and $E_1$ transform. This information is detailed in Table 5.7 (where we replace $p_x$ by $e_1$ or $E_1$ and $p_y$ by $e_2$ or $E_2$). We use it to obtain Table 10.4.

Multiplication by the $A_1$ characters and adding, in the usual projection operator method, leads to the conclusion that

$$\psi(A_1) = \frac{1}{\sqrt{2}}(e_1E_1 + e_2E_2)$$

Table 10.4

| E | $C_4^+$ | $C_4^-$ | $C_2$ | $\sigma_v(1)$ | $\sigma_v(2)$ | $\sigma_v'(1)$ | $\sigma_v'(2)$ |
|---|---|---|---|---|---|---|---|
| $e_1$ | $-e_2$ | $e_2$ | $-e_1$ | $-e_1$ | $e_1$ | $e_2$ | $-e_2$ |
| $E_1$ | $-E_2$ | $E_2$ | $-E_1$ | $-E_1$ | $E_1$ | $E_2$ | $-E_2$ |
| $e_1E_1$ | $e_2E_2$ | $e_2E_2$ | $e_1E_1$ | $e_1E_1$ | $e_1E_1$ | $e_2E_2$ | $e_2E_2$ |

**Problem 10.5** Use Table 10.4 to show that

$$\psi(B_1) = \frac{1}{\sqrt{2}}(e_1E_1 - e_2E_2)$$

**Problem 10.6** Derive a table similar to Table 10.4 but appropriate to the function $e_1E_2$. Use it to show that

$$\psi(A_2) = \frac{1}{\sqrt{2}}(e_1E_2 - e_2E_1)$$

and

$$\psi(B_2) = \frac{1}{\sqrt{2}}(e_1E_2 + e_2E_1)$$

In the above discussion, for convenience we have talked of electronic wavefunctions. The method, however, is not limited to such wavefunctions. Thus, the pairs $(e_1, e_2)$ and $(E_1, E_2)$ could equally have been vibrational wavefunctions, in which case the product wavefunctions would have been the ones relevant to a discussion of combination bands in a vibrational spectrum (vibrational excitations in which two different vibrations are excited by a single quantum of energy). Notice that we have not allowed the two pairs to be equal — $(e_1, e_2)$ and $(E_1, E_2)$ — although, in the vibrational spectrum, overtone bands arise from double excitations where $(e_1, e_2)$ is combined with $(e_1, e_2)$. This is because we can only distinguish *three* product functions in such cases. We can excite two quanta of vibrational energy in $e_1$ or $e_2$ or excite one quantum in each so that the distinguishable excited states are

$$e_1e_1$$

$$e_2e_2$$

$$e_1e_2$$

The concept of a direct product can be developed further to deal with this problem but this development is not particularly simple and is too specialized to be relevant to the general reader. We therefore will not include it in this book. The interested reader will find the electronic case developed in Ballhausen[1] and the vibrational one in Wilson, Decius and Cross.[2] The two derivations, not surprisingly, are closely related.

There is another simple application of direct products that we should consider. As we have seen, if, in $C_{2v}$ symmetry, we have in a molecule the electron configuration $a_2{}^1b_2{}^1$ (or, as we have expressed it previously, the product wavefunction is $\phi_1(A_2)\,\phi_2(B_2)$), then we say that this configuration gives rise to a *state* of $B_1$ symmetry (or, in the form that we stated it earlier, there is a product wavefunction $\psi(B_1)$), the direct product $A_2 \times B_2$ being $B_1$. Note that throughout the present discussion we shall neglect electron spin, although this would normally be indicated by a pre-superscript; thus a triplet spin state would be $^3B_1$, a singlet $^1B_1$, and so on. It is possible to extend the group theoretical concepts included in this book to include electron spin, but this is an advanced topic; faced with the choice of including or excluding it we take the latter road and exclude spin from our discussion. This neglect of spin is formally expressed by saying that we are only concerned with orbital states. Thus, in $C_{2v}$ the orbital configuration $a_2{}^1b_2{}^1$ gives rise to the orbital state $B_1$. Similarly, for the $C_{4v}$ example which we considered above, the electron configuration $e^1E^1$ gives rise to the states $A_1$, $A_2$, $B_1$ and $B_2$. More correctly, and following the notation used in earlier diagrams in this book, we say that the electron configuration $1e^12e^1$ gives rise to the states $A_1$, $A_2$, $B_1$ and $B_2$.

It is obvious that when a singly degenerate orbital is occupied by two electrons the product wavefunction describing this situation is totally symmetric because the direct product is totally symmetric (see, for example, Table 10.2 and the table given in Problem 10.4). It is not so easy to see that the same result follows when a set of degenerate orbitals is completely occupied by electrons because simply forming direct products leads, apparently, to a large number of states. However, the Pauli exclusion principle eliminates all but one of these states. There is only one way of filling all orbitals of a degenerate set and that is putting two electrons in each orbital. There is, then, only a single wavefunction and so we must have a singly degenerate state. This simple argument does not tell us whether or not this state is totally symmetric. It seems intuitively likely that it will be, and this is confirmed by following the transformations of the product wavefunction under the operations of the point group in the way that we did for $e_1E_1$ above.

We have, then, a general and very valuable conclusion:

| Closed shells of electrons are invariably totally symmetric. |
|---|

Here, by 'closed shell' we mean configurations like $a_1{}^2$, $b_{1u}{}^2$, $e^4$, $t_{2g}{}^6$, and so on. This conclusion means that when we have a many-electron molecule we can obtain the possible states arising from a configuration by simply considering those orbitals which are partially filled and ignore those that are totally filled — unless these are the only ones present, in which case we have a totally symmetric orbital state.

**Problem 10.7** Show that in the $C_{4v}$ group the electron configuration $e^4$ gives rise to a state of $A_1$ symmetry. (*Hint:* it may be helpful to write this configuration, using the notation adopted earlier in this chapter, as $e_1{}^1e_1{}^1e_2{}^1e_2{}^1$. The table constructed as part of Problem 10.6 can then be modified to be used in the present problem.)

It might be thought that, having determined that a totally symmetric state results from a closed shell, i.e. that the many-electron wavefunction is totally

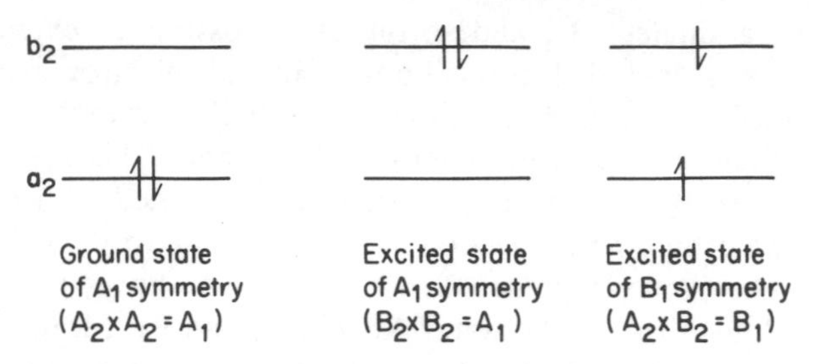


Figure 10.1 Ground and two excited configurations for a $C_{2v}$ molecule

symmetric, that this would be the end of the matter. This is not the case. Consider the situation shown in Figure 10.1, in which for a $C_{2v}$ molecule in addition to a filled $A_2$ orbital there is an empty $B_2$ orbital at higher energy. Both the configurations $a_2^2$ and the configuration in which both electrons are promoted into the $b_2$ orbitals, $b_2^2$, have orbital symmetry $A_1$. In general, it is found by detailed calculations that although the ground state wavefunction is well represented as being derived solely from a configuration such as $a_2^2$, this wavefunction is improved if there is mixed in with it a contribution from the excited state configuration $b_2^2$, which also gives rise to a state of orbital symmetry $A_1$. Such *configuration interaction* is an important step in most detailed calculations of molecular properties, although more than one excited state is usually involved in mixing with the state arising from the ground state configuration. Thus, if, in the present example, the ground state configuration were one in which a doubly occupied orbital of $A_1$ symmetry had above it a doubly occupied orbital of $A_2$ symmetry followed by empty orbitals of $B_2$ and $B_1$ symmetries (Figure 10.2) then we might expect configuration interac-

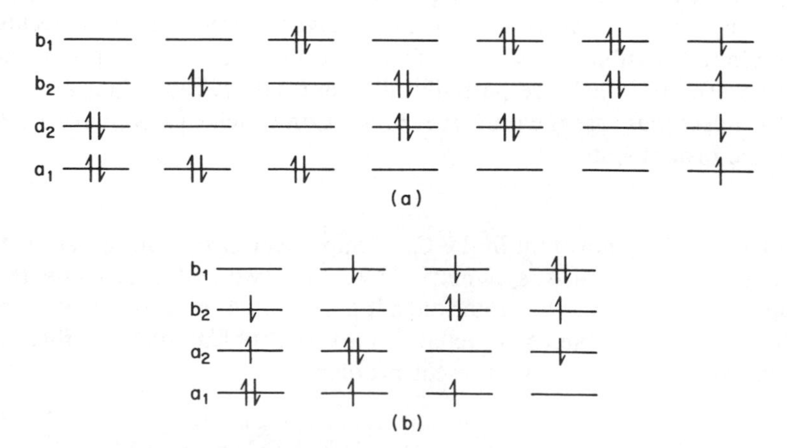


Figure 10.2 (a) Ground and excited configurations all of $A_1$ symmetry and (b) excited configurations all of $B_1$ symmetry

tion between the $A_1$ states $a_1^2a_2^2$, $a_1^2b_2^2$, $a_1^2b_1^2$, $a_2^2b_2^2$, $a_2^2b_1^2$, $b_2^2b_1^2$ and $a_1^1a_2^1b_2^1b_1^1$, and also configuration interaction between the excited $B_1$ states arising from the configurations $a_1^2a_2^1b_2^1$, $a_1^1a_2^2b_1^1$, $a_1^1b_2^2b_1^1$ and $a_2b_2b_1^2$.

**Problem 10.8** Check that the (first) set of seven configurations given above all give rise to states of $A_1$ symmetry and that the (second) set of four all give rise to $B_1$ states.

The inclusion of configuration interaction is usually a very important step in carrying out accurate calculations on the electronic structure of molecules — for instance, in obtaining those results which we have used at several points in this book. Two points should be made. Firstly, just as for orbital interactions, so too for configuration interactions; only states of the same symmetry species interact (they also have to be of the same spin multiplicity). Secondly, it is evident that as the number of orbitals included in a molecular problem — and so the number of configurations that arise — increases, so the number of states of a given symmetry species which may interact under configuration interaction increases. In that the improvement that results in the description of the ground state (and, usually, the lowest excited states) is often considerable, an upper limit on the improvement is usually set by the capacity of the computer available.

## 10.3 DIRECT PRODUCTS AND QUANTUM MECHANICAL INTEGRALS

We now come to a very important section of this book. In it lies the origin of the main value of symmetry theory to chemical problems: it enables a considerable reduction in the number of integrals that arise in quantum chemistry — for instance those that occur in the general mathematical statement of spectroscopic selection rules.

It is all too easy for quantum mechanics to appear formidable because of the large number of rather unpleasant looking integrals which it seems to involve. In practice, these integrals are found to be rather less objectionable because if they cannot be evaluated algebraically they can be evaluated numerically, either by hand or by a digital computer. Even so, we can save ourselves a great deal of work by the intelligent use of group theory, as we shall see. Let us first consider what is meant by an integration over all space (which is the integration which is usually involved in quantum mechanics). Integration may be pictorially regarded as the adding together of an infinite number of infinitesimally small fragments. As a consequence of this it is sometimes possible to see the result of an integration without actually carrying out the calculation. Consider the $p_z$ orbital shown in Figure 10.3. What is the value of the integral over all space of the $p_z$ orbital? That is, what is the value

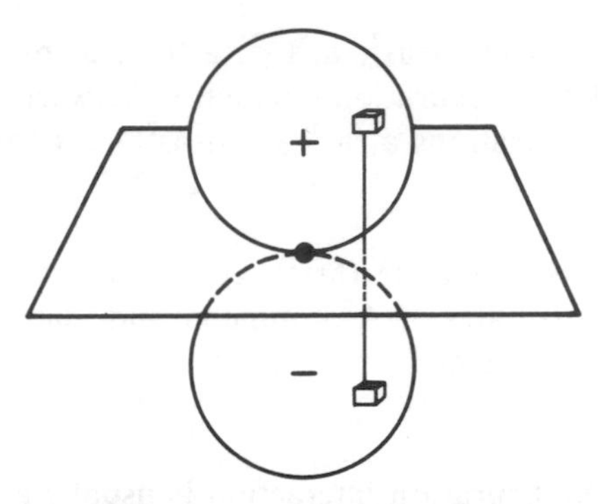


Figure 10.3 There is an exact cancellation of the contributions from the two boxes (at equivalent positions in the lobes of the $p_z$ orbital shown) to an integral over all space of the $p_z$ orbital

of $\int p_z \, dv$ where $dv$ is an infinitesimally small volume element? We treat this integral as $\Sigma p_z \, dv$, where the summation is over an infinity of minute volume elements. In order to perform this summation — this integration — one has to collect into one box, as it were, all of the infinitesimally small fragments which comprise this wavefunction. In carrying out the integration we must pay due regard to the signs of the fragments, i.e. we shall be adding fragments from that part of space in which the wavefunction has a positive amplitude to those coming from that part in which it has a negative amplitude. From the shape of the orbital it is evident that for every volume element that makes a positive contribution to the integral, there is a corresponding volume element which makes a negative contribution, so that when added together the two cancel, to give a nett contribution of zero. A pair of such mutually cancelling volumes is shown in Figure 10.3, the positive contribution from the top volume being cancelled by the negative contribution from the bottom. By adding together pairs of points in this way until the whole space is included we see that the value of the integral $\int p_z \, dv$ is zero — even though we have not explicitly evaluated it! It is the fact that arguments such as this can be cast, very simply, in the language of symmetry that makes group theory so valuable. Thus, an alternative way of stating the arguments that we have just developed is to recognize that the 'top' and 'bottom' of the $p_z$ orbital are

(a) symmetry-related by reflection in the mirror plane shown in Figure 10.3 and
(b) of opposite phase.

These two facts, taken together, establish that the integral must be zero. What we now have to do is to search for a general rule to replace the two specific points made above, which are relevant only to the $p_z$ orbital (and those functions that behave similarly).

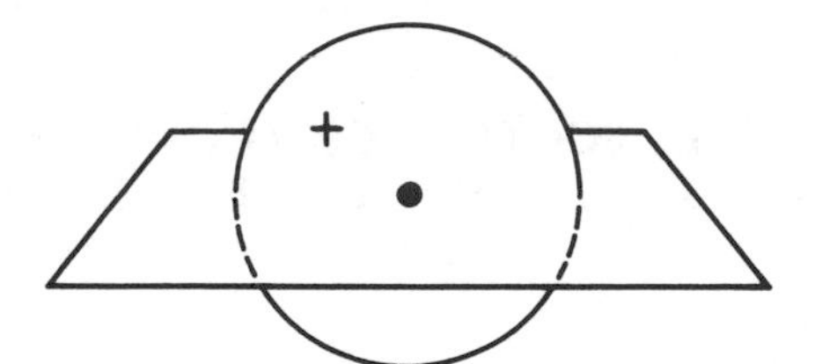


Figure 10.4 Integration over all space of an s orbital type quantity is non-zero

It will help to establish the general rule if we, qualitatively, consider the corresponding integral involving an s orbital:

$$\int s \, d\nu$$

It is clear, from Figure 10.4, that reflection in the mirror plane now interrelates two volume elements which make identical contributions to the integral. The character of the s orbital under the mirror plane reflection operation is 1 and so the contribution to the integral coming from volume elements related by this operation do not cancel. This is in contrast to the $p_z$ orbital which has a character of $-1$. Clearly, the integral over all space of a function which transforms as an irreducible representation which has all its characters equal to $+1$ will not be equal to zero by symmetry. That is, integrals over all space of functions transforming under the totally symmetric irreducible representation of a point group may be non-zero.

Does the contrary rule hold? Can we say, in general, that the integral over all space of a function transforming as a non-totally symmetric irreducible representation must be zero? We can, as the following argument shows.

In Chapter 5 we met some theorems about character tables. Let us look again at Theorem 3:

> Take any two different irreducible representations and multiply together the two characters associated with each class. Then, in each case, multiply the product by the number of operations in the class. Finally, add the answers together. The result is always zero.

Consider the case in which one of the irreducible representations chosen is the totally symmetric one. Multiplication by its elements is, then, always to multiply by the number 1. The second irreducible representation cannot be a totally symmetric one because we are working with two different irreducible representations. For this second irreducible representation, the theorem tells us that the sum of the products of the character in a class multiplied by the number of operations in that class is equal to zero. As an example consider the $T_{2u}$ irreducible representation of the $O_h$ point group discussed in Chapter

7, given in Table 7.2:

| | $O_h$: | E | $8C_3$ | $6C_4$ | $3C_2$ | $6C_2'$ | i | $8S_6$ | $6S_4$ | $3\sigma_h$ | $6\sigma_d$ |
|---|---|---|---|---|---|---|---|---|---|---|---|
| | $T_{2u}$: | 3 | 0 | −1 | −1 | 1 | −3 | 0 | 1 | 1 | −1 |
| Product of $T_{2u}$ characters with the number of operations in the class | | 3 | 0 | −6 | −3 | 6 | −3 | 0 | 6 | 3 | −6 |

It is easy to see that the sum of the numbers in the final row is zero. This means that an integral over all space of a set of $T_{2u}$ functions is zero — the negative contributions exactly cancel the positive.

**Problem 10.9** Select any one of the character tables we have discussed in a previous chapter. Select a non-totally symmetric irreducible representation and show that the sum of all (character multiplied by the number of operations in the corresponding class) is zero. The case of the $T_{2u}$ irreducible representation of the $O_h$ group considered above provides an example.

It follows that we may replace our earlier statement about non-zero integrals by the stronger one:

| Only integrals of functions transforming under the totally symmetric irreducible representation of a point group may give rise to non-zero integrals over all space. |
|---|

It is this theorem which leads to the simplifications introduced by group theory; one knows immediately which integrals must be zero without ever having to actually evaluate them. As we shall see, this as the basis for, amongst other things, all spectroscopic selection rules.

The reader may have noted that we used the word 'may' in the generalization presented above, rather than the stronger 'will'. The reason for this will become clear from a comparison of Figures 10.3 and 10.5. The latter shows the $p_z$ orbital of Figure 10.3 but now in $C_{2v}$ symmetry. Unlike Figure 10.3, none of the operations corresponding to the symmetry elements of Figure 10.5 interrelate the bottom and top of the $p_z$ orbital — the mirror plane of Figure 10.3 is not a symmetry element of $C_{2v}$. As a consequence, and as we first met in Chapter 2, the $p_z$ orbital is totally symmetric in $C_{2v}$. This does not alter the fact that the integral

$$\int p_z \, dv$$

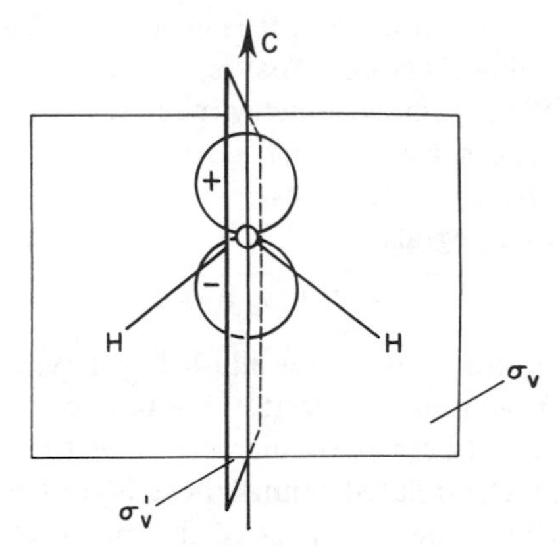


Figure 10.5 An integration over all space of $p_z$ is zero, even when it transforms as the totally symmetric irreducible representation (here, in $C_{2v}$)

remains equal to zero. We see, then, that integrals over all space of functions of $A_1$ symmetry *can* be equal to zero, and so we must say 'may' rather than 'will'.

The discussion on non-zero integrals which we have just developed in the context of simple functions, like $p_z$, can immediately be extended to product functions using the arguments developed earlier in this chapter. Just as for simple functions, only product functions which are totally symmetric may give non-zero integrals over all space. We know how to obtain the symmetry species of such product functions from the symmetries of their component functions — one simply forms the appropriate direct product to determine if it gives rise to the totally symmetric irreducible representation and, if necessary, apply the projection operator method if the explicit form of the totally symmetry product function is needed.

As an example of this application we now justify the assertion made in Section 3.5 when discussing the water molecule that 'interactions between orbitals of different symmetry species are always zero'. 'Interactions' in this context means, specifically, that both the overlap integral

$$\int \psi_a \psi_b \, dv$$

and the energy integral

$$\int \psi_a \mathcal{H} \psi_b \, dv$$

(where $\mathcal{H}$ is the so-called Hamiltonian operator for the system; fortunately we do not need to have a detailed expression for $\mathcal{H}$ in the present context) are only non-zero when $\psi_a$ and $\psi_b$ are of the same symmetry species. Let us consider the overlap integral first. We have to form direct products of the symmetry species of $\psi_a$ and $\psi_b$ and select those which give rise to the totally symmetric irreducible representation ($A_1$ in $C_{2v}$). It is clear from Table 10.2 that $A_1$ only appears along the leading diagonal. That is, $A_1$ only results, in the $C_{2v}$ group, when $\psi_a$ and $\psi_b$ are of the same symmetry species. It follows that only overlap between orbitals of the same symmetry species may give rise to non-zero overlap integrals in the $C_{2v}$ group.

What of the energy integrals

$$\int \psi_a \mathcal{H} \psi_b \, d\nu?$$

The Hamiltonian operator $\mathcal{H}$ expresses all of the energies — be they attractive, repulsive, kinetic or potential – present in the molecule. At a particular point in the molecule there will be a particular blend of the corresponding forces. However, at all symmetry-related points these blends must be equivalent. It follows that $\mathcal{H}$ must have the symmetry of the molecule. That is, $\mathcal{H}$ is totally symmetric, $A_1$, and inclusion of this irreducible representation in a direct product will not change the answer; therefore, whether or not the above energy integral is symmetry-required to be zero is solely determined by the direct product of the symmetry species of $\psi_a$ and $\psi_b$. However, we have already dealt with this problem when we discussed the overlap integral between these orbitals. The energy integral will also be zero unless $\psi_a$ and $\psi_b$ have the same symmetries. The validity of the statement made in Chapter 3 that 'interactions between orbitals of different symmetry species are always zero' at once follows.

The above discussion leads us to have a particular interest in the occurrence of the totally symmetric irreducible representations in tables of direct products. The general conclusion is implicit in our preceding discussion. In Table 10.2 and also in the table given in Problem 10.4 we see that the totally symmetric irreducible representation occurs along the leading diagonals and nowhere else. Further, it occurs in *every* entry along the leading diagonals. It is found that these two statements are true for all points groups:

> The totally symmetric irreducible representation always occurs when the direct product is formed between a particular irreducible representation and itself. It never appears when a direct product is formed between two different irreducible representations.

If nothing else, this generalization will save us some work when determining whether or not an integral is required to be zero. If, for instance, we have an integral over three functions of different symmetries (different irreducible representations) then we need only form the direct product of *two* of these irreducible representations and then compare the irreducible representation(s) contained in this direct product with that of the third function. If there is a matching, the integral may be non-zero. If there is no matching, then it certainly is zero. In this argument we quickly switched from the word 'functions' to 'irreducible representations'. This is because the general argument holds no matter what the form of the functions involved. Indeed, we have just met a case where we did not use an ordinary function — the case of the energy integrals when we were concerned with the operator $\mathcal{H}$. However, we were able to put a symmetry label on this operator. This is important because the general form of almost all the integrals of quantum mechanics is

$$\int(\text{wavefunction})_2(\text{operator})(\text{wavefunction})_1 \, dv$$

where the wavefunctions may be one-electron wavefunctions, many-electron wavefunctions, vibrational wavefunctions or many others. A wide range of operators occurs in quantum mechanics and the above discussion applies to all of them, but in the next section we shall restrict our discussion to those pertinent to the spectroscopic properties of molecules.

## 10.4 SPECTROSCOPIC SELECTION RULES

It is a fundamental postulate of quantum mechanics that corresponding to every observable associated with a system there is a corresponding operator, a postulate justified by the fact that it has always been found to work! Thus, if we are interested in the electronic spectrum of a molecule, we have to insert the appropriate ground and excited state electronic wavefunctions and the appropriate operator into the general expression given at the end of the previous section. If the transition represented by the integral is allowed then the integral will be non-zero. Since we know how to associate symmetry labels with the two wavefunctions, if we can determine the symmetry species of the appropriate operator then it will be a matter of simple group theory to determine the allowed-ness of the transition. However, if we wished to predict the actual intensity of the transition we would have to evaluate the integral properly.

Fortunately, the task of working out the symmetry species of the operators appropriate to a particular form of spectroscopy is a very simple task compared with that of determining the detailed form of the operators themselves. In most forms of spectroscopy a beam of electromagnetic radiation is allowed to interact with the system under study (we use the term 'electromagnetic radiation' rather than the word 'light' because the wavelength of the radiation may be far from the visible region of the spectrum). Integrals such as those above then describe the consequences of this interaction between radiation and matter.

In the simplest (Maxwell) picture, electromagnetic radiation is regarded as being composed of two mutually perpendicular oscillating fields, one an electric field and the other magnetic; both fields are perpendicular to the direction of propagation of the radiation. The most evident way, then, in which such radiation can interact with matter is by virtue of either or both of the electric and magnetic fields associated with it, so that the most common spectroscopic observations are of transitions which are either 'electric dipole' allowed or 'magnetic dipole' allowed. In order to determine the symmetries of the corresponding operators, then, it is only necessary to determine the symmetry species associated with an electric dipole or a magnetic dipole.

In order to obtain an electric dipole it is necessary to separate charges of opposite sign along an axis. In our three-dimensional world there are only three independent directions in which one may bring about such a charge separation and so there are just three electric dipole operators, one corresponding to the x, one to the y and one to the z axes. Further, because the Cartesian axes are dipolar — they have + and − regions — the transformations of the electric dipole operators mimic — are isomorphous to — those of the Cartesian axes of a molecule. Equally, they are isomorphous to the translations of the molecule along x, y or z axes. This isomorphism of x, y and z with $T_x$, $T_y$ and $T_z$ was noted in Section 9.3 when we discussed them in connection with vibrational analyses. We have now found a second use for them — they give the transformational properties of the three electric dipole operators — and a second reason why one or other set (and sometimes both) are included in character tables is evident.

Just as an electric dipole corresponds to a separation of electric charge along an axis, so a magnetic dipole corresponds to a rotation of charge about an axis. There are three magnetic dipole operators, one for each of the three Cartesian axes. The symmetry species of the magnetic dipole operators will be the same as those of the rotations about these axes. We met and used these rotations (usually denoted $R_x$, $R_y$ and $R_z$ in a character table) when discussing molecular vibrations in Section 9.3. The entries $R_x$, $R_y$ and $R_z$ at the right-hand side of a character table tell us how the three magnetic dipole operators transform.

We are now in a position to make some general statements about whether or not an integral related to the intensity of a transition is required to be zero; i.e. to state general selection rules for electric dipole and magnetic dipole allowed processes. This rule is derived from the integral given towards the end of the previous section by replacing wavefunctions and operator with the appropriate symmetry species. A transition is electric dipole allowed if the triple direct product of the symmetry species of the initial and final wavefunctions with that of the symmetry of a translation contains the totally symmetric irreducible representation. Similarly, a transition is magnetic dipole allowed if the triple direct product of a rotation with the symmetry species of the initial and final wavefunctions contains the totally symmetric irreducible representation. Magnetic dipole allowed transitions are those of importance in nuclear magnetic resonance and electron paramagnetic resonance spectroscopies.

These rules are rather a mouthful and triple direct products can be rather tedious to work out, so it is convenient to recast them into a simpler form by making use of the fact that the totally symmetric irreducible representation only arises in the direct product of an irreducible representation with itself. In compiling our triple direct product, let us first form the direct product of the symmetry species of the initial and final wavefunctions. The transition will only be allowed if this direct product contains within it the same symmetry species as that of the operator. This is a particularly useful way to state the selection rule because, as we have seen, there are commonly several alternative operators — dipole moment operators corresponding to $T_x$, $T_y$ and $T_z$, for instance — and in this form we are not required to choose between the alternatives until the last step. We simply see if we can match the irreducible representation(s) arising from the direct product of wavefunction symmetries with any of those of the appropriate operators. If we can, the transition is allowed. In summary, then, the general spectroscopic selection rule — of which all others are particular cases — is:

> A transition is allowed only if the direct product symmetry species of the initial and final wavefunctions contains the symmetry species of the appropriate operator.

All that we need in order to make use of this rule is a list of those spectroscopic processes which normally arise as a result of electric dipole transitions and those which normally arise from magnetic dipole transitions. Such a list is given in Table 10.5 which also lists the simple functions usually contained in character tables which are of the same symmetry species as the operators relevant to Raman spectroscopy.

The quadratic form of operator for the Raman process arises because in it one wavelength of light is incident on a molecule but a different wavelength is emitted. The symmetries of the operators relevant to Raman spectroscopy are therefore the same as those of products like $T_xT_y$, but as these are never given in character tables an equivalent form — such as xy — is used instead.

**Problem 10.10** Confirm that the following transitions are electric dipole allowed:

| Point group | Ground state symmetry | Excited state symmetry |
|---|---|---|
| $C_{2v}$ | $A_1$ | $B_2$ |
| $C_{2v}$ | $B_1$ | $B_1$ |
| $C_{4v}$ | $A_2$ | E |
| $C_{4v}$ | E | E |

Confirm that the following electric dipole transitions are forbidden:

| Point group | Ground state symmetry | Excited state symmetry |
|---|---|---|
| $C_{2v}$ | $B_2$ | $B_1$ |
| $C_{4v}$ | $A_1$ | $B_1$ |
| $C_{4v}$ | $B_1$ | $B_2$ |
| $C_{4v}$ | $A_2$ | $A_1$ |

(*Hint:* these problems anticipate the discussion of the next section of the text.)

Table 10.5

| Form of spectroscopy and spectral region | Form of operator | Symmetry properties of the operator are the same as those of |
|---|---|---|
| Electronic (visible and ultraviolet)<br>Vibrational (infrared) | Electric dipole | $T_x$, $T_y$, $T_z$ (or more simply x, y, z) |
| Rotational (microwave)<br>NMR (radiofrequency)<br>ESR (microwave) | Magnetic dipole | $R_x$, $R_y$, $R_z$ |
| Raman (visible) | Polarizability (this resembles 'electric quadrupole' but is a bit wider) | $x^2$, $y^2$, $z^2$, xy, yz, zx (or combinations of these) |

We shall conclude this chapter by two illustrations of the application of the general selection rule to vibrational spectroscopy.

Let us return to the example discussed in Chapter 9 — that of the vibrational spectrum of $CH_3Cl$. The general principle will be sufficiently well illustrated if we confine our discussion to the C—H stretching vibrations, which, as we have seen in Section 9.4, have $A_1 + E$ symmetry in the molecular point group $C_{3v}$. Molecular vibrational spectroscopy almost invariably admits of a further simplification to the selection rule problem. This is because the ground vibrational state can be assumed to be one in which no vibrations are excited. It is thus a totally symmetric state. In the excited state the symmetry is that of the vibration excited. Our general selection rule therefore reduces to a very simple one:

> A vibration will be spectroscopically active if that vibration has the symmetry species of the relevant operator.

The character table for $C_{3v}$ is repeated in Table 10.6. From Table 10.6 we see that a vibration of $A_1$ symmetry is infrared allowed (because $T_z$ — or equivalently z — has $A_1$ symmetry). It is also a Raman allowed vibration because $z^2$ and, for that matter, the sum $x^2 + y^2$ transform as $A_1$. Similarly, because ($T_x$, $T_y$) or, equivalently, (x, y) transform as E, an E vibration is in-

Table 10.6

| $C_{3v}$ | E | $2C_3$ | $3\sigma_v$ | |
|---|---|---|---|---|
| $A_1$ | 1 | 1 | 1 | $T_z$, z, $z^2$, $x^2 + y^2$ |
| $A_2$ | 1 | 1 | −1 | |
| E | 2 | −1 | 0 | ($T_x$, $T_y$), (x, y), (zx, yz), ($x^2 - y^2$, xy) |

frared allowed. It is also Raman allowed because products of coordinate axes also transform as E. We conclude that the $CH_3Cl$ molecule is expected to have two infrared peaks and two coincident Raman peaks in the carbon–hydrogen stretching region of the spectrum. Apart from additional complications caused by, for example, the low moment of inertia about the $C_3$ axis, this is precisely what is seen.

The spectral prediction which we have just made is specific to $C_{3v}$ geometry. If, for instance, $CH_3Cl$ were a planar molecule with $C_{2v}$ symmetry we would expect three infrared and Raman-coincident peaks in the C—H stretching region of the spectrum.

**Problem 10.11** If $CH_3Cl$ were a planar molecule it could have $C_{2v}$ symmetry provided that one hydrogen, the carbon and chlorine were colinear. Use the $C_{2v}$ character table to derive the spectral predictions given above.

| $C_{2v}$ | E | $C_2$ | $\sigma_v$ | $\sigma_v'$ | |
|---|---|---|---|---|---|
| $A_1$ | 1 | 1 | 1 | 1 | $T_z$, $x^2$, $y^2$, $z^2$ |
| $A_2$ | 1 | 1 | −1 | −1 | xy |
| $B_1$ | 1 | −1 | 1 | −1 | $T_y$, yz |
| $B_2$ | 1 | −1 | −1 | 1 | $T_x$, xz |

There is an important subtlety hidden in our discussion of selection rules; we will find it convenient to discuss it in the context of the vibrations of $CH_3Cl$ but the ideas which we shall introduce have a general validity.

When talking of electric dipole selection rules we became interested in the transformation properties of quantities like $T_x$, $T_y$, $T_z$. Effectively, we considered an isolated molecule and ignored its relationship with its environment: $T_x$, $T_y$ and $T_z$ referred to *molecular* axes. In solution, the molecule will be tumbling and so these molecular axes will bear no fixed relationship to the axes within which we must, perforce, work — the laboratory-fixed axes. However, if we can bring the molecular and laboratory axes into some sort of fixed relationship with each other, then we can obtain additional spectroscopic information. First we must bring the axes of all molecules into alignment and then hold them like this. If we can do this, we can hope to relate these common molecular axes to the laboratory ones. There are several techniques for aligning axes; in some cases the application of a strong electric field produces an appreciable alignment while alternatively, molecules can be incorporated into a thin sheet of some transparent plastic which is then stretched. However, the simplest and most powerful technique is to crystalize the material. It sometimes happens that molecular axes persist within a crystal. For instance, it is possible that the threefold axis of a molecule such as $CH_3Cl$ would lead to — and persist in — a crystal with a threefold axis. Such an axis can be identified by either microscopic or X-ray examination of the crystal. Suppose that such a crystal has been obtained and characterized. We now use it for an infrared experiment in which we polarize the infrared light (infrared polarizers are readily available and can be used with any infrared spectrometer). Let us call the direction of polarization of the light — that in which the electric vector lies — the direction p. In laboratory axes we are working with an operator which behaves like $T_p$. We now insert our crystal into the beam of polarized infrared radiation so that, within the crystal — and so for molecular axes — p is coincident with the threefold axis (for the crystal we have previously discussed). This means that in terms of molecular axes the operator $T_p$ is to be identified with the operator $T_z$, and *only vibrations which are $T_z$ allowed* will be seen in the spectrum. That is, if we were to run a spectrum, only $A_1$ bands would be seen (to return to our $CH_3Cl$ example, ignoring all the obvious experimental difficulties). There would be no absorption at the frequencies of E bands. Conversely, rotating our crystal so that p is perpendicular to the crystal threefold axis, $T_p$, would, in the crystal, be $T_x$ (or $T_y$ — they transform as a pair so no distinction between them can be made). In this case only the E bands would be seen in the spectrum.

This discussion has been somewhat idealized — molecular axes do not usually persist in a crystal (although the discussion can be modified to cover this case unless vibrational coupling occurs between the individual molecules in the crystal, when a rather different method of analysis has to be used). Even when molecular axes persist, real crystals are not perfect, alignment is not perfect, polarization is not perfect and so bands which, according to the above arguments, should not appear, in fact usually do. They do so, however, with very much reduced intensity and so studies such as those described above would, almost certainly, enable the determination of the actual symmetry species of a molecular vibration.

## 10.5 SUMMARY

This is an important chapter, one in which the main reason for the importance of group theory in chemical problems has been developed. Group theory is important because it simply and reliably gives answers of zero (page 202). Knowledge of which integrals are zero simplifies discussions of molecular bonding and spectra (page 208).

The development falls into two parts: firstly, direct products and, secondly, molecular integrals. With direct products it was important to discover that only the direct product of an irreducible representation with itself gives the totally symmetric representation (page 206). With molecular integrals we found that only integrals over all space of totally symmetric functions are non-zero (page 205). These two parts come together in the recognition that all integrals of any importance in quantum mechanics are integrals involving product functions (page 206).

## REFERENCES

1. C. J. Ballhausen, *Introduction to Ligand Field Theory*, p. 48, McGraw Hill, New York, 1962.
2. Wilson, Decius and Cross, *Molecular Vibrations*, p. 152, McGraw Hill, New York, 1955.

# CHAPTER 11

# *π-Electron systems*

## 11.1 SQUARE CYCLOBUTADIENE AND THE $C_4$ GROUP

One of the areas of chemistry in which relatively simple quantum mechanical ideas have had a very important impact has been in the field of unsaturated organic molecules. The idea that when a molecule contains a system of alternate single and double carbon–carbon bonds then those electrons involved in $\pi$ bonding could be considered on their own — the $\sigma$ electrons could be almost completely ignored — and that it is these $\pi$ electrons which largely determine the chemistry of such molecules, has brought a large measure of rationalization not only to the chemical stability and reactions of these molecules but also to their spectroscopic properties. The distinction between $\sigma$ and $\pi$ orbitals was made in Section 4.4. It is important to recognize that when a molecule contains a series of atoms linked by alternate single and double bonds then on *each* atom in the series there is an orbital involved in the $\pi$ bonding. It is usually the case that this orbital is a p orbital.

Recent detailed numerical calculations on many simple organic molecules have shown that the idea of $\sigma$–$\pi$ separability rests on less secure foundations than was once held to be the case. The orbital symmetry distinctions persist but configuration interaction of the type outlined in the last chapter serves to mix different electron configurations. Nonetheless, there is no doubt that the predictions made by the simple theory are rather good, even if a detailed and general justification for this is not available. It is when the results of the simple model are symmetry-determined that the most evident justification occurs, and it is with such applications that we shall be largely concerned in this chapter.

The symmetry aspects of Hückel theory, as the best known $\pi$-electron model is called, are most readily seen from an example. We shall choose a simple example which, however, illustrates all of the main points of the theory. The molecule which we shall consider is a very unstable and fugitive one — cyclobutadiene, $C_4H_4$ — which we shall take to be a planar molecule with its four carbon atoms arranged at the corners of a square. The carbon atoms are known to have this arrangement when the molecule is stabilized by complexing with a transition metal atom, as in the molecule $C_4H_4Fe(CO)_3$.

Figure 11.1 shows square cyclobutadiene together with the four $2p_\pi$ orbitals that will be our concern (we suppose that the carbon 2s and other carbon 2p

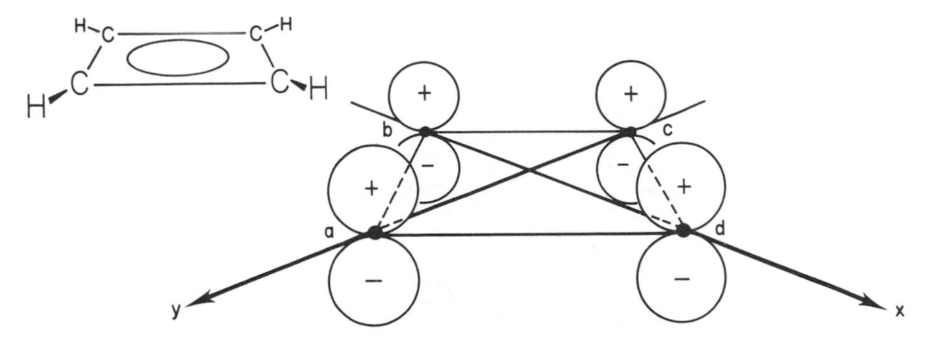


Figure 11.1 The four carbon $2p_\pi$ orbitals in cyclobutadiene, $C_4H_4$

orbitals are involved in the bonding of the $\sigma$ framework). The molecular symmetry is $D_{4h}$ and so this is the obvious group in which to work. However, we shall not. Although it is not particularly obvious from the way that the $D_{4h}$ character table is usually written (Appendix 3), the $D_{4h}$ group is the direct product of $C_{4v} \times C_s$. That is, add a $\sigma_h$ mirror plane to $C_{4v}$ and other symmetry elements are at once generated so that the group becomes $D_{4h}$. The problem that we are now considering immediately defines the effect of this $\sigma_h$ mirror plane. We are only interested in the $p_\pi$ orbitals shown in Figure 11.1 and these, and anything we may derive from them, are antisymmetric with respect to reflection in the $\sigma_h$ mirror plane. Therefore, we might well find it simpler to work in the $C_{4v}$ point group and, at the end, move to $D_{4h}$ by recognizing the $\sigma_h$ antisymmetry. It is probable that most workers would be content to stop here and work in $C_{4v}$, but we shall press on!

**Problem 11.1**

(a) Using Appendix 3 and Figure 11.1 show that square planar cyclobutadiene has $D_{4h}$ symmetry.

(b) Using Appendix 3, show that the $D_{4h}$ group is the direct product of $C_{4v}$ and $C_s$.

The $C_{4v}$ group possesses two sorts of $\sigma_v$ mirror planes: $2\sigma_v$ and $2\sigma_v'$. Either the $\sigma_v$ or $\sigma_v'$ mirror planes (it does not matter which we choose) cut vertically through our $p_\pi$ orbitals. They therefore relate one side of each lobe of this orbital to the other side (Figure 11.2). However, these sides must be of the same phase. The operation of reflection in these mirror planes thus gives us no new information. We can discard these operations, but if we do so we must also discard the other mirror planes — we either have both $2\sigma_v$ and $2\sigma_v'$ or none. The most sensible thing to do would be to play safe and keep them — after all, not much additional work is involved. We, shall be more daring, however, and eliminate them because this will give us the opportunity to work in what

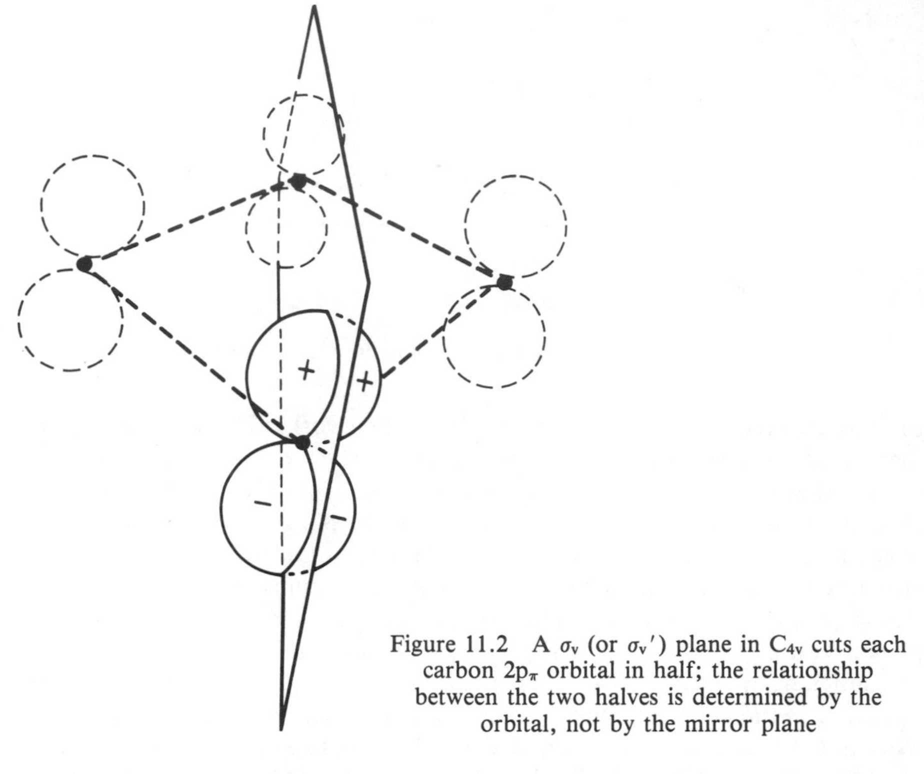


Figure 11.2 A $\sigma_v$ (or $\sigma_v'$) plane in $C_{4v}$ cuts each carbon $2p_\pi$ orbital in half; the relationship between the two halves is determined by the orbital, not by the mirror plane

seems a rather strange group — the $C_4$ point group. It is unusual to discuss a group of pure rotations such as the $C_4$ group in a text at the level of the present one and as a consequence these groups tend to be regarded as rather strange and difficult. We deliberately choose to work in the $C_4$ group in the hope that the reader will not share these misapprehensions. It must be admitted, however, that the discussion which results is a bit more difficult than would have been the case had we worked in an 'easier' group. As the reader may check (Problem 11.9), we shall ultimately obtain the same answers as we would have done in $C_{4v}$ (or $D_{4h}$)!

The $C_4$ character table is given in Table 11.1. Note that it is an Abelian group — there is only one operation in each class. In particular, note that $C_4$ and $C_4^3$ (the $C_4$ operation carried out three times in the same sense) are in different classes. In Chapter 5 we mentioned the importance of the definition

Table 11.1

| $C_4$ | E | $C_4$ | $C_2$ | $C_4^3$ | |
|---|---|---|---|---|---|
| A | 1 | 1 | 1 | 1 | |
| B | 1 | −1 | 1 | −1 | |
| E | $\left\{\begin{matrix}1\\1\end{matrix}\right.$ | $\begin{matrix}i\\-i\end{matrix}$ | $\begin{matrix}-1\\-1\end{matrix}$ | $\left.\begin{matrix}-i\\i\end{matrix}\right\}$ | $i=\sqrt{-1}$ |

of classes and this definition is given in Appendix 1. The proof that $C_4$ and $C_4{}^3$ are in different classes in the $C_4$ group is explicitly given in that appendix.

There are two odd things about the $C_4$ character table: the appearance of i $(=\sqrt{-1})$ and the failure of the number 2 to appear against the E irreducible representation under the identity column. Note that *if* the number 2 were to appear we would violate Theorem 2 of Chapter 5. The sum of squares of characters in the identity column would no longer be equal to the order of the group.

We shall look at the E irreducible representation in some detail in this chapter — the sole reason for working the $C_4$ point group was to give us the opportunity to do this. For the moment we remind the reader that the complex conjugate of $(a + ib)$, a and b being ordinary numbers, is $(a - ib)$. They have a special relationship to each other because when they are multiplied together a real number results $(a + ib)(a - ib) = a(a - ib) + ib(a - ib) = a^2 - iab + iab - (i)^2b^2 = a^2 + b^2$ because $-(i)^2 = -(-1) = 1$. In contrast, neither $(a + ib)^2$ nor $(a - ib)^2$ are free from i. We note that where, in the E irreducible representation, one component contains i, the other contains $-i$. These are complex conjugates (set $a = 0$, $b = 1$ in the expressions earlier in this paragraph). This hints at what is, in fact, correct.

In the previous chapter, in particular, we sometimes multiplied an irreducible representation by itself (when forming direct products). Earlier in the book we did the same when decomposing reducible representations into their irreducible components. We shall do the same when working with the $C_4$ point group for all irreducible representations except the E. For such mutiplications involving the E we shall multiply complex conjugates. That is, we shall multiply the first component of this doubly degenerate representation by the second, and vice versa.

The E irreducible representation of the $C_4$ point group is said to be a 'separable degenerate' representation. Some purists object to this name, holding that it is self-contradictory, but it is the name commonly used and we persist with it. The word 'degenerate' is used because functions transforming as this representation have the same energy — we shall meet an example shortly. 'Separable' is used because it is possible to design an experiment on a molecule of $C_4$ symmetry which shows that all functions transforming as the E irreducible representations are not necessarily quite equivalent. In order to illustrate this we shall digress to give a brief discussion of optical activity.

## 11.2 OPTICAL ACTIVITY

Classically, a molecule is optically active when, in an electronic transition, there is a helical movement of charge density. Just as a left-hand screw thread is not superimposable on a right-hand thread, so there is an optical rotation difference between molecules in which otherwise identical charge displacements follow right-hand and left-hand helical paths. A characteristic of a helix is that it corresponds to a simultaneous translation and rotation and so, following the discussion of Section 10.4, optically active molecules are those in which a transition is simultaneously both electric dipole (charge translation) and magnetic dipole (charge rotation) allowed. That is:

> Molecules* may be optically active when they have a symmetry such that $T_\alpha$ and $R_\alpha$ ($\alpha = x$, y or z) transform as the same irreducible representation.

Comparison of this rule with the data given on the right-hand side of the character tables in Appendix 3 confirms the applicability of the commonly stated criteria for optical activity. Optically active molecules possess neither a centre of symmetry nor a mirror, do not have any improper rotation operations or, as an alternative general statement, do not have any $S_n$ axis, where $n$ can assume any value ($n = 1$ corresponds to a mirror plane and $n = 2$ to a centre of symmetry).

**Problem 11.2** The separation of the cobalt complex ion, $[Co(en)_3]^{3+}$, into optical isomers is a common undergraduate experiment. The complex is, essentially, octahedral and en is the bidentate ligand ethylenediamine, $NH_2 . CH_2 . CH_2 . NH_2$, which is bonded to the cobalt through the nitrogen atoms on adjacent (cis) coordination sites. Determine the symmetry of this molecule and thus show that it has no $S_n$ axis. (*Hint:* the discussion of Section 7.5 should be helpful.)

In the particular case of the $C_4$ point group, $T_z$ and $R_z$ both transform as A and the complex combination $T_x + iT_y$ transforms in the same way as $R_x + iR_y$. Similarly, $T_x - iT_y$ transforms isomorphically with $R_x - iR_y$. The complex form of these latter combinations is a bit off-putting, although it should be less so by the end of this chapter. Ignoring this, it is clear that $T_\alpha$ and $R_\alpha$ transform isomorphically in the $C_4$ group so that a molecule of $C_4$ symmetry is potentially optically active. A beam of polarized light incident on such a molecule down the fourfold axis might suffer a rotation. Clearly, this is not compatible with the isotropy which one normally associates with degeneracy in the xy plane.

**Problem 11.3** Despite the discussion of optical activity in the context of cyclobutadiene in the text, it is believed that cyclobutadiene is not optically active. Why?

### 11.3 WORKING WITH COMPLEX CHARACTERS

All of the character tables we met in earlier chapters in this book contained simple integers as characters. Most people approach complex characters with

*Note the word 'molecules' in this statement. It does not apply to crystals which, under some circumstances, can contain mirror planes of symmetry and yet be optically active.

some apprehension, expecting some strange twists. This apprehension is justified! We can see this by recalling the statement we made towards the end of Section 10.3 that 'The totally symmetric irreducible representation always occurs when the direct product is formed between a particular irreducible representation and itself.' This statement is true for the $C_4$ point group but needs some elaboration. Consider the direct product of the first of the E irreducible representations of Table 11.1 with itself:

| | E | $C_4$ | $C_2$ | $C_4^3$ |
|---|---|---|---|---|
| E(1): | 1 | i | −1 | −i |
| E(1)×E(1): | 1 | −1 | 1 | −1 |

This direct product is the B irreducible representation, not the A. To obtain the A, we have to take the direct product of E(1) with its complex conjugate, E(2):

| | | | | |
|---|---|---|---|---|
| E(1): | 1 | i | −1 | −i |
| E(2): | 1 | −i | −1 | i |
| E(1) × E(2): | 1 | 1 | 1 | 1 |

That is, when working with a separately degenerate representation, one has to elaborate statements about direct products made in Chapter 10. The way to proceed is reasonably straightforward. Thus, the general expression for an overlap integral given in texts on quantum mechanics is

$$S_{ab} = \int \psi_a^* \psi_b \, dv$$

where the asteristic on $\psi_a^*$ indicates the complex conjugate of $\psi_a$. If $\psi_a$ and $\psi_b$ are not complex this reduces to the simple form considered in Section 10.1:

$$S_{ab} = \int \psi_a \psi_b \, dv$$

However, when $\psi_a$ and $\psi_b$ are both complex we have to use the more general form. In such a case we have to use complex conjugate irreducible representations when carrying out the associated group theory. We shall meet an explicit example of this in the next section.

**Problem 11.4** Modify the discussion of selection rules in Section 10.4 so that it covers the case where the wavefunctions are complex.

## 11.4 THE $\pi$ ORBITALS OF CYLCOBUTADIENE

We now return to the problem of the $\pi$ electrons of cyclobutadiene. We know that these $\pi$ electrons interact with each other — they form $\pi$ bonds of some sort — and so our problem is that of finding the $\pi$ molecular orbitals which they occupy. We shall tackle this in two stages. First, we determine the irreducible representations generated by the transformations of the four carbon

$p_\pi$ orbitals and then derive the symmetry-adapted combinations of these orbitals. Finally, we shall approximately determine their energies.

It is easy to show that the transformations of the four carbon $p_\pi$ orbitals of cyclobutadiene in the $C_4$ point group (Figure 11.1) generate the reducible representation

| E | $C_4$ | $C_2$ | $C_4^3$ |
|---|---|---|---|
| 4 | 0 | 0 | 0 |

and that this gives rise to A + B + E irreducible components. (Reducible representations like this one — in which the number which is the order of the group appears in the identity operation column with all other entries zero — are called 'the regular representation' (of the particular point group). They always span each and every irreducible representation once. The regular representation plays a part in the proof of some theorems of group theory.)

The determination of the symmetry-adapted combinations is straightforward and follows the projection operator procedure detailed in Chapter 4 very closely. Using the labels shown in Figure 11.1 for the four $p_\pi$ orbitals and neglecting overlap between these orbitals, we obtain, using the two E components in Table 11.1 separately, the following linear combinations:

$$\psi(A) = \tfrac{1}{2}(a + b + c + d)$$
$$\psi(B) = \tfrac{1}{2}(a - b + c - d)$$
$$\psi_1(E) = \tfrac{1}{2}(a - ib - c + id)$$
$$\psi_2(E) = \tfrac{1}{2}(a + ib - c - id)$$

**Problem 11.5** Use the projection operator technique to obtain the above linear combinations. The normalization of the E functions will be discussed in the text below. (*Hint:* the derivation is similar to that detailed in Section 4.6.)

As indicated in the above problem, the only difficult point in this derivation concerns the two E functions. Firstly, there is a hidden catch. In using the projection operator technique in order to generate a function transforming as a component of a separably degenerate representation one should, strictly, use its complex conjugate in the derivation. Thus, using the second E component in Table 11.1, we obtain the function listed above as $\psi_1(E)$ in an unnormalized form:

$$a - ib - c + id = \psi, \text{ say}$$

It is easy to show that this procedure has given the correct answer. As Table 11.1 shows, the effect of a $C_4$ rotation on a function transforming as the first E component is to multiply it by i. Now this rotation permutes our $p_\pi$ orbitals thus:

$$\begin{array}{ccc} a & \rightarrow & b \\ \uparrow & & \downarrow \\ d & \leftarrow & c \end{array}$$

so that it turns $\psi$ into

$$b - ic - d + ia$$

which is $i(a - ib - c + id) = i\psi$, as expected for the first E component. We now have to normalize $\psi$, i.e. multiply it by a coefficient such that it satisfies

$$\int \phi^* \phi \, dv = 1$$

where $\phi^*$ is the complex conjugate of $\phi$. The complex conjugate of a function is obtained by replacing i by $-i$ within it, so the complex conjugate of $\psi$ is

$$a + ib - c - id = \psi^*$$

We have met this function before; it is $\psi_2(E)$. We see that $\psi_1(E)$ and $\psi_2(E)$ are complex conjugates of each other.

It follows that the overlap integral of $\psi$ with itself has the value

$$\int \psi^* \psi \, dv = \int (a + ib - c - id)(a - ib - c + id) \, dv = 4$$

where, as mentioned earlier, we have assumed that the functions a, b, c and d do not overlap each other. From this value of the overlap integral it follows that the normalization constant for $\psi_1(E)$ — and, equally, $\psi_2(E)$ — must be $\frac{1}{2}$.

**Problem 11.6** Show that $\int \psi^* \psi \, dv = 4$. (*Hint:* expand the explicit form of $\psi^* \psi$ given above in the text. The fact that a and b, for example, do not overlap each other means that the integral $\int ab \, dv$ is equal to zero. The fact that a and b are normalized means that $\int aa \, dv = \int bb \, dv = 1$.)

## 11.5 THE ENERGIES OF THE $\pi$ ORBITALS OF CYCLOBUTADIENE IN THE HÜCKEL APPROXIMATION

We have now reached the limit to which simple group theory can help our discussion. To proceed we have to add chemical knowledge or, failing that, chemical intuition! In practice, this means that the next step involves using some model which provides us with a recipe for obtaining relative orbital energies. We have used such a model in earlier chapters of this book when we used a nodal criterion to obtain orbital energies — the more nodes that an orbital contained the higher we expected its energy to be. This model was augmented by an overlap one — the greater the overlap between two orbitals, the larger the energetic consequences of the interaction between them. The latter part of the discussion of Section 7.2 provides a good example of the augmentation of symmetry arguments by these models.

In the present section we shall use the nodal plane argument in a more mathematical form. (Because when we normalized the functions obtained in the previous section we assumed the overlap between $p_\pi$ orbitals on adjacent carbon atoms in cyclobutadiene to be zero it would scarcely be convincing to use an overlap model!) This mathematical form is that contained in Hückel theory; this is the simplest of all mathematical models of chemical bonding and one that is particularly appropriate to unsaturated organic molecules. It will be recalled that in Section 10.3 we encountered energy integrals

$$\int \psi_a \mathcal{H} \psi_b \, dv$$

and it is these that are important in Hückel theory. In this application, the orbitals $\psi_a$ and $\psi_b$ are $p_\pi$ orbitals and so, in cyclobutadiene, they are the orbitals a, b, c and d of Figure 11.1. The energy of each of these orbitals, before each is involved in any interaction with its partners, is the same. This energy is conventionally designated $\alpha$. For the orbital a we have, then,

$$\int a \mathcal{H} a \, dv = \alpha$$

with similar expressions for b, c and d. The energy of interaction between adjacent $p_\pi$ orbitals is called $\beta$. That is, the interaction between a and b is

$$\int a \mathcal{H} b \, dv = \int b \mathcal{H} a \, dv = \beta$$

with similar expressions for the pairs b/c, c/d and d/a. Those $p_\pi$ orbitals which are not adjacent are assumed not to interact so that, for instance,

$$\int a \mathcal{H} c \, dv = \int c \mathcal{H} a \, dv = 0$$

and similarly for b/d.

To obtain the energy, within the Hückel model, of the A combination

$$\psi(A) = \tfrac{1}{2}(a + b + c + d)$$

we simply have to evaluate

$$\int \mathcal{H}(A) \mathcal{H} \psi(A) \, dv = \tfrac{1}{4} \int (a + b + c + d) \mathcal{H} (a + b + c + d) \, dv$$

Expansion of the right-hand side of this expression and substitution of $\alpha$, $\beta$ and 0 for the resulting integrals gives the energy of $\psi(A)$ as

$$E[\psi(A)] = \alpha + 2\beta$$

**Problem 11.7**

(a) Show that the energy of $\psi(A)$ is $\alpha + 2\beta$.
(b) Show that the energy of $\psi(B)$ is $\alpha - 2\beta$.

It is a simple matter to show that the energy of the $\psi(B)$ orbital is

$$E[\psi(B)] = \alpha - 2\beta$$

but that of $\psi_1(E)$ is a bit more difficult. This is because we have to use the form of the energy expression appropriate to complex functions. This is

$$\int \psi_a^* \mathcal{H} \psi_b \, dv$$

In our case, if we take $\psi_b$ to be $\psi_1(E)$ then $\psi_a^*$ is its complex conjugate, i.e. $\psi_2(E)$. It follows that we have to evaluate

$$E[\psi_1(E)] = \tfrac{1}{4}\int (a + ib - c - id)\mathcal{H}(a - ib - c + id) \, dv$$

On expansion of this expression all of the complex quantities disappear.

**Problem 11.8** Show that the energy of $\psi_1(E)$ is $\alpha$.

The energy of $\psi_2(E)$, which is given by

$$E[\psi_2(E)] = \tfrac{1}{4}\int (a - ib - c + id)\mathcal{H}(a + ib - c - id) \, dv$$

will be the same as that for $\psi_1(E)$ because the right-hand side of this expression on expansion is identical to that for $\psi_1(E)$. We have, then,

$$E[\psi_1(E)] = E[\psi_2(E)] = \alpha$$

Evidently, for the present problem at least, it is entirely reasonable that $\psi_1(E)$ and $\psi_2(E)$ should be called 'degenerate'. Actually, this degeneracy between them is general — the actual algebraic expressions that we obtain for their energies were identical and so the degeneracy did not result from the Hückel approximations.

Because the interaction between two of the $p_\pi$ orbitals is one that leads to a stablization —it requires more energy to ionize an electron from a stabilized orbital than from an isolated $p_\pi$ orbital — the energy $\beta$ is negative (as too is $\alpha$, but, as its contribution to all of the energy levels is the same, its value does not affect the relative order of orbital energies). We conclude that the relative energies of the $\pi$ molecular orbitals of cyclobutadiene are those given in Figure 11.3. There are four $p_\pi$ electrons — one from each carbon atom — located in these orbitals and so we conclude that in the most stable arrangement they will be distributed as shown in Figure 11.3, the degenerate E orbitals containing one electron each; these two electrons will, in the ground state, have parallel spins (the maximum spin multiplicity principle). The total $\pi$-electron stabilization, compared to four carbon $p_\pi$ orbitals of energy $\alpha$, is $4\beta$ ($2\beta$ from each electron in the A orbital).

Suppose that, instead of a delocalized $\pi$ system, we had two localized, non-interacting, $\pi$ bonds — i.e. suppose cyclobutadiene to be rectangular rather than square:

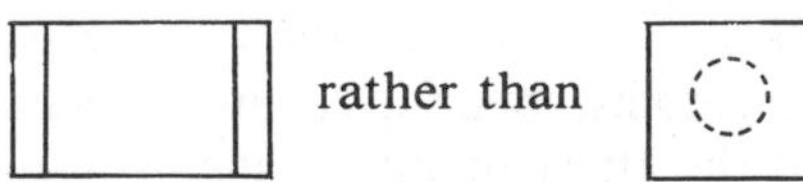

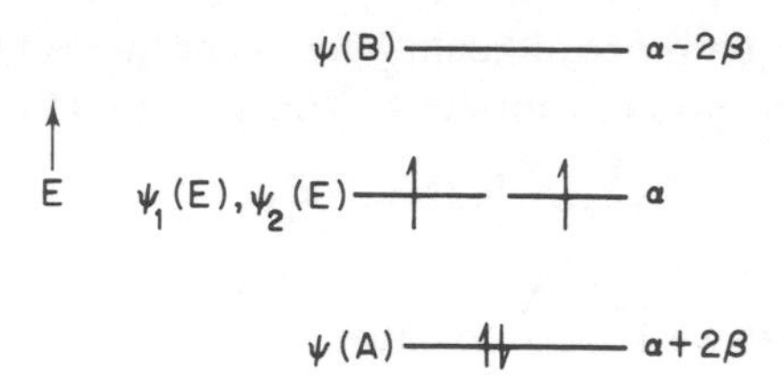


Figure 11.3 Relative energies of the $\pi$ molecular orbitals of cyclobutadiene in the Hückel approximation

Each of the two isolated $\pi$ bonds will have the form we derived in Section 4.4 for ethylene:

$$\frac{1}{\sqrt{2}}(e + f)$$

The energy of this function is given by

$$\tfrac{1}{2}\int(e + f)\mathcal{H}(e + f)\ d\nu$$

which, on expansion and substitution of the Hückel values for the integrals, leads to an energy of

$$\alpha + \beta$$

There would be two such $\pi$-bonding orbitals, each doubly occupied, so that the total $\pi$-electron stabilization would, again, be $4\beta$. We conclude that, as far as the $\pi$ electrons are concerned, at this level of approximation there is nothing to choose between rectangular and square cyclobutadiene. Cyclobutadiene is a very reactive compound — it readily dimerizes (a reaction that can be discussed by the methods of the next section) — but it has been prepared at 35 K in an argon matrix. In its ground state it seems almost certain that it is a planar molecule with no unpaired electrons. Whether it is rectangular or square is not known (for the latter the electron configuration $a^2e^2$ gives rise to both triplet and singlet spin states so, notwithstanding the predictions of Hückel theory, the possibility of a square molecule is not completely excluded by the observation of a singlet spin state).

Of the $\pi$-electron wavefunctions which we obtained working in the $C_4$ point group, two, $\psi(A)$ and $\psi(B)$, are identical in form to two that we would have obtained working in either $C_{4v}$ or $D_{4h}$ (the symmetry labels would have been different, of course). On the other hand, the $\psi(E_1)$ and $\psi(E_2)$ wavefunctions are different. The reason for this can be traced back to the existence of operations in $C_{4v}$ and $D_{4h}$ which have the effect of either mixing or interchanging the two degenerate functions. If we took the x and y axes as shown in Figure 11.1, through the carbon atoms, then the $2\sigma_v$ mirror planes we chose to discard in $C_{4v}$ would have had the effect of interchanging x and y (and so also any functions transforming like them). Hence, they transformed 'as a pair'. In contrast, there is in $C_4$ no operation which will interchange or mix $\psi_1(E)$ and $\psi_2(E)$; they are separate functions, although degenerate. The only way that we can, in $C_4$, obtain those E functions which would have been obtained in $C_{4v}$ is to mix $\psi_1(E)$ and $\psi_2(E)$, although, of course, this is not permissible in the $C_4$ group itself. Taking the sum and difference of $\psi_1(E)$ and $\psi_2(E)$ we obtain

$$\psi_1(E) + \psi_2(E) = \tfrac{1}{2}(2a - 2c)$$

or, renormalizing,

$$\psi_1'(E) = \frac{1}{\sqrt{2}}(a - c)$$

and
$$\psi_1(E) - \psi_2(E) = \tfrac{1}{2}(2ib - 2id)$$

or, renormalizing — remembering that the complex conjugate of $(ib - id)$ is $(-ib + id)$ — we obtain

$$\psi_2'(E) = \frac{1}{\sqrt{2}}(b - d)$$

These functions are those E functions we would have obtained working in $C_{4v}$ (or $D_{4h}$).

**Problem 11.9** Work in either the point group $D_{4h}$ or $C_{4v}$ and
(a) obtain the explicit forms of the four $p_\pi$ molecular orbitals of cyclobutadiene;
(b) check that the double degenerate functions obtained have an energy of $\alpha$ within the Hückel approximations.

Before we finally leave the $C_4$ group there is one further point that we should make. In deriving the $C_{2v}$ character table in Chapter 2 we asserted that there is no other set of characters other than those we considered which, when substituted for the operations of the $C_{2v}$ group in the group multiplication table, would give a table which is arithmetically correct. We did not explore the possibility of complex characters such as those which occur in the $C_4$ character table although we gave a hint of their existence in Problem 2.3. However, it is clear that a set of characters such as

$$1 \qquad i \qquad -1 \qquad -i$$

would not lead to a multiplication table which is arithmetically correct (because, for instance, when i multiplies i it gives $-1$ on the leading diagonal rather than 1). It is evident from this discussion, and can be readily checked, that the multiplication tables of $C_4$ and $C_{2v}$ are not isomorphous. Any operation in $C_{2v}$ carried out twice leads to E, whereas in $C_4$ the $C_4$ and $C_4{}^3$ operations have to be carried out four times to give E. This, incidentally, explains the appearance of i in the $C_4$ character table. Because

$$C_2 \times C_2 = E$$

only characters of either 1 or $-1$ for the $C_2$ operation are possible for a singly degenerate irreducible representation. In the $C_4$ group we also have that

$$C_4 \times C_4 = C_2$$

In other words, the character for the $C_4$ operation, squared, must give the character of the $C_2$. This presents no problems when the character for $C_2$ is 1, because that for the $C_4$ can then be either $+1$ or $-1$ (leading to the A and B irreducible representations of $C_4$). When the character for $C_2$ is $-1$, however, the only possibilities are that the character for the $C_4$ operation is either i or $-i$ (either of these squared gives $-1$), leading to the two E irreducible representations.

We have given just one example of the application of symmetry to the energy levels of $\pi$-electron systems. There are many others, but, having established the principles and procedures involved, we shall not pursue the subject further in detail. Suffice to say that the concept of aromaticity in organic chemistry is closely related to the type of stabilization arguments that we used when comparing square and rectangular cyclobutadiene. Roughly speaking, aromatic systems are those for which the delocalized system is more stable than any corresponding localized one.

## 11.6 SYMMETRY AND CHEMICAL REACTIONS

There have been many attempts to apply symmetry concepts to molecular reactions. This is a difficult area; often the symmetry is low and so of little help. Further, large molecular distortions are usually involved in chemical reactions; i.e. the molecules involved are vibrationally very excited. This has two consequences. Firstly, the analysis we gave of vibrations in Chapter 9 evidently needs modification for large amplitude vibrations; when there are several symmetry-related atoms in a molecule the evidence is that one bond breaks before the others, whereas the discussion of Chapter 9 would lead us to expect several bonds to break simultaneously. This is akin to the failure of simple molecular orbital theory at large internuclear distances (it predicts a mixture of dissociation products), a failure which is also of relevance. Secondly, as we saw at the end of Chapter 9, the actual group with which one is working will usually differ from the point group which one is, formally, using.

By far the most fruitful of the applications of symmetry to molecular reactivity has been the symmetry correlation method introduced by Woodward and Hoffmann and which is applicable to many organic reactions. We shall consider a simple example of the application of their approach, although the formalism which we shall use was introduced by other workers.

Consider the possible reaction of two ethylene molecules to give cyclobutane, a molecule which, for simplicity, we shall assume to be planar:

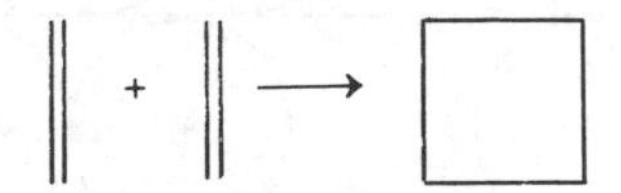

Written like this, it seems a feasible reaction, yet it is not one that readily occurs. Why? The answer is not difficult to find. Let us place two ethylene molecules close together so that they are just about to react. The 'before' and 'after' reaction bonding arrangements are shown in Figure 11.4. The actual symmetry shown in Figure 11.4 is $D_{2h}$, but it is common to work in $C_{2v}$, so that the geometrical constraints on the molecular arrangement are not as rigid as required by $D_{2h}$ symmetry. Choosing the $C_2$ axis as shown in Figure 11.4, it is easy to show that the symmetry species subtended by the two $\pi$-bonding molecular orbitals in the two ethylene molecules shown in Figure 11.4(a) is $2A_1$. It is these two $\pi$ orbitals that are involved in the reaction and that are assumed to smoothly become the two new C—C $\sigma$ bonds as the reaction takes place. These two new $\sigma$ bonds give rise to the symmetry species $A_1 + B_1$ (Figure 11.4(b)). This is not the same as those generated by the $\pi$ orbitals. There is a discontinuity; the $\pi$ bonds cannot smoothly become the new $\sigma$ bonds and so a ready reaction is not to be expected.

Let us look at this further by asking whether there exist any $B_1$ orbitals in the two ethylene molecules. The answer is 'yes'. There are two of them and

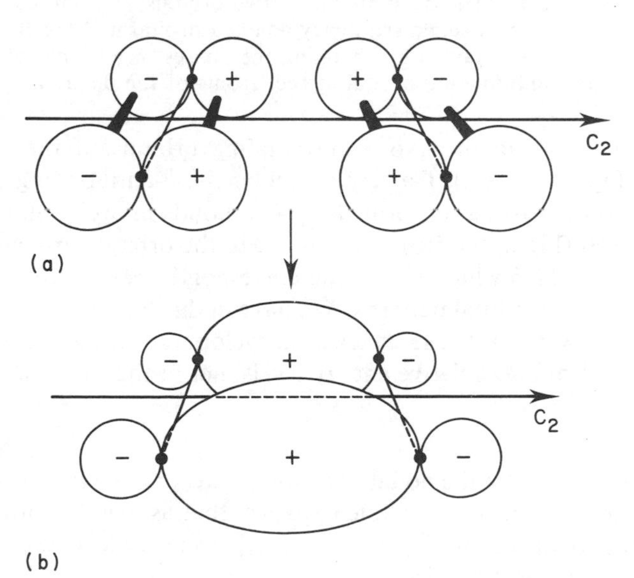


Figure 11.4 (a) Each $\pi$-bonding orbital of each $C_2H_4$ molecule transforms in $C_{2v}$ as $A_1$. (b) The two new $\sigma$-bonding orbitals in $C_4H_8$ transform together together in $C_{2v}$ as $A_1 + B_1$.

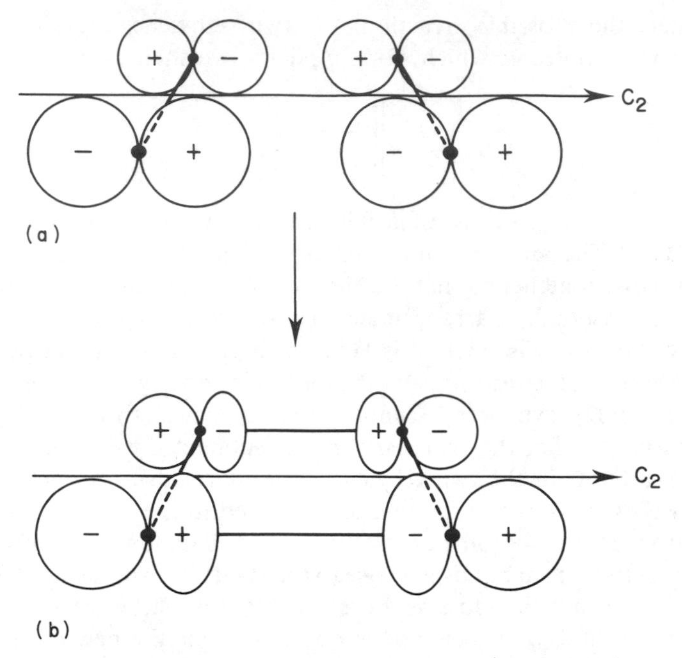


Figure 11.5 (a) Each $\pi$-antibonding orbital of each $C_2H_4$ molecule transforms in $C_{2v}$ as $B_1$. (b) The two new $\sigma$-antibonding orbitals in $C_4H_8$ transform together in $C_{2v}$ as $A_1$ and $B_1$. (The diagram shows *two* orbitals; if regarded, however, as a single symmetry-adapted orbital it is the $B_1$. The $A_1$ is obtained by changing the phases of all lobes of the $\sigma$-antibonding orbital at the 'front' of the diagram.)

they are derived from the two $\pi$-antibonding orbitals of the two ethylene molecules (Figure 11.5(a)). Correspondingly, the $\sigma$-antibonding orbitals corresponding to the two newly formed C—C $\sigma$ bonds in cyclobutane have symmetries $A_1 + B_1$ (Figure 11.5(b)). We are led to the orbital correlation diagram shown in Figure 11.6 which shows the correspondences between the 'before' and 'after' reaction orbital patterns. We have in this figure obtained the detailed pattern of new $\sigma$ orbital energy levels in cyclobutane using the nodal pattern method of determining relative energy levels met in the early chapters of this book.

**Problem 11.10** Use the nodal criterion (used for the example in Section 4.7) to show that it is reasonable to expect that as reaction proceeds, the less stable $a_1$ orbital in Figure 11.6 ($2a_1$) becomes less stable than the more stable $b_1$ orbital ($1b_1$).

It is clear from Figure 11.6 that there will be no strong driving force by which cyclobutane may be formed from two ethylene molecules. As the reac-

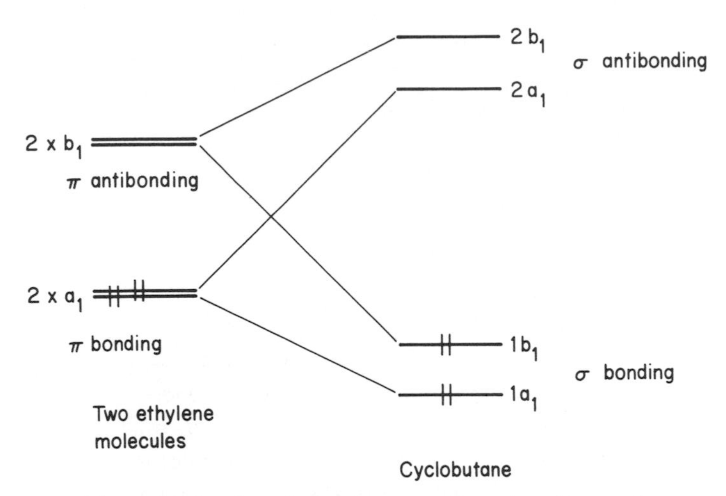


Figure 11.6 Correlation between the $\pi$-bonding and antibonding orbitals of two ethylene molecules and the $\sigma$ bonding and antibonding orbitals of cyclobutane

tion proceeds the lowest energy orbitals would be expected to be filled. Up to the energy level crossing-over point in the middle of Figure 11.6 this means that the two $A_1$ orbitals will be filled. However, whilst the stability of one increases somewhat (as a $\pi$ bond becomes a $\sigma$ bond) that of the other will decrease rapidly (as a $\pi$ bond becomes a $\sigma$-antibonding orbital), so that no reaction is to be expected.

There is an alternative approach to this problem, an approach which is based on states rather than orbitals. The ground state electronic configuration of two ethylene molecules is $(a_1)^2(a_1)^2$, a configuration which gives rise to a $^1A_1$ state (we shall be concerned with spin singlet states throughout the following discussion). A configuration such as $(a_1)^2(b_1)^2$ is an excited state configuration but also gives rise to a $^1A_1$ state (as is readily seen since the quadruple direct product $A_1 \times A_1 \times B_1 \times B_1 = A_1$). In cyclobutane the situation is reversed. The ground state configuration (considering only the newly formed $\sigma$ orbitals) is $(1a_1)^2(1b_1)^2$ and $(1a_1)^2(2a_1)^2$ is an excited state configuration, although both give rise to $^1A_1$ states. The appropriate *state* correlation diagram is shown in Figure 11.7, where we have invoked the non-crossing rule (states of the same symmetry species only cross in very rare circumstances). Physically, this application of the non-crossing rule in the present example arises because electron repulsion favours electrons being as spatially separated as possible and the energy gained from this can contribute the energy apparently required to promote an electron to a higher orbital. Figure 11.7 demonstrates rather more clearly than does Figure 11.6 that we should not expect ethylene to spontaneously dimerize to cyclobutane.

Even here, our discussion is somewhat simplified, but it does correctly

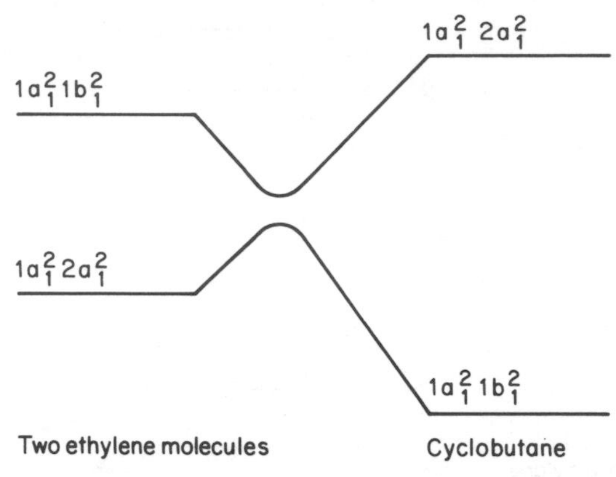


Figure 11.7 Correlation between $^1A_1$ states in the dimerization of two ethylene molecules to give cyclobutane

indicate that one can sometimes be misled by a simple 'filling of the lowest orbitals' approach to chemical bonding. We might remind the reader that the complicating effects of repulsive forces on simple pictures of chemical bonding were also encountered in the first chapter of this book.

The discussion which we have presented above can readily be extended to photochemically induced reactions (i.e. reactions involving electronically excited molecules) but this, and a host of other applications and examples, are beyond the scope of the present text. Many very readable accounts of the topic have been written but these tend to use symmetry arguments in a rather less formal manner than the present text. Commonly, particular symmetry operations are selected and orbital behaviour classified as either A (antisymmetric) or S (symmetric) under these operations; these labels are equivalent to the characters $-1$ and 1 used in the present volume.

One final cautionary note can be made. In this discussion we have been concerned with a single reaction mechanism. Other mechanisms may exist which provide an alternative, and more accessible, route to a particular product. Thus, although ethylene does not dimerize to cyclobutane, reaction between the ethylene derivatives $CH_3(H)C{=}C(H)OC_2H_5$ and $(CN)_2C{=}C(CN)_2$ proceeds smoothly at room temperature to give the corresponding cyclobutane derivatives. In this case there is evidence that a zwitterion intermediate, $(CN)_2C^{\ominus}{-}C(CN)_2{-}CH(CH_3){-}C^{\oplus}H(OC_2H_5)$, is formed. Symmetry arguments are powerful, but nature may be yet more cunning!

## 11.7 SUMMARY

Discussion of the $\pi$ orbitals of cyclobutadiene has provided a relatively simple example of the use of groups containing complex quantities in their character tables (page 216). It is necessary in such cases to work with complex conjugate basis functions and an example was provided in deriving the Hückel energies of cyclobutadiene (page 220). The fact that the $C_4$ group contains no $S_n$ operations enabled a discussion of optical activity (page 217). Molecules having symmetries without such axes are, in principle, optically active (page 218).

Finally, it was shown that symmetry correlations can give insight into some chemical reactions (page 226). Correlations between molecular states may be preferable to simple orbital correlations (page 229).

# APPENDIX 1

# *Groups and classes: definitions and examples*

## A1.1 GROUPS

In Chapter 2 we gave a definition of a group which was adequate for our purposes but which was incomplete; we did not detail all of the requirements. The first object of this appendix is to remedy this deficiency and to accurately define the word 'group'. At some points in this appendix we shall implicitly assume that the group about which we are talking does not contain an infinite number of elements. This excludes $C_{\infty v}$ and $D_{\infty h}$ but all general statements we make can be shown to apply to these two groups also.

Suppose we have a collection of elements (we shall give some examples shortly which will help to indicate the breadth of the term 'element'). The set of these elements, A, B, C, . . . , form group G, written as

$$G = \{A, B, C, \ldots\}$$

if the following conditions apply.

1. There is some *law of combination* which relates the elements one to another. No matter what the precise nature of the operation of combination we shall call it 'multiplication'. So, if A combined with B gives C we would write

$$AB = C \qquad \text{(A1.1)}$$

At the end of this section we shall illustrate this with several different laws of multiplication.

*Note:* (a) Whenever we specify a group it is, formally, necessary to also specify the law of combination. (b) The order in which elements multiply is important. There is *no* general requirement that, for instance,

$$AB = BA$$

so we must assume that, in general,

$$AB \neq BA \qquad \text{(A1.2)}$$

A more detailed consideration of this inequality will lead us to the concept of 'class' later in this appendix.

2. Multiplication is closed (the 'closure requirement'). That is, the product of any two elements within a group is an element within the group.
*Note:* 'an element' here means a *single* element. Multiplication is single valued; there can never be any ambiguity about the outcome of a multiplication. Thus, it can happen that AB = C and AB = D if and only if

$$C = D.$$

3. Multiplication is *associative*. One might think that once one has defined how to multiply two elements there would be no problem about multiplying any number together. This is not the case. Consider the triple product

$$ABC$$

Because we only know how to multiply pairs of elements we have to select a pair from this trio and multiply them first. We have a choice between

$$(AB)C \text{ and } A(BC)$$

But suppose AB = C (as we have said above) and BC = D. Then our products are

$$CC \text{ and } AD$$

It is by no means evident that these are equal unless we introduce this equality as a requirement. This is just what the statement 'multiplication is associative' does. It means that it must be true that

$$(AB)C = A(BC)$$

for the elements to form a group.
*Note:* this means that we can now multiply a string of elements together. Thus, (AB)CD = A(BC)D = AB(CD).

4. The group contains a *unit element* (often denoted E or I). This unit element plays a role which in some ways resembles that of the number 1 in ordinary arithmetic. Thus, when it multiplies any other element of the group, A, say, the product is A; i.e.

$$EA = AE = A \tag{A1.3}$$

5. For each element in the group there is a unique element which is its *inverse*. Loosely speaking, the inverse of an element 'undoes' the effect of that element. Thus, $C_3^-$ is the inverse of $C_3^+$ (Chapter 6). The inverse of the element A is usually written $A^{-1}$ (in ordinary arithmetic think of multiplying by, say, the number 7. This multiplication can be cancelled out by multiplying again, this time by the number $7^{-1} = \frac{1}{7}$); i.e.

$$AA^{-1} = A^{-1}A = E \tag{A1.4}$$

*Note:* The element which we have here called $A^{-1}$ would normally have another label within the group; it could be B, for instance, or it could be A itself if A were self-inverse. The label $A^{-1}$ is here used in preference to, say, B, because the latter label does not reveal the special relationship between A and $A^{-1}$ given by the equation above.

**Problem A1.1** Apply the relationships given above to the elements of the $C_{2v}$ group $\{E, C_2, \sigma_v, \sigma_v'\}$ and thus, formally, show that they comprise a group.

## A1.2 SOME EXAMPLES OF GROUPS

*Permutation groups* were briefly encountered towards the end of Chapter 9. The groups formed by the operations permuting $n$ objects form a fascinating subject for study. The character table for the so-called 'symmetric group' (permutation group) with $n = 2$ is isomorphic to that of $C_2$, that for $n = 3$ is isomorphic to $C_{3v}$ and that for $n = 4$ is isomorphic to $T_d$. The groups with $n \geqslant 5$ are not isomorphic to any point group. The symmetric groups are of potential importance when identical particles are of interest. In chemistry these particles could be identical nuclei but more frequently they are electrons.

The symmetric group with $n = 3$ has the following six operations, the three particles being labelled a, b and c:

| (a)(b)(c) | (a b c) | (a c b) | (a b)(c) | (a c)(b) | (a)(b c) |
|---|---|---|---|---|---|
| E | $P_1$ | $P_2$ | $X_1$ | $X_2$ | $X_3$ |

Problem 9.4 gives an example of the combination of permutation operations of this type. Using the shorthand symbols indicated (P = *P*ermutation; X = e*X*change) the following group multiplication table is obtained:

| | | First operation | | | | | |
|---|---|---|---|---|---|---|---|
| | | E | $P_1$ | $P_2$ | $X_1$ | $X_2$ | $X_3$ |
| Second operation | E | E | $P_1$ | $P_2$ | $X_1$ | $X_2$ | $X_3$ |
| | $P_1$ | $P_1$ | $P_2$ | E | $X_2$ | $X_3$ | $X_1$ |
| | $P_2$ | $P_2$ | E | $P_1$ | $X_3$ | $X_1$ | $X_2$ |
| | $X_1$ | $X_1$ | $X_3$ | $X_2$ | E | $P_1$ | $P_2$ |
| | $X_2$ | $X_2$ | $X_1$ | $X_3$ | $P_2$ | E | $P_1$ |
| | $X_3$ | $X_3$ | $X_2$ | $X_1$ | $P_1$ | $P_2$ | E |

This multiplication table is isomorphic to that for $C_{3v}$, given in Table 8.2 (substitute systematically $C_3^+$ for $P_1$, $C_3^-$ for $P_2$, $X_1$ for $\sigma_v(1)$, etc.).

**Problem A1.2** Check that the above multiplication table is correct.

*Substitution groups* are fun but do not seem to have any general application. Consider the six functions

$$E = x \qquad P = \frac{1}{1 - x} \qquad Q = \frac{x - 1}{x}$$

$$R = \frac{1}{x} \qquad S = 1 - x \qquad T = \frac{x}{x - 1}$$

These form a group when the law of combination is substitution as function of a function. Thus,

$$SR = S\left(\frac{1}{x}\right) = 1 - \left(\frac{1}{x}\right) = \frac{x - 1}{x} = Q$$

and

$$PT = P\left(\frac{x}{x - 1}\right) = \frac{1}{1 - x/(x - 1)} = 1 - x = S$$

The multiplication table, given below, is also isomorphic to that of the $C_{3v}$ group (Table 8.2). The reader should check this, but should be warned. The table isomorphic to $C_{3v}$ given in the previous example was set out in a way that should have made the isomorphism self-evident. This is not true for that given below; more work will be required to demonstrate the isomorphism.

| | | First operation | | | | | |
|---|---|---|---|---|---|---|---|
| | | E | P | Q | R | S | T |
| Second operation | E | E | P | Q | R | S | T |
| | P | P | Q | E | T | R | S |
| | Q | Q | E | P | S | T | R |
| | R | R | S | T | E | P | Q |
| | S | S | T | R | Q | E | P |
| | T | T | R | S | P | Q | E |

**Problem A1.3** Check that the above multiplication table is correct.

An example of a two-colour group has been given in the discussion associated with Figure 2.5. Here, the changing of a colour has been introduced as a component of a symmetry operation.

*Colour groups* are of some importance in chemistry in the context of space groups (the groups of symmetry operations of crystalline materials). Many of the operations of space groups have the effect of relating one molecule in a crystal lattice to another. However, what if the molecules are not quite identical? For instance, the molecules could be atomically identical but have opposite magnetic properties (because the electron spins are arranged in opposite ways, for example). In this case the operation — put colloquially — of 'turn the magnet over' is similar to our 'change the colour' operation; it forms a composite with another symmetry operation to relate not-quite-identical objects.

'Two-colour' space groups are also known as 'black and white' groups or 'Shubnikov' groups. Polychromatic groups also exist!

With the exception of those groups indicated above, all of the point groups discussed in this book related to our three-dimensional world. However, it is possible to add to this picture that of electron spin. This has the effect of doubling the number of operations in the group compared to the corresponding point group without spin. Hence, these are called *double groups*. They have the rather unexpected property that the identity operation corresponds to a rotation of 720°. The double groups are of importance in some areas of chemistry — in particular, in the theory of transition metal ions.

## A1.3 THE CLASSES OF A GROUP

When in the previous section we started to detail the definition of a group we quickly found it necessary to recognize that the multiplication of any two elements, A and B, of a group could not be assumed to be *commutative*. That is, it is not generally true that

$$AB = BA$$

(When either A or B is the identity E, the equation is always true — it is equation A1.3). This equation may hold for some pairs of operations within a group but not others (for example, it is true for all pairs of $\sigma_v$ operations in the $C_{3v}$ point group, but is untrue when a $C_3$ is combined with a $\sigma_v$; see Table 8.2). Groups for which it is true for all pairs of elements are *Abelian point groups*. $C_{2v}$ (Chapters 2 and 3), $D_{2h}$ (Chapter 4) and $C_4$ (Chapter 11) are examples of Abelian point groups. As we have seen, in Abelian point groups there are never two elements in the same class. Non-Abelian point groups may have more than one element in each class and so, in giving a more precise meaning to the word 'class', we shall start with equation (A1.2) since this applies to at least some of the elements of non-Abelian groups:

$$AB \neq BA$$

Multiply each side of this equation, on the right, by the element $A^{-1}$. We obtain

$$ABA^{-1} \neq BAA^{-1}$$

But $AA^{-1} = E$ (equation A1.4) and so $BAA^{-1} = BE = B$ (by equation A1.3). That is,

$$ABA^{-1} \neq B$$

The product $ABA^{-1}$ must be equivalent to a single element in the group. To be general, let us call this single element D. That is,

$$ABA^{-1} = D \tag{A1.5}$$

There is a hidden symmetry in equation (A1.5). To see this, multiply on the left of each side of the equation by $A^{-1}$ and on the right of each side by A. We obtain

$$A^{-1}(ABA^{-1})A = A^{-1}(D)A$$

Because multiplication is associative we can write this as

$$(A^{-1}A)B(A^{-1}A) = A^{-1}DA$$

Which, by equation (A1.4), becomes

$$B = A^{-1}DA \tag{A1.6}$$

which is to be compared with equation (A1.5). Because of this relationship between B and D they are said to be *conjugate* elements of the group.

However, A was picked at random in the above development — we placed no restrictions on it. Suppose we choose a different element, C say, in its place. We have no theorem which would require that because (equation A1.5)

$$ABA^{-1} = D$$

then

$$CBC^{-1} = D$$

No, rather, we must assume that $CBC^{-1}$ gives yet another element (even if, sometimes, it does not). Let us consider the case where it does not give D but another element which we will call F. That is,

$$CBC^{-1} = F \tag{A1.7}$$

We can parallel the arguments leading up to equation (A1.6) above with a similar development to show from equation (A1.7) that

$$B = C^{-1}FC \tag{A1.8}$$

That is, B is conjugate with F as well as with D. We now show that this sequence requires that F and D are also conjugate elements.

We do this by combining equations (A1.6) and (A1.8):

$$A^{-1}DA = B = C^{-1}FC$$

Consider the two outer expressions and multiply each on the left by A and on the right by $A^{-1}$:

$$(AA^{-1})D(AA^{-1}) = (AC^{-1})F(CA^{-1})$$

That is,

$$D = (AC^{-1})F(CA^{-1}) \tag{A1.9}$$

Equation (A1.9) is of the form we seek (i.e. analogous to A1.5, A1.6, A1.7 and A1.8) provided that we can show that $(AC^{-1})$ and $(CA^{-1})$ are inverses of each other. If they are inverses then they satisfy (A1.4) and so all we have to do is to multiply them together and see whether their product is E. We have

$$
\begin{aligned}
(AC^{-1})(CA^{-1}) &= AC^{-1}CA^{-1} \\
&= A(C^{-1}C)A^{-1} \\
&= A(E)A^{-1} \\
&= AA^{-1} \\
&= E
\end{aligned}
$$

That is, $(AC^{-1})$ and $(CA^{-1})$ are, indeed, inverses. Now $AC^{-1}$ must be equal to a single element of the group; let us call it H. $CA^{-1}$ must then be $H^{-1}$ so that equation (A1.9) becomes

$$D = HFH^{-1} \tag{A1.10}$$

which is of the form we have been seeking. We conclude that B, D and F are all conjugate elements and comprise a subset of the set of all the group elements. Each set of conjugate elements in a group forms a *class* of the group.

Formally, then, in order to find all members of a group which are of the same class as B we let each element of the group in turn (including B) become A in the expression (see equation A1.5)

$$ABA^{-1}$$

and collect together all the products. They comprise all the elements which fall in the same class as B.

As an example we will consider the substitution group given in the previous section and use its multiplication table (page 235). First, from the table we identify the inverse of each element:

| Element | Inverse |
|---|---|
| E | E |
| P | Q |
| Q | P |
| R | R |
| S | S |
| T | T |

To obtain all elements in the same class as P we work down this list forming the products, obtaining the results given below:

$$
\begin{aligned}
EPE &= P \\
PPQ &= P \\
QPP &= P
\end{aligned}
$$

$$\begin{aligned} RPR &= Q \\ SPS &= Q \\ TPT &= Q \end{aligned}$$

and we conclude that P and Q are in the same class (a result which we could have anticipated because they are isomorphous with the $C_3^+$ and $C_3^-$ operations of $C_{3v}$).

**Problem A1.4** Check the above argument.

As a second example we consider the problem encountered in Chapter 11 — that $C_4$ and $C_4^3$ are in different classes. The group multiplication table for the $C_4$ group is (note its diagonal symmetry):

| $C_4$ | E | $C_4$ | $C_2$ | $C_4^3$ |
|---|---|---|---|---|
| E | E | $C_4$ | $C_2$ | $C_4^3$ |
| $C_4$ | $C_4$ | $C_2$ | $C_4^3$ | E |
| $C_2$ | $C_2$ | $C_4^3$ | E | $C_4$ |
| $C_4^3$ | $C_4^3$ | E | $C_4$ | $C_2$ |

from which we deduce that the inverses are:

| Element | Inverse |
|---|---|
| E | E |
| $C_4$ | $C_4^3$ |
| $C_2$ | $C_2$ |
| $C_4^3$ | $C_4$ |

In the class containing $C_4$ there will be

$$\begin{aligned} EC_4E &= C_4 \\ C_4C_4C_4^3 &= C_4 \\ C_2C_4C_2 &= C_4 \\ C_4^3C_4C_4 &= C_4 \end{aligned}$$

That is, it is in a class of its own. It is easy to similarly show that $C_4^3$ is in a class of its own, as too is $C_2$. This shows that $C_4^3$ is an Abelian group.

**Problem A1.5** Demonstrate that $C_4^3$ is in a class of its own.

**Problem A1.6** Show that the $C_{2v}$ group is an Abelian group.

## A1.4 CLASS ALGEBRA

When we introduced the $C_{3v}$ character table in Chapter 6 we did so in the form:

| $C_{3v}$ | E | $2C_3$ | $3\sigma_v$ |
|---|---|---|---|
| $A_1$ | 1 | 1 | 1 |
| $A_2$ | 1 | 1 | −1 |
| E | 2 | −1 | 0 |

and this, and the character tables of all other non-Abelian groups, are given in this form in Appendix 3. Why? Why put elements in the same class together? Why not write this character table as follows?

| $C_{3v}$ | E | $C_3^+$ | $C_3^-$ | $\sigma_v(1)$ | $\sigma_v(2)$ | $\sigma_v(3)$ |
|---|---|---|---|---|---|---|
| $A_1$ | 1 | 1 | 1 | 1 | 1 | 1 |
| $A_2$ | 1 | 1 | 1 | −1 | −1 | −1 |
| E | 2 | −1 | −1 | 0 | 0 | 0 |

After all, this is the form in which, effectively, we used it in the projection operator method (see, for instance, page 121). Firstly, we note that not all of the orthonormality relationships (Section 5.3) would remain true (some columns in the extended character table are identical). There is, however, another and fundamental reason. This is that there exists a *class algebra*. Take the $C_{3v}$ group as an example. It contains three classes with elements:

Class 1: E

Class 1: $C_3^+$ $C_3^-$

Class 3: $\sigma_v(1)$ $\sigma_v(2)$ $\sigma_v(3)$

Express this mathematically, thus:

$$\mathcal{C}_1 = E$$
$$\mathcal{C}_2 = \tfrac{1}{2}(C_3^+ + C_3^-)$$
$$\mathcal{C}_3 = \tfrac{1}{3}[\sigma_v(1) + \sigma_v(2) + \sigma_v(3)]$$

We can multiply these classes together. Thus,

$$\mathcal{C}_2\mathcal{C}_2 = \tfrac{1}{4}(C_3^+ + C_3^-)(C_3^+ + C_3^-)$$
$$= \tfrac{1}{4}(C_3^+C_3^+ + C_3^+C_3^- + C_3^-C_3^+ + C_3^-C_3^-)$$

which, from Table 8.2, is equal to

$$= \tfrac{1}{4}(C_3^- + E + E + C_3^+)$$
$$= \tfrac{1}{2}E + \tfrac{1}{4}(C_3^+ + C_3^-)$$
$$= \tfrac{1}{2}(\mathcal{C}_1 + \mathcal{C}_2)$$

We can thus form a class multiplication table which is easily shown to be:

| $C_{3v}$ | $\mathcal{C}_1$ | $\mathcal{C}_2$ | $\mathcal{C}_3$ |
|---|---|---|---|
| $\mathcal{C}_1$ | $\mathcal{C}_1$ | $\mathcal{C}_2$ | $\mathcal{C}_3$ |
| $\mathcal{C}_2$ | $\mathcal{C}_2$ | $\frac{1}{2}(\mathcal{C}_1 + \mathcal{C}_2)$ | $\mathcal{C}_3$ |
| $\mathcal{C}_3$ | $\mathcal{C}_3$ | $\mathcal{C}_3$ | $\frac{1}{3}(\mathcal{C}_1 + 2\mathcal{C}_2)$ |

**Problem A1.7** Check that the class multiplication table shown above is correct.

**Problem A1.8** Show that the above classes do not form a group under the operation of class multiplication. (*Hint:* refer to the relationships used to define a group at the beginning of this appendix.)

Only the classes of Abelian groups form groups under class multiplication but this is trivial because the classes are isomorphic to the elements of the Abelian group itself.

**Problem A1.9** Check the truth of the above assertion by reference to the $C_{2v}$ point group.

From the class multiplication table given above we see that, in general, we must expect the product of multiplying two classes together to be of the form

$$\mathcal{C}_j\mathcal{C}_i = \sum_k c_k\mathcal{C}_k$$

where the sum $k$ is over all classes and $c_k$ is a coefficient. We now ask what may appear a rather strange question. Is it possible to obtain a linear sum of the classes of the form

$$\mathcal{E} = \sum_j a_j\mathcal{C}_j$$

which has the property that when multiplied by any class, $\mathcal{C}_i$ say, it satisfies an equation of the form

$$\mathcal{C}_i\mathcal{E} = \lambda\mathcal{E}$$

where $\lambda$ is a number (possibly complex)?

Those with some knowledge of quantum mechanics will recognize this as an eigenvalue equation. The eigenvalues, $\lambda$, lead directly to the characters in the character table (these characters are not the $\lambda$'s but are related to them by simple, well-defined, numerical coefficients). That is, the characters are related to the classes and it is for this reason that character tables are given in the way that they are.

Clearly, the mathematics given above can be developed to provide a method for the calculation of character tables. We shall not give this development here but the interested reader will find a very readable account in a book by G. G. Hall, *Applied Group Theory* (Longmans, 1967).

APPENDIX 2

# *The matrix representations of groups*

In this book we have given a non-mathematical treatment of what, in fact, is a mathematical subject. This appendix goes some way towards reinstating the mathematics. However, it cannot claim to be comprehensive — if it were, its length would be very much greater.

## A2.1 MATRIX ALGEBRA AND SYMMETRY OPERATIONS

An array of quantities — often numbers — such as those given below is called a matrix

$$\begin{bmatrix} 3 & 2 \\ 4 & -1 \\ 0 & 2 \end{bmatrix} \quad \text{and} \quad \begin{bmatrix} 3 & 2 & -2 \\ 4 & -1 & 0 \\ 0 & 2 & 3 \end{bmatrix}$$

Clearly, matrices can be square — contain the same number of rows as they have columns — or they may be rectangular — the number of rows may be greater or less than the number of columns. Each number — or other quantity — appearing in a matrix is referred to as a 'matrix element'. If represented by an algebraic symbol a matrix element is often given suffixes to indicate in which row and which column it lies in the matrix.

Matrices of the same size may added; this is done by adding together the corresponding entries (elements). We illustrate this by adding two matrices; as an aid to clarity the elements of one matrix are given as letters:

$$\begin{bmatrix} 3 & 2 & -2 \\ 4 & -1 & 0 \\ 0 & 2 & 3 \end{bmatrix} + \begin{bmatrix} a & b & c \\ d & e & f \\ g & h & i \end{bmatrix} = \begin{bmatrix} (3+a) & (2+b) & (-2+c) \\ (4+d) & (-1+e) & f \\ g & (2+h) & (3+i) \end{bmatrix}$$

**Problem A2.1**

Fill in the missing quantities in the following matrix equation:

$$\begin{bmatrix} \sin^2\theta & \frac{1}{\sqrt{2}} & 3 \\ \cdot & \cdot & \cdot \\ 3 & 1 & -\sin^2\theta \end{bmatrix} + \begin{bmatrix} \cdot & \cdot & 0 \\ 1 & \cos^2\phi & -1 \\ -1 & 5 & \cdot \end{bmatrix} = \begin{bmatrix} 1 & \sqrt{2} & \cdot \\ 1 & 0 & -4 \\ \cdot & \cdot & \cos 2\theta \end{bmatrix}$$

The application of matrix algebra to the theory of groups is relatively limited and we shall have no occasion to add or subtract matrices. Key to our use of them, however, is the multiplication of matrices. Matrix multiplication does *not* parallel matrix addition; one does *not* simply multiply corresponding pairs of elements together. Although pairs of elements are, indeed, multiplied, each element in a complete row is multiplied by the corresponding element in a complete column — so that the row and column have to be of equal length — and the products are added together. It is this sum of products that is an element in the product matrix. To obtain the entry in the mth row and the nth column of the product matrix the elements in the mth row of the first matrix are multiplied by those in the nth column of the second.

Consider the two matrices which were added above. We will now multiply them instead. The entry at the top left-hand corner of the product matrix, the one in the first row (m = 1) and first column (n = 1) is given by:

$$\begin{matrix}\text{First} \\ \text{row}\end{matrix}\rightarrow\begin{bmatrix} 3 & 2 & -2 \\ \cdot & \cdot & \cdot \\ \cdot & \cdot & \cdot \end{bmatrix} \times \begin{bmatrix} a & \cdot & \cdot \\ d & \cdot & \cdot \\ g & \cdot & \cdot \end{bmatrix} = \begin{bmatrix} (3a + 2d - 2g) & \cdot & \cdot \\ \cdot & \cdot & \cdot \\ \cdot & \cdot & \cdot \end{bmatrix}$$

↑
First column

The complete product matrix is

$$\begin{bmatrix} (3a + 2d - 2g) & (3b + 2e - 2h) & (3c + 2f - 2i) \\ (4a - d) & (4b - e) & (4c - f) \\ (2d + 3g) & (2e + 3h) & (2f + 3i) \end{bmatrix}$$

and the reader who is unfamiliar with matrix multiplication should check several of the elements of this product.

**Problem A2.2** Fill in the blanks in the following matrix equation:

$$\begin{bmatrix} 3 & -1 \\ 2 & \cdot \end{bmatrix} \times \begin{bmatrix} 2 & -1 \\ \cdot & 3 \end{bmatrix} = \begin{bmatrix} 0 & \cdot \\ 10 & \cdot \end{bmatrix}$$

The multiplication of two matrices may be expressed algebraically. If the product of the matrices A and B (A being on the left) is denoted AB, then

$$(AB)_{mn} = \sum_{t} A_{mt}B_{tn} \tag{A2.1}$$

where m and n carry the meanings given above and t is simply a convenient running label which enables us to distinguish the individual matrix element products which have to be added together to give the element in the mth row and nth column of the product matrix AB.

In the example above, the two matrices which were multiplied together were square but this is not a requirement; the sole restriction is the obvious one — that the number of elements in each row of the matrix on the left equals the number of elements in each column of the matrix on the right.

The relevance of this to molecular symmetry can be seen by reference to Figure 3.1. This shows the transformation of the hydrogen 1s orbitals, $h_1$ and $h_2$, under the symmetry operations of the $C_{2v}$ point group and was discussed in Section 3.1.

Figure 3.1 shows that under the identity operation, E, $h_1$ and $h_2$ remain unchanged. This can be expressed by the matrix product

$$\begin{bmatrix} 1 & 0 \\ 0 & 1 \end{bmatrix} \begin{bmatrix} h_1 \\ h_2 \end{bmatrix} = \begin{bmatrix} h_1 \\ h_2 \end{bmatrix} \tag{A2.2}$$

where, following convention, we have omitted the $\times$ sign between the two matrices multiplied together. Writing them side by side in the way we have done is taken as meaning that they are to be multiplied. The reader should check that, arithmetically, equation (A2.2) is correct. It may be correct, but what does it mean? On the left-hand side the hydrogen 1s orbitals are written as the elements of a column matrix. The order in which they are written is, ultimately, unimportant but that used is clearly the more natural. When the matrix multiplication is carried out this column matrix is regenerated, unchanged.

That is, multiplication by the matrix $\begin{bmatrix} 1 & 0 \\ 0 & 1 \end{bmatrix}$ has a similar effect on $h_1$ and $h_2$ as the identity operation. However, had a different matrix been used to multiply the $h_1$, $h_2$ column matrix a different result would have been obtained.

Figure 3.1 shows that the $C_2$ rotation interchanges $h_1$ and $h_2$. The reader can readily show that this is expressed by the matrix product

$$\begin{bmatrix} 0 & 1 \\ 1 & 0 \end{bmatrix} \begin{bmatrix} h_1 \\ h_2 \end{bmatrix} = \begin{bmatrix} h_2 \\ h_1 \end{bmatrix} \tag{A2.3}$$

Here, the matrix $\begin{bmatrix} 0 & 1 \\ 1 & 0 \end{bmatrix}$ has a similar effect on $\begin{bmatrix} h_1 \\ h_2 \end{bmatrix}$ as the $C_2$ operation has on $h_1$ and $h_2$; $h_1$ and $h_2$ are interchanged.

It is left as an exercise for the reader to show that $\sigma_v$ and $\sigma_v'$ operations find parallels in the matrix products:

$$\sigma_v: \quad \begin{bmatrix} 1 & 0 \\ 0 & 1 \end{bmatrix} \begin{bmatrix} h_1 \\ h_2 \end{bmatrix} = \begin{bmatrix} h_1 \\ h_2 \end{bmatrix} \tag{A2.4}$$

$$\sigma_v': \quad \begin{bmatrix} 0 & 1 \\ 1 & 0 \end{bmatrix} \begin{bmatrix} h_1 \\ h_2 \end{bmatrix} = \begin{bmatrix} h_2 \\ h_1 \end{bmatrix} \tag{A2.5}$$

**Problem A2.3** Show, by expansion and comparison with Chapter 3, that equations (A2.4) and (A2.5) correctly describe the action of $\sigma_v$ and $\sigma_v'$ on $h_1$ and $h_2$.

In Chapter 2 it was shown that sets of numbers such as 1, 1, $-1$, $-1$ multiply in a manner which is isomorphic to the multiplication of the operations of the $C_{2v}$ point group (see Table 2.3 for example). The important thing about the square matrices in equations (A2.2) to (A2.5) is that when multiplied under the rule of matrix multiplication they, too, multiply isomorphically to the $C_{2v}$ operations. The multiplication of these $2 \times 2$ matrices is given in Table A2.1.

Table A2.1

| | | E | $C_2$ | $\sigma_v$ | $\sigma_v'$ |
|---|---|---|---|---|---|
| | | Right-hand matrix in the product | | | |
| | | $\begin{bmatrix} 1 & 0 \\ 0 & 1 \end{bmatrix}$ | $\begin{bmatrix} 0 & 1 \\ 1 & 0 \end{bmatrix}$ | $\begin{bmatrix} 1 & 0 \\ 0 & 1 \end{bmatrix}$ | $\begin{bmatrix} 0 & 1 \\ 1 & 0 \end{bmatrix}$ |
| Left-hand matrix in the product | $\begin{bmatrix} 1 & 0 \\ 0 & 1 \end{bmatrix}$ | $\begin{bmatrix} 1 & 0 \\ 0 & 1 \end{bmatrix}$ | $\begin{bmatrix} 0 & 1 \\ 1 & 0 \end{bmatrix}$ | $\begin{bmatrix} 1 & 0 \\ 0 & 1 \end{bmatrix}$ | $\begin{bmatrix} 0 & 1 \\ 1 & 0 \end{bmatrix}$ |
| | $\begin{bmatrix} 0 & 1 \\ 1 & 0 \end{bmatrix}$ | $\begin{bmatrix} 0 & 1 \\ 1 & 0 \end{bmatrix}$ | $\begin{bmatrix} 1 & 0 \\ 0 & 1 \end{bmatrix}$ | $\begin{bmatrix} 0 & 1 \\ 1 & 0 \end{bmatrix}$ | $\begin{bmatrix} 1 & 0 \\ 0 & 1 \end{bmatrix}$ |
| | $\begin{bmatrix} 1 & 0 \\ 0 & 1 \end{bmatrix}$ | $\begin{bmatrix} 1 & 0 \\ 0 & 1 \end{bmatrix}$ | $\begin{bmatrix} 0 & 1 \\ 1 & 0 \end{bmatrix}$ | $\begin{bmatrix} 1 & 0 \\ 0 & 1 \end{bmatrix}$ | $\begin{bmatrix} 0 & 1 \\ 1 & 0 \end{bmatrix}$ |
| | $\begin{bmatrix} 0 & 1 \\ 1 & 0 \end{bmatrix}$ | $\begin{bmatrix} 0 & 1 \\ 1 & 0 \end{bmatrix}$ | $\begin{bmatrix} 1 & 0 \\ 0 & 1 \end{bmatrix}$ | $\begin{bmatrix} 0 & 1 \\ 1 & 0 \end{bmatrix}$ | $\begin{bmatrix} 1 & 0 \\ 0 & 1 \end{bmatrix}$ |

**Problem A2.4** Check that Table A2.1 is correct

Table A2.1 should be compared with Table 2.1. Each matrix in Table A2.1 will be found to transform isomorphically to the operation associated with it.

Is this property limited to $2 \times 2$ matrices? No, provided that they are square matrices, matrices of any order can be found which multiply isomorphically to the operations of the $C_{2v}$ point group. Indeed, the numbers which behaved like this in Chapter 2 may be regarded as $1 \times 1$ matrices!

As an example of this, the following four matrices describe the transformations of the hydrogen atoms

$$\begin{bmatrix} H_a \\ H_b \\ H_c \\ H_d \end{bmatrix}$$

of Figure 2.16.

$$E: \begin{bmatrix} 1 & 0 & 0 & 0 \\ 0 & 1 & 0 & 0 \\ 0 & 0 & 1 & 0 \\ 0 & 0 & 0 & 1 \end{bmatrix}$$

$$C_2: \begin{bmatrix} 0 & 0 & 1 & 0 \\ 0 & 0 & 0 & 1 \\ 1 & 0 & 0 & 0 \\ 0 & 1 & 0 & 0 \end{bmatrix}$$

$$\sigma_v: \begin{bmatrix} 0 & 1 & 0 & 0 \\ 1 & 0 & 0 & 0 \\ 0 & 0 & 0 & 1 \\ 0 & 0 & 1 & 0 \end{bmatrix}$$

$$\sigma_v': \begin{bmatrix} 0 & 0 & 0 & 1 \\ 0 & 0 & 1 & 0 \\ 0 & 1 & 0 & 0 \\ 1 & 0 & 0 & 0 \end{bmatrix}$$

Further, the multiplication of these matrices is isomorphic to that of the corresponding operations of the $C_{2v}$ point group.

**Problem A2.5** Show that the above matrices do, indeed, describe the transformations of the hydrogen atoms of Figure 2.16

**Problem A2.6** Show that the multiplication of the above matrices is isomorphous to that of the operations of the $C_{2v}$ point group (it may be helpful to use Figure 2.17 as a check).

Matrix multiplication, then, provides us with a method of describing in detail the transformation of several objects under the operations of a point group. We have already — in Section 3.2 — described something similar. In Section 3.2 we first used the transformation of several objects under the operations of a point group to obtain reducible representations. Not surprisingly, the two methods are connected. In Section 3.2 we described a method of obtaining the characters of reducible representations. The bridge between the two methods appears when it is recognized that 'character' is the name given to the arithmetic sum of all of the elements on the leading diagonal (top left to bottom right) of a square matrix. If we apply this definition to the four matrices given immediately above we find that their characters are:

| | Matrix associated with | | | |
|---|---|---|---|---|
| | E | $C_2$ | $\sigma_v$ | $\sigma_v'$ |
| Character of the matrix | 4 | 0 | 0 | 0 |

This set of characters is simply that obtained for the reducible representation generated by the transformations of the four hydrogen atoms in Figure 2.16.

**Problem A2.7** (cf. Problem 3.2) Check that the transformations of the hydrogen atoms of Figure 2.16 lead to the above representation. (*Note:* the representation which has the number which is equal to the order of the group (here, 4) under the E operation and zeros elsewhere is called 'the regular representation'. It is of importance because it is used in the proof of some group theoretical theorems (but none which are included in this book). It is generated by a basis set which is not associated with any symmetry element. Thus, here, the hydrogen atoms are in general positions — they do not lie on a mirror plane or symmetry axis.)

The rules for the generation of characters given in boxes in Section 3.2 are now seen as arising from the definition of the character of a matrix and the fact that it is only when its transformation is described by an entry on the leading diagonal that an object remains unmoved under a symmetry operation. (A word of caution: this last statement will need some modification shortly when we encounter fractions on the diagonal.)

Just as one distinguishes between reducible and irreducible representations, so we distinguish reducible from irreducible matrix representations. We shall meet irreducible matrix representations later in this section and the connection between reducible and irreducible representations in Section A2.4. Both the $2 \times 2$ and $4 \times 4$ matrices given above were sets of reducible matrices.

Those functions whose transformations are described by matrices in the way that we have described are called basis functions; those basis functions given at the right-hand side of character tables (Appendix 3) are ultimately related to the transformation of the x, y and z coordinate axes. It is therefore important to consider the transformation of a set of coordinate axes under typical group symmetry operations. This is not a difficult problem. For example, the inversion operation, i, is described by

$$\text{i:} \quad \begin{bmatrix} -1 & 0 & 0 \\ 0 & -1 & 0 \\ 0 & 0 & -1 \end{bmatrix} \begin{bmatrix} x \\ y \\ z \end{bmatrix} = \begin{bmatrix} -x \\ -y \\ -z \end{bmatrix}$$

Reflection in a mirror plane (let us chose the yz plane as the mirror plane) is

$$\sigma(\mathrm{yz}): \quad \begin{bmatrix} -1 & 0 & 0 \\ 0 & 1 & 0 \\ 0 & 0 & 1 \end{bmatrix} \begin{bmatrix} \mathrm{x} \\ \mathrm{y} \\ \mathrm{z} \end{bmatrix} = \begin{bmatrix} -\mathrm{x} \\ \mathrm{y} \\ \mathrm{z} \end{bmatrix}$$

We discussed the problem of the rotation of axes in Chapter 6 (see Figure 6.4 and the related discussion). When the axes, x, y are rotated by an angle $\alpha$ around the z axis and are relabelled x′, y′, then as Figure 6.4 shows

$$\mathrm{x}' = \mathrm{x} \cos \alpha + \mathrm{y} \sin \alpha$$

Similarly,

$$\mathrm{y}' = -\mathrm{x} \sin \alpha + \mathrm{y} \cos \alpha$$

We can add, trivially,

$$\mathrm{z}' = \mathrm{z}$$

Therefore, the matrix describing the effect of a rotation, $R_z(\alpha)$, of $\alpha$ around the z axis on the coordinate axes is the $3 \times 3$ matrix in the middle of the following equation:

$$R_z(\alpha) \begin{bmatrix} \mathrm{x} \\ \mathrm{y} \\ \mathrm{z} \end{bmatrix} = \begin{bmatrix} \cos \alpha & \sin \alpha & 0 \\ -\sin \alpha & \cos \alpha & 0 \\ 0 & 0 & 1 \end{bmatrix} \begin{bmatrix} \mathrm{x} \\ \mathrm{y} \\ \mathrm{z} \end{bmatrix} = \begin{bmatrix} \mathrm{x}' \\ \mathrm{y}' \\ \mathrm{z}' \end{bmatrix} \qquad \text{(A2.6)}$$

A study of the elements on the leading diagonal of this matrix — those that contribute to the character — will show the basis for the rule given at the end of Section 6.1:

> When an axis is rotated by an angle $\alpha$ its contribution to the character for that operation is $\cos \alpha$.

Similarly, this relationship enables us to study in more detail the rotation of x and y axes by 45° shown in Figures 4.5 and 5.5 and also the rotation to give the more general set in Figure 5.6; the following discussion is based on these figures.

It is evident that the character generated by the x and y axes under a $C_4$ operation is identical for either choice of x and y axes shown in Figures 5.4 and 5.5 (the character is 0 because x and y directions are transposed by the operation). It should also be evident that the same character under this operation is obtained for the more general x and y axes of Figure 5.6 (if it is not evident, use equation A2.6 — suitably modified).

Less evident is the fact that the general axis set gives the same character as the other sets under improper rotations. Consider the operation of reflection in the $\sigma_v(2)$ mirror plane of Figure 5.3, a mirror plane which is the xz plane in Figure 5.4. All three axis sets give a character of 1 for the z axis. For the

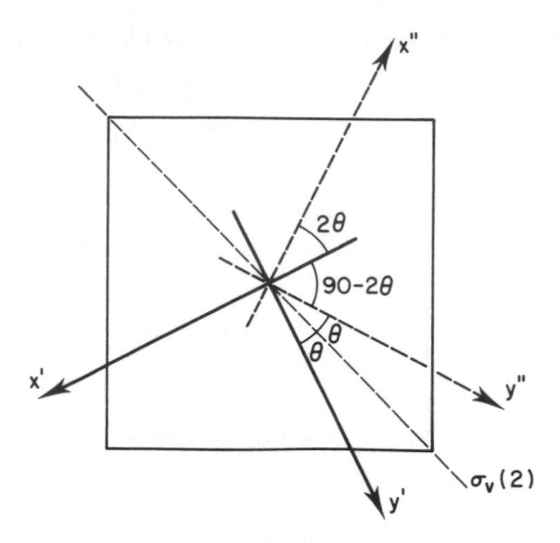


Figure A2.1 The effect of a mirror plane reflection ($\sigma_v(2)$) on x′ and y′ of Figure 5.6

(x, y) axes sets of Figures 5.4 and 5.5 we obtain characters of 0 for this reflection operation, but what of the axis set of Figure 5.6? If the angle between y′ and the adjacent Br—F bond axis contained in the $\sigma_v(2)$ mirror plane is denoted $\theta$ then the relationship between x′, y′ and their images x″, y″ is found to be (Figure A2.1)

$$x'' = -x' \cos 2\theta - y' \sin 2\theta$$
$$y'' = -x' \sin 2\theta + y' \cos 2\theta$$

That is,

$$\begin{bmatrix} -\cos 2\theta & -\sin 2\theta \\ -\sin 2\theta & \cos 2\theta \end{bmatrix} \begin{bmatrix} x' \\ y' \end{bmatrix} = \begin{bmatrix} x'' \\ y'' \end{bmatrix}$$

Clearly, the character of the $2 \times 2$ transformation matrix is 0, just as was the case for the axis choice of Figures 5.4 and 5.5.

So far, in all of the axis transformations that we have considered the z axis has remained unmoved. It is easy to see that rotation by three independent angles about coordinate axes is necessary to describe the relationship between two sets of arbitrarily orientated Cartesian axes. If we first rotate about z, then x′ and y′ must remain in the original xy plane. If we now rotate about x′ then this axis will remain in the original xy plane and does not assume a general position. Two rotations are not sufficient to place all three original axes in general positions; a third is needed.

The general transformation is shown in Figure A2.2. A rotation by $\phi$ about z is followed by one of $\theta$ about x′. Under this rotation z becomes z′ and y′ becomes y″. The final rotation is one of $\psi$ about z′, whereupon x′ becomes x″ and y″ becomes y‴. Mathematically, equation (A2.6) is applied to each of

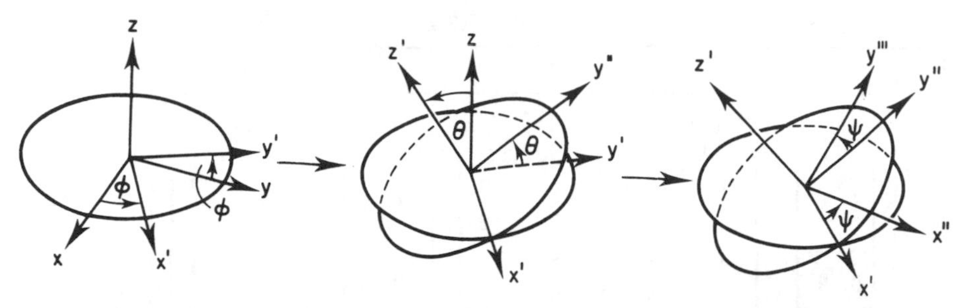


Figure A2.2 The interconversion of two sets of axes, (x, y, z) and (x″, y‴, z′), which are in a general relationship the one set to the other

these transformations. The final result is

$$\begin{bmatrix} \cos\psi\cos\phi - \cos\theta\sin\phi\sin\psi & \cos\psi\sin\phi + \cos\theta\cos\phi\sin\psi & \sin\psi\sin\theta \\ -\sin\psi\cos\phi - \cos\theta\sin\phi\cos\psi & -\sin\psi\sin\phi + \cos\theta\cos\phi\cos\psi & \cos\psi\sin\theta \\ \sin\theta\sin\phi & -\sin\theta\cos\phi & \cos\theta \end{bmatrix} \begin{bmatrix} x \\ y \\ z \end{bmatrix} = \begin{bmatrix} x'' \\ y''' \\ z' \end{bmatrix} \tag{A2.7}$$

Because a set of p orbitals transforms in the same way as the coordinate axes, this relationship is needed to answer the problem left unresolved in Section 3.2 — the transformation of a complete set of p orbitals referred to arbitrary axes under the operations of the $C_{2v}$ point group.

**Problem A2.8** Derive equation (A2.7).

**Problem A2.9** Consider the following conversion of axis set x, y, z into the general positions occupied by the set x′, y′, z′.

Take z and rotate it by an angle $\alpha$ such that it becomes coincident with z′. Now rotate the other two axes by an angle $\beta$ so that they coincide with x′ and y′ (x′ and y′ are in the plane perpendicular to z′). Does this sequence mean that it is possible to relate the two axis sets by just two angles, $\alpha$ and $\beta$, rather than the three of equation (A2.7)? If not, why not?

In the section above, we have been particularly concerned with sets of matrices which usually are reducible representations. However, similar considerations apply to irreducible representations. That is, there exist sets of irreducible matrices for each group. As we shall see in Section A2.4, it is always possible to manipulate a set of matrices which form a basis for a reducible representation in such a way that they can be rewritten as a sum of the irreducible matrices.

Table A2.2

| $C_{3v}$ | E | $C_3^+$ | $C_3^-$ | $\sigma_v(1)$ | $\sigma_v(2)$ | $\sigma_v(3)$ |
|---|---|---|---|---|---|---|
| $A_1$ | [1] | [1] | [1] | [1] | [1] | [1] |
| $A_2$ | [1] | [1] | [1] | [−1] | [−1] | [−1] |
| E | $\begin{bmatrix} 1 & 0 \\ 0 & 1 \end{bmatrix}$ | $\begin{bmatrix} -\frac{1}{2} & -\frac{\sqrt{3}}{2} \\ \frac{\sqrt{3}}{2} & -\frac{1}{2} \end{bmatrix}$ | $\begin{bmatrix} -\frac{1}{2} & \frac{\sqrt{3}}{2} \\ -\frac{\sqrt{3}}{2} & -\frac{1}{2} \end{bmatrix}$ | $\begin{bmatrix} -1 & 0 \\ 0 & 1 \end{bmatrix}$ | $\begin{bmatrix} \frac{1}{2} & -\frac{\sqrt{3}}{2} \\ -\frac{\sqrt{3}}{2} & -\frac{1}{2} \end{bmatrix}$ | $\begin{bmatrix} \frac{1}{2} & \frac{\sqrt{3}}{2} \\ \frac{\sqrt{3}}{2} & -\frac{1}{2} \end{bmatrix}$ |

As an example of the irreducible matrix representations of a group we give those for the $C_{3v}$ point group in Table A2.1. This table should be compared with the $C_{3v}$ character table given in Table 6.1 (and also in Appendix 3).

Comparison of Table A2.2 with the $C_{3v}$ character table reveals two important things. Firstly, whereas individual operations are listed separately in Table A2.2 in a character table they are grouped into classes. Secondly, for a given irreducible representation, the irreducible matrices of all operations in any one class have the same character, and this is the character listed for the class in the character table. The rapproachment between the ‘individual operation’ and ‘classes’ presentations is provided by the class algebra which was introduced in Appendix A1.4. There are some applications of group theory where it is necessary to use the complete matrix representations of groups — for instance, in some applications of the direct product. We shall meet such applications in the next section although we shall not discuss them in detail in that section.

## A2.2 DIRECT PRODUCTS

In the main text we met three different uses of the phrase ‘direct product’. Firstly, the operations of some groups were said to be the direct product of the operations of two other groups. An example is the $D_{2h}$ group, discussed in Section 4.3, for which each individual $D_{2h}$ operation may be regarded as a product of an individual operation of the $D_2$ group with an individual operation of the $C_i$ group. For such cases the character table of the product group was also said to be the direct product of those of the other two groups. The phrase ‘direct product’ was also used to describe the multiplication together of two representations of a group, a topic which was discussed as some length in Chapter 10. Clearly, the concept of a direct product is one of wide applicability in group theory; it is also an important one.

At the end of the previous section it was stated that a close connection exists between the characters in a character table and sets of irreducible matrices. Like the case of the (reducible) matrix representations which we discussed in that section in some detail, the multiplication of irreducible matrices is isomorphic to the multiplication of the group operations. Because of this isomorphism and because we can form direct products of group operations, we would expect there to be, correspondingly, a direct product of matrices. At the beginning of the previous section we met one way of multiplying two matrices, but this was such that the size of the product matrix was often the same as the size of the matrices from which it was formed (e.g. when square matrices are multiplied together). A characteristic of direct products is that an *increase* in size is the norm — the $D_{2h}$ group is larger than $D_2$ and $C_2$. This argument suggests that the direct product of the matrices involves a second form of matrix multiplication. This is not a unique situation. For instance, there are two different ways of combining — multiplying — vectors together.

The direct product of two matrices is obtained by individually and separately multiplying every element of each of the two matrices together. Thus, the direct product of the matrices

$$\begin{bmatrix} 3 & 2 \\ 4 & -1 \end{bmatrix} \quad \text{and} \quad \begin{bmatrix} a & b \\ d & e \end{bmatrix}$$

is

$$\begin{bmatrix} 3a & 3b & 2a & 2b \\ 3d & 3e & 2d & 2e \\ 4a & 4b & -a & -b \\ 4d & 4e & -d & -e \end{bmatrix}$$

If a general element of the matrix A is $a_{ij}$ (i labelling the row and j the column in which $a_{ij}$ occurs in A) and a typical element of B is $b_{km}$ (kth row, mth column), then the general element of the matrix C which is the direct product of A and B is

$$a_{ij}\, b_{km} = c_{ij,km} \qquad \text{(A2.8)}$$

For the matrices themselves,

$$A \otimes B = C \qquad \text{(A2.9)}$$

where $\otimes$ indicates that a direct product is being formed.

There are several ways in which the matrix C may be written; a convenient one is

$$C = \begin{bmatrix} a_{11}B & a_{12}B & . & \cdots \\ a_{21}B & a_{22}B & . & \cdots \\ a_{31}B & a_{32}B & . & \cdots \\ \vdots & \vdots & \vdots & \vdots \\ \vdots & \vdots & \vdots & \vdots \end{bmatrix} \qquad \text{(A2.10)}$$

where $a_{11}B$ means that each element of the matrix B is multiplied, in order, by $a_{11}$.

**Problem A2.10** Fill in the blanks in the following matrix equation (it may be helpful to regard the $4 \times 4$ matrix as consisting of four $2 \times 2$ matrices, corresponding to B above:

$$\begin{bmatrix} 1 & 2 \\ 0 & \cdot \end{bmatrix} \begin{bmatrix} \cdot & \cdot \\ -1 & \cdot \end{bmatrix} = \begin{bmatrix} 3 & 6 & 0 & 0 \\ \cdot & -3 & 0 & 0 \\ \cdot & \cdot & -4 & -8 \\ \cdot & \cdot & \cdot & \cdot \end{bmatrix}$$

The elements of the group $D_{3d}$ are formed as the direct product of the elements of the $C_{3v}$ group with the elements of the $C_i$ group. The relationship between the operations of $D_{3d}$, $C_{3v}$ and $C_i$ is indicated in the table below in the form:

| | Operations of $C_{3v}$ |
|---|---|
| Operations of $C_i$ | Operations of $D_{3d}$ |

| | E | $C_3^+$ | $C_3^+$ | $\sigma_v(1)$ | $\sigma_v(2)$ | $\sigma_v(3)$ |
|---|---|---|---|---|---|---|
| E | E | $C_3^+$ | $C_3^-$ | $\sigma_v(1)$ | $\sigma_v(2)$ | $\sigma_v(3)$ |
| i | i | $S_6^-$ | $S_6^+$ | $C_2(1)$ | $C_2(2)$ | $C_2(3)$ |

A precisely parallel relationship exists between the irreducible matrix representations of the $D_{3d}$, $C_{3v}$ and $C_i$ groups, a relationship detailed below. First, however, we note that the group $D_{3d}$ is also the direct product of $D_3$ with $C_i$, and it is this latter product which is conventionally taken to determine the labels of the irreducible representations of $D_{3d}$.

**Problem A2.11** Show that $D_{3d} = D_3 \otimes C_i$.

The matrix representations of the $C_i$ group are

| $C_i$ | E | i |
|---|---|---|
| $A_g$ | [1] | [1] |
| $A_u$ | [1] | [$-1$] |

so that the direct product with the $C_{3v}$ matrix representations given in Table A2.2 leads to the irreducible matrix representations for the $D_{3d}$ group given in Table A2.3. That this isomorphism between multiplication of operations and multiplication of matrices persists in this application of the direct product is evident when it is recalled that the isomorphism exists in each of the individual groups involved in the direct product.

Table A2.3

| $D_{3d}$ | E | $C_3^+$ | $C_3^-$ | $\sigma_v(1)$ | $\sigma_v(2)$ | $\sigma_v(3)$ | i | $S_6^-$ | $S_6^+$ | $C_2(1)$ | $C_2(2)$ | $C_2(3)$ |
|---|---|---|---|---|---|---|---|---|---|---|---|---|
| $A_{1g}$ | [1] | [1] | [1] | [1] | [1] | [1] | [1] | [1] | [1] | [1] | [1] | [1] |
| $A_{2g}$ | [1] | [1] | [1] | [−1] | [−1] | [−1] | [1] | [1] | [1] | [−1] | [−1] | [−1] |
| $E_g$ | $\begin{bmatrix} 1 & 0 \\ 0 & 1 \end{bmatrix}$ | $\begin{bmatrix} -\frac{1}{2} & -\frac{\sqrt{3}}{2} \\ \frac{\sqrt{3}}{2} & -\frac{1}{2} \end{bmatrix}$ | $\begin{bmatrix} -\frac{1}{2} & \frac{\sqrt{3}}{2} \\ -\frac{\sqrt{3}}{2} & -\frac{1}{2} \end{bmatrix}$ | $\begin{bmatrix} -1 & 0 \\ 0 & 1 \end{bmatrix}$ | $\begin{bmatrix} \frac{1}{2} & -\frac{\sqrt{3}}{2} \\ -\frac{\sqrt{3}}{2} & -\frac{1}{2} \end{bmatrix}$ | $\begin{bmatrix} \frac{1}{2} & \frac{\sqrt{3}}{2} \\ \frac{\sqrt{3}}{2} & -\frac{1}{2} \end{bmatrix}$ | $\begin{bmatrix} 1 & 0 \\ 0 & 1 \end{bmatrix}$ | $\begin{bmatrix} -\frac{1}{2} & -\frac{\sqrt{3}}{2} \\ \frac{\sqrt{3}}{2} & -\frac{1}{2} \end{bmatrix}$ | $\begin{bmatrix} -\frac{1}{2} & \frac{\sqrt{3}}{2} \\ -\frac{\sqrt{3}}{2} & -\frac{1}{2} \end{bmatrix}$ | $\begin{bmatrix} -1 & 0 \\ 0 & 1 \end{bmatrix}$ | $\begin{bmatrix} \frac{1}{2} & -\frac{\sqrt{3}}{2} \\ -\frac{\sqrt{3}}{2} & -\frac{1}{2} \end{bmatrix}$ | $\begin{bmatrix} \frac{1}{2} & \frac{\sqrt{3}}{2} \\ \frac{\sqrt{3}}{2} & -\frac{1}{2} \end{bmatrix}$ |
| $A_{2u}$ | [1] | [1] | [1] | [1] | [1] | [1] | [−1] | [−1] | [−1] | [−1] | [−1] | [−1] |
| $A_{1u}$ | [1] | [1] | [1] | [−1] | [−1] | [−1] | [−1] | [−1] | [−1] | [1] | [1] | [1] |
| $E_u$ | $\begin{bmatrix} 1 & 0 \\ 0 & 1 \end{bmatrix}$ | $\begin{bmatrix} -\frac{1}{2} & -\frac{\sqrt{3}}{2} \\ \frac{\sqrt{3}}{2} & -\frac{1}{2} \end{bmatrix}$ | $\begin{bmatrix} -\frac{1}{2} & \frac{\sqrt{3}}{2} \\ -\frac{\sqrt{3}}{2} & -\frac{1}{2} \end{bmatrix}$ | $\begin{bmatrix} -1 & 0 \\ 0 & 1 \end{bmatrix}$ | $\begin{bmatrix} \frac{1}{2} & -\frac{\sqrt{3}}{2} \\ -\frac{\sqrt{3}}{2} & -\frac{1}{2} \end{bmatrix}$ | $\begin{bmatrix} \frac{1}{2} & \frac{\sqrt{3}}{2} \\ \frac{\sqrt{3}}{2} & -\frac{1}{2} \end{bmatrix}$ | $\begin{bmatrix} -1 & 0 \\ 0 & -1 \end{bmatrix}$ | $\begin{bmatrix} \frac{1}{2} & \frac{\sqrt{3}}{2} \\ -\frac{\sqrt{3}}{2} & \frac{1}{2} \end{bmatrix}$ | $\begin{bmatrix} \frac{1}{2} & -\frac{\sqrt{3}}{2} \\ \frac{\sqrt{3}}{2} & \frac{1}{2} \end{bmatrix}$ | $\begin{bmatrix} 1 & 0 \\ 0 & -1 \end{bmatrix}$ | $\begin{bmatrix} -\frac{1}{2} & \frac{\sqrt{3}}{2} \\ \frac{\sqrt{3}}{2} & \frac{1}{2} \end{bmatrix}$ | $\begin{bmatrix} -\frac{1}{2} & -\frac{\sqrt{3}}{2} \\ -\frac{\sqrt{3}}{2} & \frac{1}{2} \end{bmatrix}$ |

**Problem A2.12** Show that $D_{3d} = C_{3v} \otimes C_i$ (see above); then check Table 2.3.

The definition of direct product given by equations (A2.8) and (A2.9) and the convention given by equation (A2.10) are, of course, also applicable to the direct products formed between two representations of the same group. Thus the direct product matrices $A_2 \otimes E$ of the $C_{3v}$ point group are (from Table A2.2)

$$A_2 \otimes E = \begin{array}{cccccc} E & C_3^{+} & C_3^{-} & \sigma_v(1) & \sigma_v(2) & \sigma_v(3) \\ \begin{bmatrix} 1 & 0 \\ 0 & 1 \end{bmatrix} & \begin{bmatrix} -\frac{1}{2} & -\frac{\sqrt{3}}{2} \\ \frac{\sqrt{3}}{2} & -\frac{1}{2} \end{bmatrix} & \begin{bmatrix} -\frac{1}{2} & \frac{\sqrt{3}}{2} \\ -\frac{\sqrt{3}}{2} & -\frac{1}{2} \end{bmatrix} & \begin{bmatrix} 1 & 0 \\ 0 & -1 \end{bmatrix} & \begin{bmatrix} -\frac{1}{2} & \frac{\sqrt{3}}{2} \\ \frac{\sqrt{3}}{2} & \frac{1}{2} \end{bmatrix} & \begin{bmatrix} -\frac{1}{2} & -\frac{\sqrt{3}}{2} \\ -\frac{\sqrt{3}}{2} & \frac{1}{2} \end{bmatrix} \end{array}$$

Examination of these matrices shows that the characters of the direct product matrices are those obtained by using a character table and multiplying the characters of the two irreducible representations together. The technique of multiplying characters to obtain direct product characters, although simple, ignores the subtle changes that have taken place in the corresponding matrices, particularly in those representing the $\sigma_v$ operations (for which, multiplying by the character 0 might well have appeared trivial).

**Problem A2.13** Using the $C_{3v}$ character table, form the direct product of the $A_2$ and E irreducible representations and compare the answer with the matrix form given above.

As a final example we consider the direct product $E \otimes E$ in $C_{3v}$ but confine our discussion to just two of the product matrices. The two direct products we will evaluate are those direct product matrices corresponding to $\sigma_v(1)$ and $C_3^{+}$ which arise from the $E \otimes E$ direct product. For the first of these, expression in the form given by equation (A2.10) leads to

$$\begin{bmatrix} -1\begin{bmatrix} -1 & 0 \\ 0 & 1 \end{bmatrix} & 0\begin{bmatrix} -1 & 0 \\ 0 & 1 \end{bmatrix} \\ 0\begin{bmatrix} -1 & 0 \\ 0 & 1 \end{bmatrix} & 1\begin{bmatrix} -1 & 0 \\ 0 & 1 \end{bmatrix} \end{bmatrix}$$

which, on expansion, gives

$$\begin{bmatrix} 1 & 0 & 0 & 0 \\ 0 & -1 & 0 & 0 \\ 0 & 0 & -1 & 0 \\ 0 & 0 & 0 & 1 \end{bmatrix}$$

a matrix with a character of 0, the same character we obtain working with the $C_{3v}$ character table.

The second direct product, that corresponding to $C_3^+$, involves more work. In the form of equation (A2.10) the product is

$$\begin{bmatrix} -\frac{1}{2}\begin{bmatrix} -\frac{1}{2} & -\frac{\sqrt{3}}{2} \\ \frac{\sqrt{3}}{2} & -\frac{1}{2} \end{bmatrix} & -\frac{\sqrt{3}}{2}\begin{bmatrix} -\frac{1}{2} & -\frac{\sqrt{3}}{2} \\ \frac{\sqrt{3}}{2} & -\frac{1}{2} \end{bmatrix} \\ \frac{\sqrt{3}}{2}\begin{bmatrix} -\frac{1}{2} & -\frac{\sqrt{3}}{2} \\ \frac{\sqrt{3}}{2} & -\frac{1}{2} \end{bmatrix} & -\frac{1}{2}\begin{bmatrix} -\frac{1}{2} & -\frac{\sqrt{3}}{2} \\ \frac{\sqrt{3}}{2} & -\frac{1}{2} \end{bmatrix} \end{bmatrix}$$

leading to

$$\begin{bmatrix} \frac{1}{4} & \frac{\sqrt{3}}{4} & \frac{\sqrt{3}}{4} & \frac{3}{4} \\ -\frac{\sqrt{3}}{4} & \frac{1}{4} & -\frac{3}{4} & \frac{\sqrt{3}}{4} \\ -\frac{\sqrt{3}}{4} & -\frac{3}{4} & \frac{1}{4} & \frac{\sqrt{3}}{4} \\ \frac{3}{4} & -\frac{\sqrt{3}}{4} & -\frac{\sqrt{3}}{4} & \frac{1}{4} \end{bmatrix}$$

and the expected character of 1.

Because these direct product matrices are $4 \times 4$ it is clear that they must describe the transformation of four quantities — basis functions — which must themselves be related to the basis functions for the E irreducible representation. The exploration and exploitation of such relationships are important aspects of advanced group theory but beyond the scope of the present text.

**Problem A2.14** Evaluate the direct product matrices of $E \otimes E$ in $C_{3v}$ for the operations $C_3^-$ and $\sigma_v(2)$.

## A2.3 THE ORTHONORMALITY RELATIONSHIPS

In Section 5.3 we discussed group orthogonality relationships in a general fashion. The object of the present section is to provide a firmer mathematical basis for these relationships. Theorems 2 to 5 of Section 5.3 were all expressed in terms of the characters of irreducible representations. However, we have seen earlier in this appendix, in Section A2.1, that these characters are derived from irreducible matrices, such as those in Table A2.2, and it is through the latter that we prove the relationships of Section 5.3. We shall first derive some orthogonality relationships which relate to the irreducible matrices and express them in a single comprehensive equation. We shall then show that the theorems of Section 5.3 derive from these matrix element orthogonality relationships. Finally, we will indicate a proof of the comprehensive equation mentioned above. In this section we will follow the arguments and use the notation used by H. Eyring, J. Walter and G. E. Kimball in Chapter 10 of their book *Quantum Chemistry* (Wiley, 1944) where the argument is more general and rigorous than that given here.

If Table A2.2 (page 252) is carefully examined it will be found that if corresponding matrix elements in the same irreducible representation are squared and added together a rather simple result is obtained. Thus, if the 1,2 (top right — first row, second column) elements of the E irreducible matrices are squared and added we obtain:

$$0+\tfrac{3}{4}+\tfrac{3}{4}+0+\tfrac{3}{4}+\tfrac{3}{4}=3$$

The answer, 3, is equal to

$$\frac{\text{The number of operations in the group (the order of the group)}}{\text{The dimension of the irreducible representation}}$$

$$=\frac{6}{2}\ \text{in the present example}$$

**Problem A2.15** Check that 3 is the result when the above procedure is applied to the 1,1, to the 2,1 and to the 2,2 elements of the E irreducible matrices in Table A2.2. Predict the answer expected for either $E_g$ or $E_u$ of Table A2.3 and check your prediction.

This result is perfectly general and may be written as

$$\sum_R \Gamma_i(R)_{mn}\Gamma_i(R)_{mn}=\frac{h}{l_i} \qquad \text{(A2.11)}$$

where i labels the irreducible representation chosen, $\Gamma_i(R)_{mn}$ indicating the mnth element of the irreducible matrix corresponding to the operation R. The order of the group is denoted by h and $l_i$ is the dimension of the ith irreducible representation. The symbol Γ is that commonly used to indicate a matrix — reducible or irreducible — which transforms isomorphically to a symmetry operation.

If, on the other hand, instead of squaring matrix elements and then adding, two *different* elements from within the same matrix are multiplied and added then the result is 0. Thus, for the 1,1 and 1,2 elements of the E matrices in Table A2.2 we have

$$0+\frac{\sqrt{3}}{4}-\frac{\sqrt{3}}{4}+0-\frac{\sqrt{3}}{4}+\frac{\sqrt{3}}{4}=0$$

In general,

$$\sum_R \Gamma_i(R)_{mn}\Gamma_i(R)_{m'n'}=0 \qquad \text{(A2.12)}$$

where either $m \neq m'$ and/or $n \neq n'$. Similarly, for any pair of matrix elements of any two *different* irreducible representations i and j (as long as the matrices are associated with the same operation, R):

$$\sum_R \Gamma_i(R)_{mn}\Gamma_j(R)_{m'n'}=0 \qquad \text{(A2.13)}$$

where $i \neq j$ and there is no restriction on mn and m′n′ as long as they exist. Thus for the 1,1 (the only!) element of the $A_2$ matrices and the 2,1 element of the E matrices in Table A2.2 we have

$$0+\frac{\sqrt{3}}{2}-\frac{\sqrt{3}}{2}+0-\frac{\sqrt{3}}{2}+\frac{\sqrt{3}}{2}=0$$

Equations (A2.11), (A2.12) and (A2.13), and the restrictions on the quantities within them, may be combined into the general — and therefore important — relationship which is often called ‘the great orthonormality theorem’:

$$\sum_R \Gamma_i(R)_{mn}\Gamma_j(R)_{m'n'}=\frac{h}{\sqrt{l_i l_j}}\,\delta_{ij}\,\delta_{mm'}\,\delta_{nn'} \qquad \text{(A2.14)}$$

where $\delta_{ij}=1$ if $i=j$ but 0 if $i \neq j$. Similarly $\delta_{mm'}=1$ if $m=m'$ but 0 if $m \neq m'$ and $\delta_{nn'}=1$ if $n=n'$ but 0 if $n \neq n'$. The appearance of a square root term in equation (A2.14) is deceptive. Because of the $\delta_{ij}$ term this term will only be of importance when $i=j$ and then $\sqrt{l_i l_j}=l_i=l_j$, so we could just as well have put one of these on the right-hand side of (A2.14). The square root is included so that (A2.14) is symmetric in i and j.

In order to relate A2.14 to the theorems of Section 5.3 we have to adapt it to refer to the characters (= sum of elements on the leading diagonal) of the irreducible matrices. Whereas $\Gamma_i(R)_{mn}$ is a general matrix element, diagonal matrix elements will have $m=n$ and so be of the form

$$\Gamma_i(R)_{mm}$$

For these diagonal elements (A2.14) assumes the form

$$\sum_R \Gamma_i(R)_{mm}\Gamma_j(R)_{m'm'}=\frac{h}{l_j}\,\delta_{ij}\,\delta_{mm'} \qquad \text{(A2.15)}$$

Because, by definition, the ith irreducible representation is of dimension $l_i$, there will be $l_i$ terms $\Gamma_i(R)_{mm}$ which have to be added to give the character of the ith irreducible matrix appropriate to the operation R. That is, m assumes values from 1 (one) through to $l_i$ (if different from 1). Similarly, m′ runs from 1 (one) to $l_j$.

Writing, as is conventionally done, $\chi_i(R)$ for the character (the sum of elements along the leading diagonal) of the ith irreducible representation under the operation R we have

$$\chi_i(R) = \sum_m \Gamma_i(R)_{mm} \tag{A2.16}$$

and

$$\chi_j(R) = \sum_{m'} \Gamma_j(R)_{m'm'} \tag{A2.17}$$

Summing each side of (A2.15) over m and m′ and substituting (A2.16) and (A2.17) we have

$$\sum_R \chi_i(R)\chi_j(R) = \frac{h}{l_j} \delta_{ij} \sum_{m=1}^{l_i} \sum_{m'=1}^{l_j} \delta_{mm'} \tag{A2.18}$$

**Problem A2.16** Derive equation (A2.18) from equations (A2.15), (A2.16) and (A2.17).

Because of the $\delta_{mm'}$ term on the right-hand side of equation (A2.18) ($\delta_{mm'} = 0$ if $m \neq m'$ but 1 if $m = m'$) there will only be a contribution from the two summations where $m = m'$ and then the contribution will be 1. The right-hand side becomes

$$\frac{h}{l_i} \delta_{ij} \sum_{m'=1}^{l_j} 1$$

The meaning of the summation in this expression is 'every term in the summation from 1 (one) through to $l_j$ contributes 1 (one) to the total'; thus the value of the sum is $l_j$. The summation is not to be confused with

$$\sum_{m'=1}^{l_j} m'$$

This means that equation (A2.18) can be rewritten as

$$\sum_R \chi_i(R)\chi_j(R) = h\ \delta_{ij} \tag{A2.19}$$

In this expression each operation appears separately. Because for a given irreducible representation the character of each matrix is the same for all operations in the same class we can add their contributions to the left-hand side of equation (A2.19) together. If there are $g_\varrho$ operations in the class $\varrho$ we can write

$$\sum_R \chi_i(R)\chi_j(R) = \sum_\varrho \chi_i(R_\varrho)\chi_j(R_\varrho)g_\varrho$$

where $\chi_i(R_\varrho)$ is the character of the ith irreducible representation under the class $R_\varrho$ (i.e. the quantity given in the group character table). Of course, $\chi_i(R_\varrho)$ is numerically equal to $\chi_i(R)$. With this substitution equation (A2.19) becomes

$$\sum_\varrho \chi_i(R_\varrho)\chi_j(R_\varrho)g_\varrho = h\ \delta_{ij} \qquad \text{(A2.20)}$$

All of the Theorems 2 to 5 given in Section 5.3 are contained within (A2.20).

**Problem A2.17** Show that Theorems 2 to 5 of Section 5.3 are contained within equation (A2.20).

Because we built our argument on Table A2.2, which contains only real quantities, equation (A2.20) contains the hidden assumption that all characters are real. As Appendix 3 (and Chapter 11) show, this assumption is not valid — some character tables contain complex quantities. Fortunately, the generalization of equation (A2.20) to include these cases is simple; the general expression is

$$\sum_\varrho \chi_i^*(R_\varrho)\chi_j(R_\varrho)g_\varrho = h\ \delta_{ij} \qquad \text{(A2.21)}$$

where the * means that one uses not the quantity given in a character table as $\chi_i(R_\varrho)$ but, instead, its complex conjugate. An example of this usage is given in some detail in Chapter 11.

**Problem A2.18** Show that in the $C_{2v}$ point group there is an infinite number of functions of the general form $x^m y^m$ which transform as the totally symmetric ($A_1$) irreducible representation.

We now proceed to the derivation of equation (A2.14). Again our derivation is closely related to that given by H. Eyring, J. Walter and G. E. Kimball in their book *Quantum Chemistry*, in Appendix VI, and uses the same notation but is less rigorous and is structured rather differently.

We first assume that a function F, which, for generality, we express as a sum:

$$F = \sum_{k=1}^{l} x_k y_k \qquad \text{(A2.22)}$$

is invariant under all operations of the group — it transforms as the totally symmetric irreducible representation. That is, any operation, R, of the group operating on F gives F:

$$RF = F$$

The variables $x_1, \ldots, x_k$ (there are usually no more than three members of the set) themselves form a basis for the representation $\Gamma_x$ of the group with matrices $\Gamma_x(R)$. Similarly, the $y_k$ form a basis for $\Gamma_y$ with matrices $\Gamma_y(R)$. It is to be noted that the functions which we here call $x_k y_k$ may be identified with the functions $x^n y^m$ of Problem A2.18 — there is nothing in the following derivation which requires that $n = m = 1$.

The explicit expression for RF is

$$RF = R\left[\sum_{k=1}^{l} x_k y_k\right] = F$$

Because R operates on each function individually, this is

$$RF = \sum_{k=1}^{l} Rx_k \cdot Ry_k = F$$

We now sum over all operations of the group and obtain

$$\sum_R RF = \sum_R \sum_{k=1}^{l} Rx_k \cdot Ry_k = hF \tag{A2.23}$$

by the assumption that $RF = F$, there being h operations in the group. Consider $Rx_k$; it is evident from equation (A2.6) that, for example,

$$Rx_k = \sum_{s=1}^{n} \Gamma_x(R)_{sk} x_s \tag{A2.24a}$$

where $\Gamma_x(R)_{sk}$ is the skth element of the matrix $\Gamma_x(R)$. Similarly,

$$Ry_k = \sum_{t=1}^{m} \Gamma_y(R)_{tk} y_t \tag{A2.24b}$$

We shall later show that the invariance of F requires that

$$\Gamma_x(R)_{ik} = \Gamma_y(R)_{ik} \tag{A2.25}$$

so we shall drop the x, y suffixes on the Γ's and set $m = n = l$ (the letter l) in equations (A2.24a) and (A2.24b).

Substituting these two expressions into equation (A2.23) gives

$$\sum_R RF = \sum_R \sum_{k=1}^{l} \sum_{s=1}^{l} \sum_{t=1}^{l} \Gamma(R)_{sk} \Gamma(R)_{tk} x_s y_t = hF \tag{A2.26}$$

**Problem A2.19** Derive equation (A2.26)

Because equation (A2.22) contains only terms $x_k y_k$ so too must equation (A2.26), i.e. $s = t$. That is in equation (A2.26), all terms with $s \neq t$ must equal zero. Thus,

$$\sum_R \Gamma(R)_{sk} \Gamma(R)_{tk} = 0 \qquad (s \neq t) \tag{A2.27}$$

Because $\sum_R RF$ in equation (A2.26) spans all the R in the group and because each R has a unique inverse, $R^{-1}$ (Section A1.1), $\sum_R R^{-1}F$ also spans all operations of the group. It follows that

$$\sum_R R^{-1}F = \sum_R RF$$

i.e. the sums over R and $R^{-1}$ merely give the same operations in a different order.

In this appendix we have not investigated the relationship between the matrices associated with R and with $R^{-1}$ but it is a simple one — when all the elements of the matrices are real, the case to which we confine ourselves, one simply interchanges columns and rows of R to obtain $R^{-1}$, and vice versa.

**Problem A2.20** Use the above technique to obtain the inverses of the matrices in Table A2.1. Check your answers by reference to Table 8.2.

Thus, for example, the inverse of the matrix associated with $R_z(\alpha)$ in equation (A2.6) is simply

$$R_z^{-1}(\alpha) = \begin{bmatrix} \cos\alpha & -\sin\alpha & 0 \\ \sin\alpha & \cos\alpha & 0 \\ 0 & 0 & 1 \end{bmatrix}$$

**Problem A2.21** Show, using the above $3 \times 3$ matrix and the one given in equation (A2.6), that

$$\Gamma(R_z(\alpha))\Gamma(R_z^{-1}(\alpha)) = \Gamma(R_z^{-1}(\alpha))\Gamma(R_z(\alpha)) = \Gamma_E$$

Note that this relationship is obtained from equation (A1.4) by the substitution of a matrix for the corresponding operation.

Returning to equation (A2.26), it is clear from the above that on the left-hand side we can replace $\sum_R RF$ by $\sum_R R^{-1}F$. However, when we make this substitution we must also replace on the right-hand side those matrix elements associated with R by those associated with $R^{-1}$. We do this, as we have seen, by interchanging rows and columns — i.e. by reversing the subscripts on the matrix elements (this causes no problems because we are dealing with square matrices). It follows that

$$\sum_R R^{-1}F = \sum_R \sum_{k=1}^{1} \sum_{s=1}^{1} \sum_{t=1}^{1} \Gamma(R)_{ks}\Gamma(R)_{kt}x_s y_t = hF \qquad \text{(A2.28)}$$

**Problem A2.22** Check equation (A2.28).

Therefore, for $s \neq t$ we have, following the discussion under equation (A2.26),

$$\sum_R \Gamma(R)_{ks}\Gamma(R)_{kt} = 0 \qquad (s \neq t) \tag{A2.29}$$

Having exploited the $s \neq t$ situation to the full we now set $s = t = j$ in equation (A2.26) and obtain

$$\sum_R RF = \sum_R \sum_{k=1}^{l} \sum_{j=1}^{l} \Gamma(R)_{jk}\Gamma(R)_{jk}x_j y_j = hF \tag{A2.30}$$

Comparing this with equation (A2.22) we deduce

$$\sum_R \sum_{k=1}^{l} \Gamma(R)_{jk}\Gamma(R)_{jk} = h \qquad (j = 1, \ldots, l)\text{ (j from one to l)} \tag{A2.31}$$

Similarly, setting $s = t = j$ in equation (A2.28) gives

$$\sum_R \sum_{k=1}^{l} \Gamma(R)_{kj}\Gamma(R)_{kj} = h \qquad (j = 1, \ldots, l) \tag{A2.32}$$

**Problem A2.23** Check the derivation of equations (A2.31) and (A2.32).

The right-hand sides of equations (A2.31) and (A2.32) are both independent of j and k, no matter in which order these dummy suffixes appear in the left-hand side expressions. It follows that each term in the summation over k makes an equal contribution to the sum. The contribution of each term is therefore given by

$$\sum_R \Gamma(R)_{jk}\Gamma(R)_{jk} = \frac{h}{l} \tag{A2.33}$$

Finally, we combine equations (A2.27), (A2.29) and (A2.33). None of these expressions contain any suffixes on the Γ's but we use equation (A2.24) to justify the requirement that the suffixes be identical, although, strictly, this should be formally proved. This combination gives, with minor changes in notation, the expression

$$\sum_R \Gamma_i(R)_{mn}\Gamma_j(R)_{m'n'} = \frac{h}{\sqrt{l_i l_j}}\,\delta_{ij}\,\delta_{mm'}\,\delta_{nn'}$$

which is identical to equation (A2.14).

**Problem 2.24** Detail the derivation of equation (A2.14) along the lines outlined in the text.

The derivation of this equation was the object of the present section and many readers may wish to stop at this point. There is, however, one loose end in our derivation in equation (A2.25), presented again as follows:

$$\Gamma_x(R)_{ik} = \Gamma_y(R)_{ik}$$

The validity of this equation was essential to our derivation of equation (A2.14) but was not proven. We now give this, somewhat lengthy, proof.

We start by operating on each side of equation (A2.22) with an operation, R, of the group. That is, we start with

$$RF = R \sum_{k=1}^{r} x_k y_k$$

where, for convenience, we have replaced the l of equation (A2.22) by r:

$$= \sum_{k=1}^{r} R(x_k y_k)$$

$$= \sum_{k=1}^{r} R(x_k)R(y_k)$$

$$= F \text{ (see the discussion associated with equation A2.22)}$$

That is,

$$F = \sum_{k=1}^{r} R(x_k)R(y_k) \qquad \text{for all R}$$

It is now convenient to write equation (A2.22) in expanded form:

$$F = \sum_{k=1}^{r} x_k y_k = x_1 y_1 + x_2 y_2 + \cdots + x_r y_r \tag{A2.34}$$

Now let R operate on just the $x_k$ in equation (A2.22); we have

$$F' = \sum_{k=1}^{r} R(x_k) y_k \tag{A2.35}$$

or, in expanded form,

$$F' = R(x_1)y_1 + R(x_2)y_2 + \cdots + R(x_r)y_r$$

Now, for an individual $x_k$,

$$Rx_k = \sum_{s=1}^{n} \Gamma_x(R)_{sk} x_s \tag{A2.36}$$

where $\Gamma_n(R)_{sk}$ is the skth element of $\Gamma_n(R)$. So equation (A2.35) becomes

$$F' = y_1 \sum_{s=1}^{n} \Gamma_x(R)_{s1}x_s + y_2 \sum_{s=1}^{n} \Gamma_x(R)_{s2}x_s + \cdots + y_r \sum_{s=1}^{n} \Gamma_x(R)_{sr}x_s \quad \text{(A2.37)}$$

Now, within each summation term, each function $x_1, x_2, \ldots, x_k$ may appear. We now regroup the terms in equation (A2.37) according to these x's; each x may be multiplied by any of the y's. We note that in each $\Gamma_x(R)_{sk}$ term in A2.37, the s goes with the x and the k with the y. The rearranged expression is

$$F' = x_1 \sum_{k=1}^{r} \Gamma_x(R)_{1k}y_k + x_2 \sum_{k=1}^{r} \Gamma_x(R)_{2k}y_k + \cdots + x_n \sum_{k=1}^{r} \Gamma_x(R)_{nk}y_k$$

If we now operate on the y's on the right-hand side of this expression with R then $F'$ becomes equal to F and we have

$$F = x_1 \sum_{k=1}^{r} \Gamma_x(R)_{1k}Ry_k + \cdots + x_n \sum_{k=1}^{r} \Gamma_x(R)_{nk}Ry_k$$

By comparison with equation (A2.34) we have

$$\sum_{k=1}^{r} \Gamma_x(R)_{ik}Ry_k = y_i \quad \text{(A2.38)}$$

The inverse of the operation R is $R^{-1}$; operating with it on both sides of equation (A2.38) gives

$$R^{-1} \sum_{k=1}^{r} \Gamma_x(R)_{ik}Ry_k = R^{-1}y_i$$

Remembering that $\Gamma_x(R)_{ik}$ is a number, this is

$$\sum_{k=1}^{r} \Gamma_x(R)_{ik}R^{-1}Ry_k = R^{-1}y_i$$

that is, by equation (A1.4),

$$\sum_{k=1}^{r} \Gamma_x(R)_{ik}y_k = R^{-1}y_i \quad \text{(A2.39)}$$

Now a general expression for $Ry_i$, analogous to equation (A2.25), is

$$Ry_i = \sum_{k=1}^{r} \Gamma_y(R)_{ki}y_k$$

so that $R^{-1}y_i$ is given by

$$R^{-1}y_i = \sum_{k=1}^{r} \Gamma_y(R)_{ik}y_k$$

Comparison with equation (A2.39) requires that

$$\sum_{k=1}^{r} \Gamma_x(R)_{ik} = \sum_{k=1}^{r} \Gamma_y(R)_{ik}$$

We have thus found a relationship between the elements of $\Gamma_x(R)$ and $\Gamma_y(R)$, but this relationship holds for all choices of R and all ik — we placed no restrictions on either. This generality only arises when corresponding individual terms in the two summations are themselves equal. That is,

$$\Gamma_x(R)_{ik} = \Gamma_y(R)_{ik}$$

which is equation (A2.25).

## A2.4 THE REDUCTION OF REDUCIBLE REPRESENTATIONS

An extremely important property of the matrices which multiply isomorphically to group operations is the fact that 'their characters are invariant to a similarity transformation'. What does the phrase in quotes mean? We shall see this shortly but first we digress.

We consider a set of matrices A, B, C, ... which form a representation of the group. From these we generate other matrices which also form a representation of the group, as we shall now show. The importance of this step is that the new matrices may be chosen so that they are more convenient to work with than the starting set. In particular, the new matrices can be cunningly engineered to be a sum of irreducible matrices.

Suppose that the multiplication of the original matrices is such that

$$AB = C$$

We have to show that a similar relationship holds for the new matrices; if we can do this then they, indeed, multiply correctly. The engineering process consists of selecting another matrix, call it M, which is not a member of the set A, B, C, ... but one which is of the same dimension. Thus, if A, B, C are all $6 \times 6$ matrices, so too must be M. For the moment we will not specify M further beyond requiring that it has an inverse, $M^{-1}$. We form the products

$$\begin{aligned} M^{-1}AM &= A' \\ M^{-1}BM &= B' \\ M^{-1}CM &= C' \\ \vdots \quad & \quad \vdots \end{aligned}$$

where the product matrices $A'$, $B'$, $C'$, ... will be of the same dimensions as A, B, C, ... The set of matrices $A'$, $B'$, $C'$, ... are a new set of representation matrices of the same group. We show this by considering the product $A'B'$:

$$A'B' = M^{-1}AM\ M^{-1}BM = M^{-1}ABM = M^{-1}CM = C'$$

That is, $A'B' = C'$ whenever $AB = C$ and so $A'$, $B'$, $C'$, ... multiply in the same way as (isomorphically with) A, B, C, ... .

The matrices A and $A'$ are said to be 'related by a similarity transformation', as are B and $B'$, C and $C'$, etc. It is the property of matrices related in this way that they have the same characters — hence the phrase 'their characters are invariant to a similarity transformation'. This relationship between the characters is not difficult to prove. The character of the matrix A′, $\chi(A')$, is given by

$$\chi(A') = \sum_i A'_{ii}$$

which, written in terms of the elements of the product $M^{-1}AM$ is

$$\chi(A') = \sum_i \sum_j \sum_k M^{-1}_{ik} A_{kj} M_{ji}$$

Each term on the right-hand side is a number and we will get the same result no matter in which order they are multiplied together. We can therefore rewrite this expression as

$$\chi(A') = \sum_i \sum_j \sum_k A_{kj} M_{ji} M^{-1}_{ik}$$

$$= \sum_j \sum_k A_{kj} \sum_i M_{ji} M^{-1}_{ik}$$

Now, the final summation is that required for matrix multiplication (see equation A2.1). Thus,

$$\chi(A') = \sum_j \sum_k A_{kj} (MM^{-1})_{jk}$$

But $MM^{-1} = E$, a unit matrix with ones on the diagonal and zeros elsewhere, so

$$\chi(A') = \sum_j \sum_k A_{kj}\, \delta_{jk} \qquad \text{where } \delta_{jk} = \begin{cases} 1 \text{ if } j = k \\ 0 \text{ if } j \neq k \end{cases}$$

$$= \sum_k A_{kk}$$

We conclude that

$$\chi(A') = \chi(A)$$

which demonstrates the invariance of the character under a similarity transformation.

The transformation of reducible matrix representations into sums of irreducible matrix representations is of vital importance. Although it is rarely necessary to go through the formal procedure, as we shall see, it is implicitly involved, for example, in the formation of the symmetry-adapted linear combinations of orbitals used in the discussions of molecular bonding in this book. The procedure involved may be seen in the schematic equation given below; by an educated choice of M, the general matrix A is converted to the matrix

A′ which has irreducible matrices (cross-hatched) strung along its leading diagonal, all other elements being zero:

$$M^{-1}[A]M = \begin{bmatrix} \blacksquare & 0 & 0 & 0 \\ 0 & \blacksquare & 0 & 0 \\ 0 & 0 & \blacksquare & \\ 0 & 0 & & \end{bmatrix} \tag{A2.40}$$

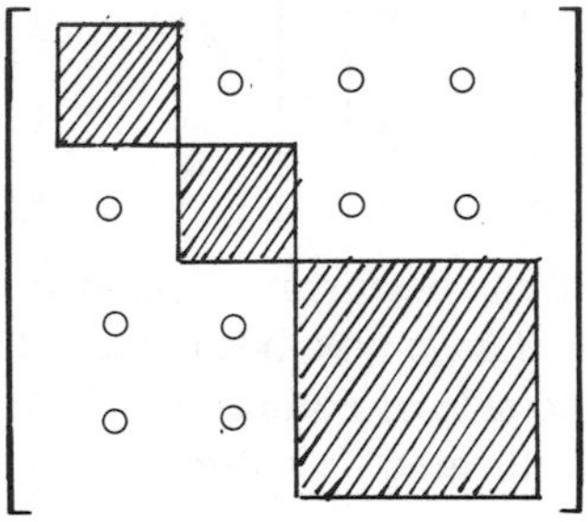

The next step in our development is to obtain matrices, M, which bring about this simplification. It is simplest to proceed by way of an example and this we do by returning to the two hydrogen atoms in the water molecule problem (Section 3.2), for which Table A2.1 shows that there are two reducible matrices to consider,

$$\begin{bmatrix} 1 & 0 \\ 0 & 1 \end{bmatrix} \quad \text{and} \quad \begin{bmatrix} 0 & 1 \\ 1 & 0 \end{bmatrix}$$

The first of these is diagonal (only zeros off the leading diagonal) so the similarity transformation has to lead to the retention of this characteristic. The second matrix is non-diagonal and has to be transformed to diagonal form. We shall shortly show how the matrix, M, is obtained but, for the moment, we simply give it and show that it leads to the desired result.

In the present case the matrix M is

$$\begin{bmatrix} \frac{1}{\sqrt{2}} & \frac{1}{\sqrt{2}} \\ \frac{1}{\sqrt{2}} & -\frac{1}{\sqrt{2}} \end{bmatrix} \text{ so that } M^{-1} \text{ is } \begin{bmatrix} \frac{1}{\sqrt{2}} & \frac{1}{\sqrt{2}} \\ \frac{1}{\sqrt{2}} & -\frac{1}{\sqrt{2}} \end{bmatrix}$$

i.e. M is self-inverse. We evaluate the two products of the form $M^{-1}AM$, working from the left. For the first we have:

$$\begin{bmatrix} \frac{1}{\sqrt{2}} & \frac{1}{\sqrt{2}} \\ \frac{1}{\sqrt{2}} & -\frac{1}{\sqrt{2}} \end{bmatrix} \begin{bmatrix} 1 & 0 \\ 0 & 1 \end{bmatrix} \begin{bmatrix} \frac{1}{\sqrt{2}} & \frac{1}{\sqrt{2}} \\ \frac{1}{\sqrt{2}} & -\frac{1}{\sqrt{2}} \end{bmatrix} = \begin{bmatrix} \frac{1}{\sqrt{2}} & \frac{1}{\sqrt{2}} \\ \frac{1}{\sqrt{2}} & -\frac{1}{\sqrt{2}} \end{bmatrix} \begin{bmatrix} \frac{1}{\sqrt{2}} & \frac{1}{\sqrt{2}} \\ \frac{1}{\sqrt{2}} & -\frac{1}{\sqrt{2}} \end{bmatrix} = \begin{bmatrix} 1 & 0 \\ 0 & 1 \end{bmatrix}$$

Since A in this case was an identity matrix (ones for all diagonal elements and zeros elsewhere) this product was really $M^{-1}EM = M^{-1}M = E$ so we could have anticipated the fact that A would be left unchanged by the similarity transformation. The second case is less trivial:

$$\begin{bmatrix} \frac{1}{\sqrt{2}} & \frac{1}{\sqrt{2}} \\ \frac{1}{\sqrt{2}} & -\frac{1}{\sqrt{2}} \end{bmatrix} \begin{bmatrix} 0 & 1 \\ 1 & 0 \end{bmatrix} \begin{bmatrix} \frac{1}{\sqrt{2}} & \frac{1}{\sqrt{2}} \\ \frac{1}{\sqrt{2}} & -\frac{1}{\sqrt{2}} \end{bmatrix} = \begin{bmatrix} \frac{1}{\sqrt{2}} & \frac{1}{\sqrt{2}} \\ -\frac{1}{\sqrt{2}} & -\frac{1}{\sqrt{2}} \end{bmatrix} \begin{bmatrix} \frac{1}{\sqrt{2}} & \frac{1}{\sqrt{2}} \\ \frac{1}{\sqrt{2}} & -\frac{1}{\sqrt{2}} \end{bmatrix} = \begin{bmatrix} 1 & 0 \\ 0 & -1 \end{bmatrix}$$

As required, this matrix has been transformed into a diagonal form. Both of our original $2 \times 2$ matrices have now been block-diagonalized into the form indicated on the right-hand side of equation (A2.40). The transformed matrices are (with the blocking indicated as submatrices)

$$\begin{bmatrix} [1] & 0 \\ 0 & [1] \end{bmatrix} \quad \text{and} \quad \begin{bmatrix} [1] & 0 \\ 0 & [-1] \end{bmatrix}$$

Remembering that, as equations (A2.2) to (A2.5) show, we need to consider each of the original matrices twice, we now write separately the matrices (here $1 \times 1$, i.e. numbers) that appear on the diagonals of the transformed matrices. We obtain

| Operation | E | $C_2$ | $\sigma_v$ | $\sigma_v'$ | |
|---|---|---|---|---|---|
| First matrix | [1] | [1] | [1] | [1] | $A_1$ |
| Second matrix | [1] | [−1] | [1] | [−1] | $B_1$ |

The characters of these matrices are equal to the matrices themselves and so we see that we have generated the $A_1$ and $B_1$ representations of Chapter 3 (Section 3.3).

**Problem A2.25** Show that the four $4 \times 4$ matrices given on page 247 are simultaneously reduced to a diagonal form by a similarity transformation using for M:

$$\begin{bmatrix} \frac{1}{2} & \frac{1}{2} & \frac{1}{2} & \frac{1}{2} \\ \frac{1}{2} & \frac{1}{2} & -\frac{1}{2} & -\frac{1}{2} \\ \frac{1}{2} & -\frac{1}{2} & \frac{1}{2} & -\frac{1}{2} \\ \frac{1}{2} & -\frac{1}{2} & -\frac{1}{2} & \frac{1}{2} \end{bmatrix}$$

We now come to the key problem; how do we obtain the matrix which reduces a set of reducible matrices to block diagonal form? The answer is indicated by comparing the matrix used in the example detailed above with the two symmetry-adapted $A_1$ and $B_1$ functions obtained in Chapter 3 (at the end of Section 3.4). These were

$$\psi(A_1) = \frac{1}{\sqrt{2}}(h_1 + h_2)$$

$$\psi(B_1) = \frac{1}{\sqrt{2}}(h_1 - h_2)$$

If we write the coefficients which multiply $h_1$ and $h_2$ in these expressions in matrix form we obtain

$$\begin{bmatrix} \frac{1}{\sqrt{2}} & \frac{1}{\sqrt{2}} \\ \frac{1}{\sqrt{2}} & -\frac{1}{\sqrt{2}} \end{bmatrix}$$

which is just the matrix M we used above. The result is general; the matrix which diagonalizes a reducible matrix is obtained by forming a matrix from the coefficients with which the basis functions appear in the symmetry-adapted functions. The method of obtaining these symmetry-adapted functions has been described in Chapter 4 (Section 4.6), extended in Chapters 5 (Section 5.5) and 6 (Section 6.2) and simplified for difficult problems in Appendix 4. There is one point at which care is needed: it is important to make sure that the listing of the basis functions used in obtaining a reducible matrix (and in real-life problems this is the way such matrices are usually obtained) is identical to that implied in the matrix obtained from the coefficients in the symmetry-adapted functions. Thus, in the example above, the functions $\psi(A_1)$ and $\psi(B_1)$ are obtained by multiplying out the product

$$\begin{bmatrix} \frac{1}{\sqrt{2}} & \frac{1}{\sqrt{2}} \\ \frac{1}{\sqrt{2}} & -\frac{1}{\sqrt{2}} \end{bmatrix} \begin{bmatrix} h_1 \\ h_2 \end{bmatrix}$$

and here the listing of $h_1$ and $h_2$ coincides with that of equations (A2.2) to (A2.5). It so happens that in this case we would have escaped penalty had we listed $h_1$ and $h_2$ in the incorrect order and obtained for M the matrix

$$\begin{bmatrix} \frac{1}{\sqrt{2}} & \frac{1}{\sqrt{2}} \\ -\frac{1}{\sqrt{2}} & \frac{1}{\sqrt{2}} \end{bmatrix}$$

but, in general, we could not expect to be so fortunate.

**Problem A2.26** Show that the alternative form of M given above leads to the same results as the form used in the text.

We conclude this appendix by deriving the algebraic expression that is the basis of the recipe frequently used in the main text to obtain the irreducible components of a reducible representation.

As the discussion in this section shows a reducible matrix representation of a group can be reduced to a matrix which contains only irreducible matrices along its leading diagonal by a suitable similarity transformation. Further, the character of the original and transformed matrices are identical. If we write $\chi(R)$ for the character of a reducible matrix under the operation R then we have

$$\chi(R) = \sum_{j=1}^{k} a_j \chi_j(R) \tag{A2.41}$$

where $\chi_j(R)$ is the character of the jth irreducible representation under the same operation (there are k irreducible representations in all) and $a_j$ is either zero or a positive integer, the number of times the jth irreducible representation occurs in the reducible representation. It is these $a_j$'s that we seek to determine. Equation (A2.19) is

$$\sum_R \chi_i(R)\chi_j(R) = h\ \delta_{ij}$$

where $\delta_{ij} = 1$ if $i = j$ but 0 if $i \neq j$. We now multiply each side of equation (A2.19) by $a_j$ and sum over all j. We thus obtain, using equation (A2.41),

$$\sum_R \chi_i(R) \sum_{j=1}^{k} a_j \chi_j(R) = \sum_R \chi_i(R)\chi(R) = ha_i \tag{A2.42}$$

Only $a_i$ appears on the right-hand side of this expression because all other $a_j$'s have been multiplied by 0 because of the $\delta_{ij}$. Rearranging equation (A2.42) we obtain

$$a_i = \frac{1}{h} \sum_R \chi(R)\chi_i(R) \tag{A2.43}$$

which is the equation implicitly used in the text.

# APPENDIX 3

# *Character tables of the more important point groups*

To the right of each character table are given two columns of bases for irreducible representations. *R*otations and *T*ranslations, given in the first column, are needed for vibrational analyses (see Chapter 9) and for some forms of spectroscopy (see Chapter 10). The second column is useful for other spectroscopies (Chapter 10) and for discussions of molecular bonding. Invariably, $x^2$, $y^2$ and $z^2$ in some combination or independently transform as the totally symmetric irreducible representation. It follows that any linear combination of the functions so transforming and, in particular, $x^2 + y^2 + z^2 = r^2$, transform under this irreducible representation. The function $r^2$, transform under this irreducible representation. The function $r^2$ is spherically symmetrical and so is associated with the s orbital of an atom.

'Note' comments have usually either been repeated where relevant or cross-references given. However, the reader encountering problems should scan the notes for related character tables, where helpful comments may well be found.

## A3.1 THE ICOSAHEDRAL GROUPS

$I_h$: The icosahedral group

| $I_h$ | E | $12C_5$ | $12C_5^2$ | $20C_3$ | $15C_2$ | i | $12S_{10}$ | $12S_{10}^3$ | $20S_6$ | $15\sigma$ | | |
|---|---|---|---|---|---|---|---|---|---|---|---|---|
| $A_g$ | 1 | 1 | 1 | 1 | 1 | 1 | 1 | 1 | 1 | 1 | | $x^2+y^2+z^2$ |
| $T_{1g}$ | 3 | $-2\cos 144^\circ$ | $-2\cos 72^\circ$ | 0 | $-1$ | 3 | $-2\cos 144^\circ$ | $-2\cos 72^\circ$ | 0 | $-1$ | $(R_x, R_y, R_z)$ | |
| $T_{2g}$ | 3 | $-2\cos 72^\circ$ | $-2\cos 144^\circ$ | 0 | $-1$ | 3 | $-2\cos 72^\circ$ | $-2\cos 144^\circ$ | 0 | $-1$ | | |
| $G_g$ | 4 | $-1$ | $-1$ | 1 | 0 | 4 | $-1$ | $-1$ | 1 | 0 | | |
| $H_g$ | 5 | 0 | 0 | $-1$ | 1 | 5 | 0 | 0 | $-1$ | 1 | | $\left\{[\frac{1}{\sqrt{6}}(2z^2-x^2-y^2), \frac{1}{\sqrt{2}}(x^2-y^2), xy, yz, zx\right.$ |
| $A_u$ | 1 | 1 | 1 | 1 | 1 | $-1$ | $-1$ | $-1$ | $-1$ | $-1$ | | |
| $T_{1u}$ | 3 | $-2\cos 144^\circ$ | $-2\cos 72^\circ$ | 0 | $-1$ | $-3$ | $2\cos 144^\circ$ | $2\cos 144^\circ$ | 0 | 1 | $(T_x, T_y, T_z)$ | $(x, y, z)$ |
| $T_{2u}$ | 3 | $-2\cos 72^\circ$ | $-2\cos 144^\circ$ | 0 | $-1$ | $-3$ | $2\cos 72^\circ$ | $2\cos 144^\circ$ | 0 | 1 | | |
| $G_u$ | 4 | $-1$ | $-1$ | 1 | 0 | $-4$ | 1 | 1 | $-1$ | 0 | | |
| $H_u$ | 5 | 0 | 0 | $-1$ | 1 | $-5$ | 0 | 0 | 1 | $-1$ | | |

*Notes:* 1. This character table is the direct product of I with $C_i$, as indicated by the dotted lines.
2. When working with groups containing a $C_5$ axis the following relationship will be found useful:

$$-2\cos 72^\circ = -0.61803 = \tfrac{1}{2}(1-5^{1/2}) = \alpha$$

$$-2\cos 144^\circ = 1.61803 = \tfrac{1}{2}(1+5^{1/2}) = \beta$$

$$\alpha^2 = 1+\alpha;\ \beta^2 = 1+\beta;\ \alpha\beta = -1;\ \alpha+\beta = 1$$

3. The icosahedral groups are the only ones which contain fourfold (G) and fivefold (H) degenerate representations.

*Example:* the icosahedron shown in Figure 7.31.

## I: The group of pure rotations of an icosahedron

| I | E | $12C_5$ | $12C_5^2$ | $20C_3$ | $15C_2$ | | |
|---|---|---|---|---|---|---|---|
| A | 1 | 1 | 1 | 1 | 1 | | $x^2+y^2+z^2$ |
| $T_1$ | 3 | $-2\cos 144^\circ$ | $-2\cos 72^\circ$ | 0 | $-1$ | $(R_x, R_y, R_z)$: $(T_x, T_y, T_z)$ | $(x, y, z)$ |
| $T_2$ | 3 | $-2\cos 72^\circ$ | $-2\cos 144^\circ$ | 0 | $-1$ | | |
| G | 4 | $-1$ | $-1$ | 1 | 0 | | |
| H | 5 | 0 | 0 | $-1$ | 1 | | $[\frac{1}{\sqrt{6}}(2z^2-x^2-y^2), \frac{1}{\sqrt{2}}(x^2-y^2), xy, yz, zx]$ |

*Note:* See comments under the $I_h$ group.

*Example:* The easiest way of obtaining a figure of I symmetry is to take one edge of the icosahedron shown in Figure 7.31 and make it a zig-zag (the $C_2$ axis passing through the mid-point of this edge must be preserved). The pure rotation operations are then used to generate this zig-zag in each of the other edges. A figure of I symmetry results. Faces that are flat in the icosahedron will become fluted under the above procedure.

## A3.2 CUBIC POINT GROUPS

The two most important cubic point groups are $O_h$ and $T_d$.

| $O_h$ | E | $8C_3$ | $3C_2$ | $6C_4$ | $6C_2'$ | i | $8S_6$ | $3\sigma_h$ | $6S_4$ | $6\sigma_d$ | | |
|---|---|---|---|---|---|---|---|---|---|---|---|---|
| $A_{1g}$ | 1 | 1 | 1 | 1 | 1 | 1 | 1 | 1 | 1 | 1 | | $x^2+y^2+z^2$ |
| $A_{2g}$ | 1 | 1 | 1 | −1 | −1 | 1 | 1 | 1 | −1 | −1 | | |
| $E_g$ | 2 | −1 | 2 | 0 | 0 | 2 | −1 | 2 | 0 | 0 | | $[\frac{1}{\sqrt{6}}(2z^2-x^2-y^2), \frac{1}{\sqrt{2}}(x^2-y^2)]$ |
| $T_{1g}$ | 3 | 0 | −1 | 1 | −1 | 3 | 0 | −1 | 1 | −1 | $(R_x, R_y, R_z)$ | |
| $T_{2g}$ | 3 | 0 | −1 | −1 | 1 | 3 | 0 | −1 | −1 | 1 | | $(xy, xz, yz)$ |
| $A_{1u}$ | 1 | 1 | 1 | 1 | 1 | −1 | −1 | −1 | −1 | −1 | | |
| $A_{2u}$ | 1 | 1 | 1 | −1 | −1 | −1 | −1 | −1 | 1 | 1 | | |
| $E_u$ | 2 | −1 | 2 | 0 | 0 | −2 | 1 | −2 | 0 | 0 | | |
| $T_{1u}$ | 3 | 0 | −1 | 1 | −1 | −3 | 0 | 1 | −1 | 1 | $(T_x, T_y, T_z)$ | $(x, y, z)$ |
| $T_{2u}$ | 3 | 0 | −1 | −1 | 1 | −3 | 0 | 1 | 1 | −1 | | |

*Notes:* 1. In some texts inorganic chemists refer to one set of the d orbitals on a transition metal atom at the centre of an octahedral complex as $d_\gamma$ (which are $e_g$ in our notation) and $d_\varepsilon$ the other (which we call $t_{2g}$).

2. The $O\eta$ group is a direct product of O and $C_i$. This is indicated by the dotted lines in the character table.

3. Although the $\sigma_h$ mirror planes also satisfy the definition for $\sigma_d$, here and later (for the $D_{nh}$ groups) the $_h$ subscript is given precedence.

4. The $1/\sqrt{6}$ and $1/\sqrt{2}$ in the expressions for $e_g$ basis functions normalize each function to unity. In the text we dropped the $1/\sqrt{2}$ on $x^2-y^2$ and therefore used $1/\sqrt{3}$ in front of $2z^2-x^2-y^2$.

*Examples:* (a) A cube, an octahedron (Figure 7.1).
(b) $SF_6$ (Figure 7.2 when M = S, L = F).

| $T_d$ | E | $8C_3$ | $3C_2$ | $6S_4$ | $6\sigma_d$ | | |
|---|---|---|---|---|---|---|---|
| $A_1$ | 1 | 1 | 1 | 1 | 1 | | $x^2+y^2+z^2$ |
| $A_2$ | 1 | 1 | 1 | −1 | −1 | | |
| E | 2 | −1 | 2 | 0 | 0 | | $[\frac{1}{\sqrt{6}}(2z^2-x^2-y^2), \frac{1}{\sqrt{2}}(x^2-y^2)]$ |
| $T_1$ | 3 | 0 | −1 | 1 | −1 | $(R_x, R_y, R_z)$ | |
| $T_2$ | 3 | 0 | −1 | −1 | 1 | $(T_x, T_y, T_z)$ | (x, y, z): (xy, xz, yz) |

*Examples:* (a) A tetrahedron (Figure 7.1).
(b) $CH_4$, $P_4$, $Ni(CO)_4$, $C(CH_3)_4$ in its most symmetrical configuration.

Other cubic point groups are O, T and $T_h$.

O: The group of the pure rotation operations of the octahedron

| O | E | $8C_3$ | $3C_2$ | $6C_4$ | $6C_2'$ | | |
|---|---|---|---|---|---|---|---|
| $A_1$ | 1 | 1 | 1 | 1 | 1 | | $x^2+y^2+z^2$ |
| $A_2$ | 1 | 1 | 1 | −1 | −1 | | |
| E | 2 | −1 | 2 | 0 | 0 | | $[\frac{1}{\sqrt{6}}(2z^2-x^2-y^2), \frac{1}{\sqrt{2}}(x^2-y^2)]$ |
| $T_1$ | 3 | 0 | −1 | 1 | −1 | $(T_x, T_y, T_z)$ $(R_x, R_y, R_z)$ | (x, y, z) |
| $T_2$ | 3 | 0 | −1 | −1 | 1 | | (xy, xz, yz) |

*Example:* To obtain a figure of O symmetry follow the instructions given under the I character table, replacing ‘Figure 7.30’ by ‘Figure 7.1’, ‘icosahedron’ by ‘octahedron’ and ‘I’ by ‘O’.

## T: The group of the pure rotation operations of the tetrahedron

| T | E | $4C_3$ | $4C_3^2$ | $3C_2$ | $\varepsilon = \exp \frac{2\pi i}{3}$ | |
|---|---|---|---|---|---|---|
| A | 1 | 1 | 1 | 1 | | $x^2+y^2+z^2$ |
| E | $\begin{cases}1\\1\end{cases}$ | $\begin{matrix}\varepsilon\\\varepsilon^2\end{matrix}$ | $\begin{matrix}\varepsilon^2\\\varepsilon\end{matrix}$ | $\begin{rcases}1\\1\end{rcases}$ | | $[\frac{1}{\sqrt{6}}(2z^2-x^2-y^2), \frac{1}{\sqrt{2}}(x^2-y^2)]$ |
| T | 3 | 0 | 0 | −1 | $(T_x, T_y, T_z)$ $(R_x, R_y, R_z)$ | $(x, y, z)$ $(xy, yz, zx)$ |

*Note:* $\varepsilon$ and $\varepsilon^2$ $(= \exp 4\pi i/3 = -2\pi i/3)$ are complex conjugates. See also Chapter 11 and the notes under the $C_3$ group below.
*Example:* To obtain a figure of T symmetry follow the instructions given under the I character table, replacing 'Figure 7.31' by 'Figure 7.1', 'icosahedron' by 'tetrahedron' and 'I' by 'T'.

## $T_h$:

| $T_h$ | E | $4C_3$ | $4C_3^2$ | $3C_2$ | i | $4S_6^5$ | $4S_6$ | $3\sigma_h$ | | |
|---|---|---|---|---|---|---|---|---|---|---|
| $A_g$ | 1 | 1 | 1 | 1 | 1 | 1 | 1 | 1 | | $x^2+y^2+z^2$ |
| $E_g$ | $\begin{cases}1\\1\end{cases}$ | $\begin{matrix}\varepsilon\\\varepsilon^2\end{matrix}$ | $\begin{matrix}\varepsilon^2\\\varepsilon\end{matrix}$ | $\begin{matrix}1\\1\end{matrix}$ | $\begin{matrix}1\\1\end{matrix}$ | $\begin{matrix}\varepsilon\\\varepsilon^2\end{matrix}$ | $\begin{matrix}\varepsilon^2\\\varepsilon\end{matrix}$ | $\begin{rcases}1\\1\end{rcases}$ | | $[\frac{1}{\sqrt{6}}(2z^2-x^2-y^2), \frac{1}{\sqrt{2}}(x^2-y^2)]$ |
| $T_g$ | 3 | 0 | 0 | −1 | 3 | 0 | 0 | −1 | $(R_x, R_y, R_z)$ | $(xy, yz, zx)$ |
| $A_u$ | 1 | 1 | 1 | 1 | −1 | −1 | −1 | −1 | | |
| $E_u$ | $\begin{cases}1\\1\end{cases}$ | $\begin{matrix}\varepsilon\\\varepsilon^2\end{matrix}$ | $\begin{matrix}\varepsilon^2\\\varepsilon\end{matrix}$ | $\begin{matrix}1\\1\end{matrix}$ | $\begin{matrix}-1\\-1\end{matrix}$ | $\begin{matrix}-\varepsilon\\-\varepsilon^2\end{matrix}$ | $\begin{matrix}-\varepsilon^2\\-\varepsilon\end{matrix}$ | $\begin{rcases}-1\\-1\end{rcases}$ | | |
| $T_u$ | 3 | 0 | 0 | −1 | −3 | 0 | 0 | 1 | $(T_x, T_y, T_z)$ | $(x, y, z)$ |

*Notes:* 1. The $T_h$ character table is the direct product of T with $C_i$. This is indicated by the dotted lines in the character table.
2. For the meaning of $\varepsilon$ see under the T character table above.

*Example:* An 'octahedral' complex $[M(OH_2)_6]$ in which all the atoms of *trans*-$H_2O$ ligands lie in a $\sigma_h$ plane.

## A3.3 THE GROUPS $D_{nh}$

A regular, planar polygon with $n$ sides has $D_{nh}$ symmetry. Thus, an equilateral triangle has $D_{3h}$ symmetry; a square has $D_{4h}$ symmetry. The label D arises because of the presence of twofold axes (*D*ihedral axes) perpendicular to a $C_n$ axis. There are $n$ of these twofold axes. The subscript 'h' means that all of the groups have a unique mirror plane perpendicular to the $C_n$ axis (if this axis is vertical then the mirror plane is horizontal). Although these groups all have $\sigma_d$ operations, to avoid possible confusion with the $D_{nd}$ groups many authors label some or all of the $\sigma_d$ mirror planes as $\sigma_v$.

| $D_{2h}$ | E | $C_2(z)$ | $C_2(y)$ | $C_2(x)$ | i | $\sigma(xy)$ | $\sigma(xz)$ | $\sigma(yz)$ | | |
|---|---|---|---|---|---|---|---|---|---|---|
| $A_g$ | 1 | 1 | 1 | 1 | 1 | 1 | 1 | 1 | | $x^2$; $y^2$; $z^2$ |
| $B_{1g}$ | 1 | 1 | −1 | −1 | 1 | 1 | −1 | −1 | $R_z$ | xy |
| $B_{2g}$ | 1 | −1 | 1 | −1 | 1 | −1 | 1 | −1 | $R_y$ | xz |
| $B_{3g}$ | 1 | −1 | −1 | 1 | 1 | −1 | −1 | 1 | $R_x$ | yz |
| $A_u$ | 1 | 1 | 1 | 1 | −1 | −1 | −1 | −1 | | |
| $B_{1u}$ | 1 | 1 | −1 | −1 | −1 | −1 | 1 | 1 | $T_z$ | z |
| $B_{2u}$ | 1 | −1 | 1 | −1 | −1 | 1 | −1 | 1 | $T_y$ | y |
| $B_{3u}$ | 1 | −1 | −1 | 1 | −1 | 1 | 1 | −1 | $T_x$ | x |

*Notes:* 1. Because there are three mutually perpendicular $C_2$ axes the choice of x, y and z is arbitary. A relabelling of these axes will lead to an interchange of the labels $B_1$, $B_2$ and $B_3$. Similarly, the h, v subscript notation on the mirror planes is unhelpful and so the mirror planes and the corresponding operations are defined by the Cartesian axes that lie in them.

2. The $D_{2h}$ group is a direct product of $D_2$ and $C_i$. This is indicated by the dotted lines in the character table.

*Examples:* $C_2H_4$; $B_2H_6$ (see Chapter 4).

| $D_{3h}$ | E | $2C_3$ | $3C_2$ | $\sigma_h$ | $2S_3$ | $3\sigma_d$ | | |
|---|---|---|---|---|---|---|---|---|
| $A_1'$ | 1 | 1 | 1 | 1 | 1 | 1 | | $x^2 + y^2$; $z^2$ |
| $A_2'$ | 1 | 1 | −1 | 1 | 1 | −1 | $R_z$ | |
| $E'$ | 2 | −1 | 0 | 2 | −1 | 0 | $(T_x, T_y)$ | $(x, y)$; $[\frac{1}{\sqrt{2}}(x^2 - y^2), xy]$ |
| $A_1''$ | 1 | 1 | 1 | −1 | −1 | −1 | | |
| $A_2''$ | 1 | 1 | −1 | −1 | −1 | 1 | $T_z$ | |
| $E''$ | 2 | −1 | 0 | −2 | 1 | 0 | $(R_x, R_y)$ | $(xz, yz)$ |

*Notes: 1.* The $D_{3h}$ group is a direct product of $D_3$ and $C_s$. This is indicated by the dotted lines in the character table. Irreducible representations symmetric with respect to reflection in the $\sigma_h$ mirror plane are denoted by primes whilst antisymmetry is denoted by a double prime.

2. The $1/\sqrt{2}$ factor on $(x^2 - y^2)$ as an $E'$ basis function means that, like xy, it is normalized to unity.

*Example:* A triangular prism (Figure A3.1).

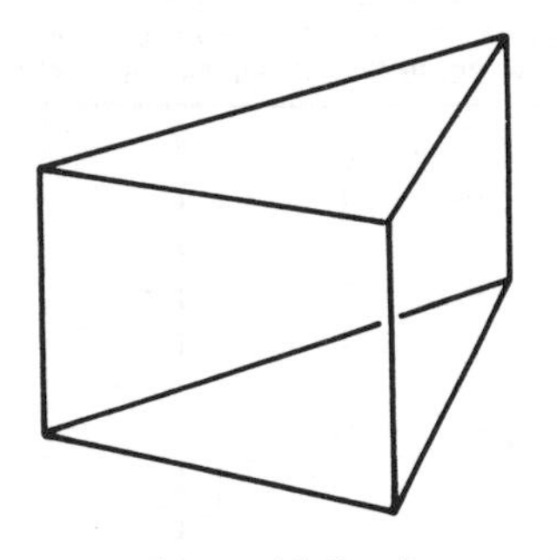

Figure A3.1 A triangular prism, of $D_{3h}$ symmetry

| $D_{4h}$ | $E$ | $2C_4$ | $C_2$ | $2C_2'$ | $2C_2''$ | $i$ | $2S_4$ | $\sigma_h$ | $2\sigma_d$ | $2\sigma_d'$ | | |
|---|---|---|---|---|---|---|---|---|---|---|---|---|
| $A_{1g}$ | 1 | 1 | 1 | 1 | 1 | 1 | 1 | 1 | 1 | 1 | | $x^2+y^2$; $z^2$ |
| $A_{2g}$ | 1 | 1 | 1 | −1 | −1 | 1 | 1 | 1 | −1 | −1 | $R_z$ | |
| $B_{1g}$ | 1 | −1 | 1 | 1 | −1 | 1 | −1 | 1 | 1 | −1 | | $x^2-y^2$ |
| $B_{2g}$ | 1 | −1 | 1 | −1 | 1 | 1 | −1 | 1 | −1 | 1 | | $xy$ |
| $E_g$ | 2 | 0 | −2 | 0 | 0 | 2 | 0 | −2 | 0 | 0 | $(R_x, R_y)$ | $(xz, yz)$ |
| $A_{1u}$ | 1 | 1 | 1 | 1 | 1 | −1 | −1 | −1 | −1 | −1 | | |
| $A_{2u}$ | 1 | 1 | 1 | −1 | −1 | −1 | −1 | −1 | 1 | 1 | $T_z$ | $z$ |
| $B_{1u}$ | 1 | −1 | 1 | 1 | −1 | −1 | 1 | −1 | −1 | 1 | | |
| $B_{2u}$ | 1 | −1 | 1 | −1 | 1 | −1 | 1 | −1 | 1 | −1 | | |
| $E_u$ | 2 | 0 | −2 | 0 | 0 | −2 | 0 | 2 | 0 | 0 | $(T_x, T_y)$ | $(x, y)$ |

*Notes:* 1. The $D_{4h}$ group is a direct product of $D_4$ and $C_i$. This is indicated by the dotted lines in the character table.
2. The choice between which pair of $C_2$ axes (and operations) are labelled $C_2'$ and which are labelled $C_2''$ is arbitrary. A redefinition will interchange $B_{1g}$ with $B_{2g}$ and $B_{1u}$ with $B_{2u}$. Similarly, the choice of the vertical planes $\sigma_d$ and $\sigma_d'$ is arbitrary but the $\sigma_d$ planes must contain the $C_2'$ axes and the $\sigma_d'$ planes must contain the $C_2''$.
3. In this character table we have followed the strict definition of $\sigma_d$ but many authors label the mirror planes containing a $C_2'$ axis as $\sigma_v$ and those containing a $C_2''$ axis as $\sigma_d$.

*Example:* A square prism (Figure A3.2).

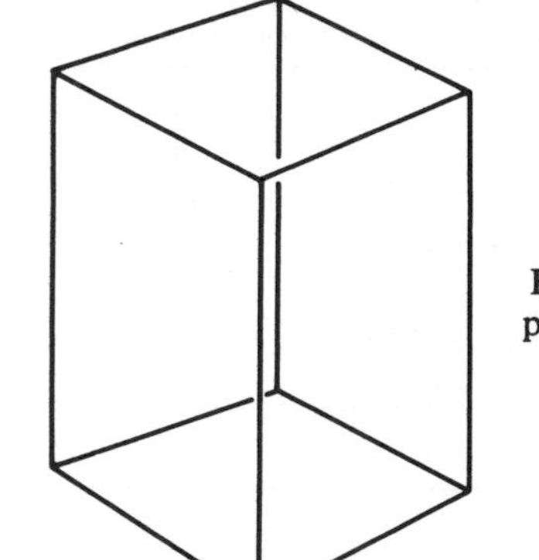

Figure A3.2 A square prism, of $D_{4h}$ symmetry

| $D_{5h}$ | E | $2C_5$ | $2C_5^2$ | $5C_2$ | $\sigma_h$ | $2S_5$ | $2S_5^3$ | $5\sigma_d$ | | |
|---|---|---|---|---|---|---|---|---|---|---|
| $A_1'$ | 1 | 1 | 1 | 1 | 1 | 1 | 1 | 1 | | $x^2+y^2$; $z^2$ |
| $A_2'$ | 1 | 1 | 1 | −1 | 1 | 1 | 1 | −1 | $R_z$ | |
| $E_1'$ | 2 | 2 cos 72° | 2 cos 144° | 0 | 2 | 2 cos 72° | 2 cos 144° | 0 | $(T_x, T_y)$ | $(x, y)$ |
| $E_2'$ | 2 | 2 cos 144° | 2 cos 72° | 0 | 2 | 2 cos 144° | 2 cos 72° | 0 | | $[\frac{1}{\sqrt{2}}(x^2-y^2), xy]$ |
| $A_1''$ | 1 | 1 | 1 | 1 | −1 | −1 | −1 | −1 | | |
| $A_2''$ | 1 | 1 | 1 | −1 | −1 | −1 | −1 | 1 | $T_z$ | z |
| $E_1''$ | 2 | 2 cos 72° | 2 cos 144° | 0 | −2 | −2 cos 72° | −2 cos 144° | 0 | $(R_x, R_y)$ | (xz, yz) |
| $E_2''$ | 2 | 2 cos 144° | 2 cos 72° | 0 | −2 | −2 cos 144° | −2 cos 72° | 0 | | |

*Notes:* 1. The $D_{5h}$ group is a direct product of $D_5$ and $C_s$. This is indicated by the dotted lines in the character table. Irreducible representations symmetric with respect to reflection in the $\sigma_h$ mirror plane are denoted by the superscript prime and antisymmetry is denoted by a double prime.
2. See the notes under the $I_h$ character table.

*Examples:* A regular pentagonal prism (Figure A3.3) and eclipsed ferrocene (Figure A3.4).

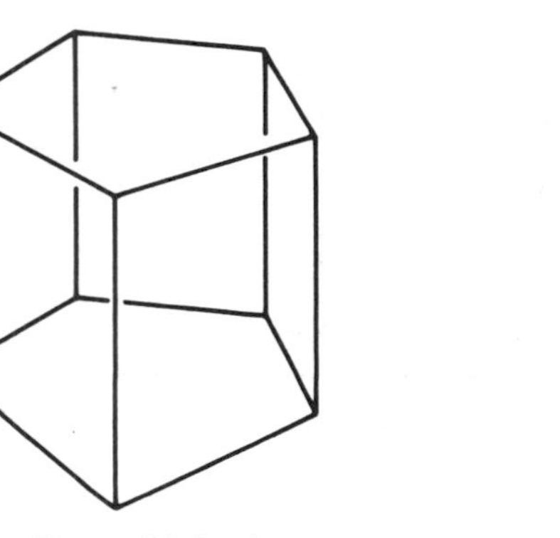

Figure A3.3 A pentagonal prism, of $D_{5h}$ symmetry

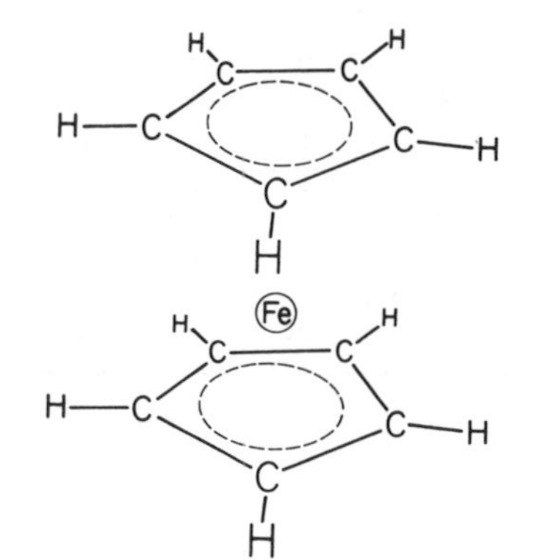


Figure A3.4 Ferrocene, $Fe(C_5H_5)_2$, in the eclipsed configuration, of $D_{5h}$ symmetry

| $D_{6h}$ | E | $2C_6$ | $2C_3$ | $C_2$ | $3C_2'$ | $3C_2''$ | i | $2S_3$ | $2S_6$ | $\sigma_h$ | $3\sigma_d$ | $3\sigma_d'$ | | |
|---|---|---|---|---|---|---|---|---|---|---|---|---|---|---|
| $A_{1g}$ | 1 | 1 | 1 | 1 | 1 | 1 | 1 | 1 | 1 | 1 | 1 | 1 | | $x^2+y^2; z^2$ |
| $A_{2g}$ | 1 | 1 | 1 | 1 | −1 | −1 | 1 | 1 | 1 | 1 | −1 | −1 | $R_z$ | |
| $B_{1g}$ | 1 | −1 | 1 | −1 | 1 | −1 | 1 | −1 | 1 | −1 | 1 | −1 | | |
| $B_{2g}$ | 1 | −1 | 1 | −1 | −1 | 1 | 1 | −1 | 1 | −1 | −1 | 1 | | |
| $E_{1g}$ | 2 | 1 | −1 | −2 | 0 | 0 | 2 | 1 | −1 | −2 | 0 | 0 | $(R_x, R_y)$ | $(xy, yz)$ |
| $E_{2g}$ | 2 | −1 | −1 | 2 | 0 | 0 | 2 | −1 | −1 | 2 | 0 | 0 | | $[\frac{1}{\sqrt{2}}(x^2-y^2), xy]$ |
| $A_{1u}$ | 1 | 1 | 1 | 1 | 1 | 1 | −1 | −1 | −1 | −1 | −1 | −1 | | |
| $A_{2u}$ | 1 | 1 | 1 | 1 | −1 | −1 | −1 | −1 | −1 | −1 | 1 | 1 | $T_z$ | z |
| $B_{1u}$ | 1 | −1 | 1 | −1 | 1 | −1 | −1 | 1 | −1 | 1 | −1 | 1 | | |
| $B_{2u}$ | 1 | −1 | −1 | −1 | −1 | 1 | −1 | 1 | −1 | 1 | 1 | −1 | | |
| $E_{1u}$ | 2 | 1 | −1 | −2 | 0 | 0 | −2 | −1 | 1 | 2 | 0 | 0 | $(T_x, T_y)$ | $(x, y)$ |
| $E_{2u}$ | 2 | −1 | −1 | 2 | 0 | 0 | −2 | 1 | 1 | −2 | 0 | 0 | | |

*Notes:* 1. The $D_{6h}$ character table is a direct product of $D_6$ and $C_i$. This is indicated by the dotted lines in the character table.
2. The choice between which pair of $C_2$ axes (and operations) is labelled $C_2'$ and which is labelled $C_2''$ is arbitrary. A redefinition will interchange $B_{1g}$ with $B_{2g}$ and $B_{1u}$ with $B_{2u}$. Similarly, the choice of the vertical planes $\sigma_d$ and $\sigma_d'$ is arbitrary but the $\sigma_d$ planes must contain the $C_2''$ axes and the $\sigma_d'$ planes must contain the $C_2'$.

*Examples:* A regular hexagonal prism (Figure A3.5) and the benzene molecule (Figure A3.6).

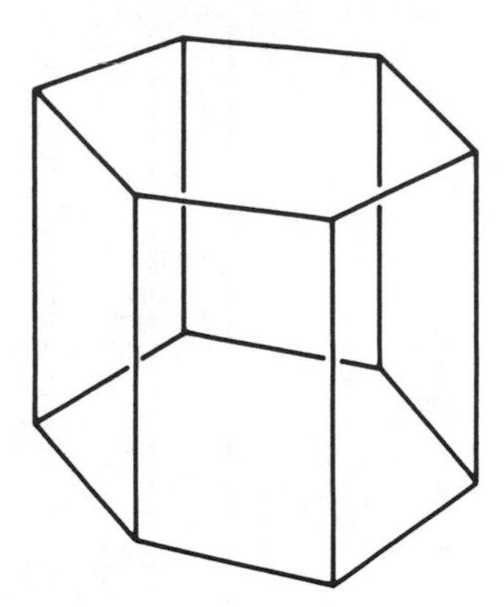

Figure A3.5 A hexagonal prism, of $D_{6h}$ symmetry

Figure A3.6 The benzene molecule, of $D_{6h}$ symmetry

## A3.4 THE GROUPS $D_{nd}$

These groups do not have the $\sigma_h$ mirror plane of the $D_{nh}$ groups. Objects with $D_{nd}$ symmetry typically have two similar halves, staggered with respect to each other. Thus solid objects of $D_{nd}$ symmetry are called 'antiprisms', a term which indicates the staggering.

When $n$ is odd, the groups $D_{nd}$ are direct products of the groups $D_n$ and $C_i$ (when $n$ is even, the direct products $D_n \times C_i$ are $D_{nh}$ groups).

| $D_{2d}$ | E | $2S_4$ | $C_2$ | $2C_2'$ | $2\sigma_d$ | | |
|---|---|---|---|---|---|---|---|
| $A_1$ | 1 | 1 | 1 | 1 | 1 | | $x^2+y^2$; $z^2$ |
| $A_2$ | 1 | 1 | 1 | −1 | −1 | $R_z$ | |
| $B_1$ | 1 | −1 | 1 | 1 | −1 | | $x^2-y^2$ |
| $B_2$ | 1 | −1 | 1 | −1 | 1 | $T_z$ | z; xy |
| E | 2 | 0 | −2 | 0 | 0 | $(T_x, T_y)(R_x, R_y)$ | (x, y); (xz, yz) |

*Notes:* 1. The x axis is taken as coincident with one $C_2'$.
2. Singly degenerate irreducible representations which are symmetric with respect to an $S_4$ operation are indicated by an A label; antisymmetry is indicated by a B label.
3. This group is sometimes called $V_d$.
4. Most people find the $2C_2'$ axes and $2\sigma_d$ mirror planes difficult to locate in this group. Time spent with the examples below would be time well spent.

*Examples:* A pentagonal dodecahedron (Figure A3.7) and the molecule spiropentane (Figure A3.8).

| $D_{3d}$ | E | $2C_3$ | $3C_2$ | i | $2S_6$ | $3\sigma_d$ | | |
|---|---|---|---|---|---|---|---|---|
| $A_{1g}$ | 1 | 1 | 1 | 1 | 1 | 1 | | $x^2+y^2$; $z^2$ |
| $A_{2g}$ | 1 | 1 | −1 | 1 | 1 | −1 | $R_z$ | |
| $E_g$ | 2 | −1 | 0 | 2 | −1 | 0 | $(R_x, R_y)$ | $[\frac{1}{\sqrt{2}}(x^2-y^2), xy]$; (xz, yz) |
| $A_{1u}$ | 1 | 1 | 1 | −1 | −1 | −1 | | |
| $A_{2u}$ | 1 | 1 | −1 | −1 | −1 | 1 | $T_z$ | z |
| $E_u$ | 2 | −1 | 0 | −2 | 1 | 0 | $(T_x, T_y)$ | (x, y) |

*Notes:* 1. This group is a direct product of $D_3$ with $C_i$, indicated by the dotted lines in the character table.
2. The $C_3$ and i operations may be considered as derived from the $S_6$ because

$$S_6^2 = C_3$$
$$S_6^3 = i$$

*Examples:* The staggered ethane molecule (Figure A3.9).

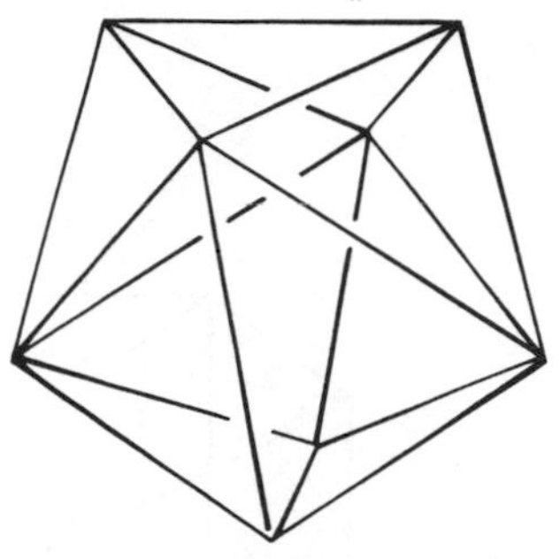

Figure A3.7 A pentagonal bipyramid, of $D_{2d}$ symmetry. Note that all of the apices of this figure lie in one of two mutually perpendicular planes

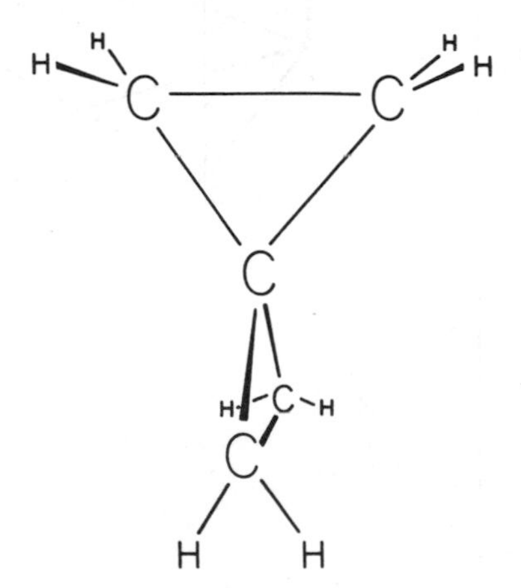


Figure A3.8 The molecule spiropentane, $C_5H_8$, of $D_{2d}$ symmetry. The carbon atoms in this figure lie at four of the eight apices of the pentagonal bipyramid shown in Figure A3.7

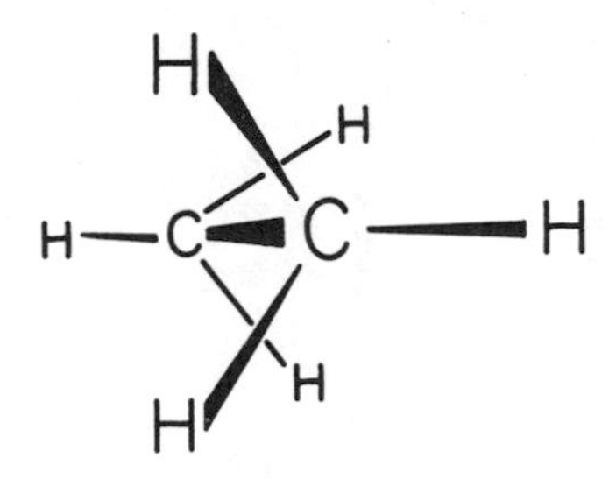


Figure A3.9 The staggered ethane molecule, of $D_{3d}$ symmetry. This symmetry is best seen if the molecule is viewed along the C—C bond but it is difficult to draw it adequately in this orientation

| $D_{4d}$ | $E$ | $2S_8$ | $2C_4$ | $2S_8^3$ | $C_2$ | $4C_2'$ | $4\sigma_d$ | | |
|---|---|---|---|---|---|---|---|---|---|
| $A_1$ | 1 | 1 | 1 | 1 | 1 | 1 | 1 | | $x^2+y^2; z^2$ |
| $A_2$ | 1 | 1 | 1 | 1 | 1 | $-1$ | $-1$ | $R_z$ | |
| $B_1$ | 1 | $-1$ | 1 | $-1$ | 1 | 1 | $-1$ | | |
| $B_2$ | 1 | $-1$ | 1 | $-1$ | 1 | $-1$ | 1 | $T_z$ | $z$ |
| $E_1$ | 2 | $\sqrt{2}$ | 0 | $-\sqrt{2}$ | $-2$ | 0 | 0 | $(T_x, T_y)$ | $(x, y)$ |
| $E_2$ | 2 | 0 | $-2$ | 0 | 2 | 0 | 0 | | $[\frac{1}{\sqrt{2}}(x^2-y^2), xy]$ |
| $E_3$ | 2 | $-\sqrt{2}$ | 0 | $\sqrt{2}$ | $-2$ | 0 | 0 | $(R_x, R_y)$ | $(xy, yz)$ |

*Note:* The unique $C_2$, the $S_8^3$ and $C_4$ operations may be considered to be derived from the $S_8$ because

$$S_8^2 = C_4$$
$$S_8^3 = S_8^3$$
$$S_8^4 = C_2$$

*Example:* The square antiprism (Figure A3.10).

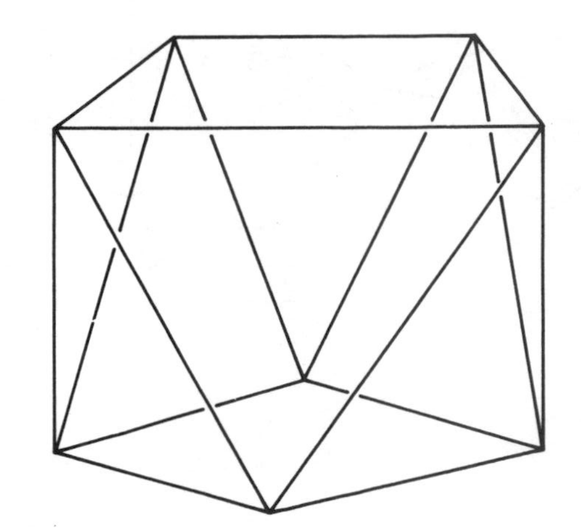

Figure A3.10 The square antiprism, of $D_{4d}$ symmetry

| $D_{5d}$ | $E$ | $2C_5$ | $2C_5^2$ | $5C_2$ | $i$ | $2S_{10}^3$ | $2S_{10}$ | $5\sigma_d$ | | |
|---|---|---|---|---|---|---|---|---|---|---|
| $A_{1g}$ | 1 | 1 | 1 | 1 | 1 | 1 | 1 | 1 | | $x^2+y^2$; $z^2$ |
| $A_{2g}$ | 1 | 1 | 1 | −1 | 1 | 1 | 1 | −1 | $R_z$ | |
| $E_{1g}$ | 2 | 2 cos 72° | 2 cos 144° | 0 | 2 | 2 cos 72° | 2 cos 144° | 0 | $(R_x, R_y)$ | $(xz, yz)$ |
| $E_{2g}$ | 2 | 2 cos 144° | 2 cos 72° | 0 | 2 | 2 cos 144° | 2 cos 72° | 0 | | $[\frac{1}{\sqrt{2}}(x^2-y^2), xy]$ |
| $A_{1u}$ | 1 | 1 | 1 | 1 | −1 | −1 | −1 | −1 | | |
| $A_{2u}$ | 1 | 1 | 1 | −1 | −1 | −1 | −1 | 1 | $T_z$ | $z$ |
| $E_{1u}$ | 2 | 2 cos 72° | 2 cos 144° | 0 | −2 | −2 cos 72° | −2 cos 144° | 0 | $(T_x, T_y)$ | $(x, y)$ |
| $E_{2u}$ | 2 | 2 cos 144° | 2 cos 72° | 0 | −2 | −2 cos 144° | −2 cos 72° | 0 | | |

*Notes:* 1. This group is the direct product of $C_5$ and $C_i$, indicated by the dotted lines in the character table.
2. Before working with this group refer to Note 2 under the $D_{5h}$ character table.
3. Many of the operations of the group may be considered to be derived from the $S_{10}$ because

$$S_{10}^2 = C_5$$
$$S_{10}^3 = S_{10}^3$$
$$S_{10}^4 = C_5^2$$
$$S_{10}^5 = i$$

*Examples:* The pentagonal antiprism (Figure A3.11) and staggered ferrocene (Figure A3.12)

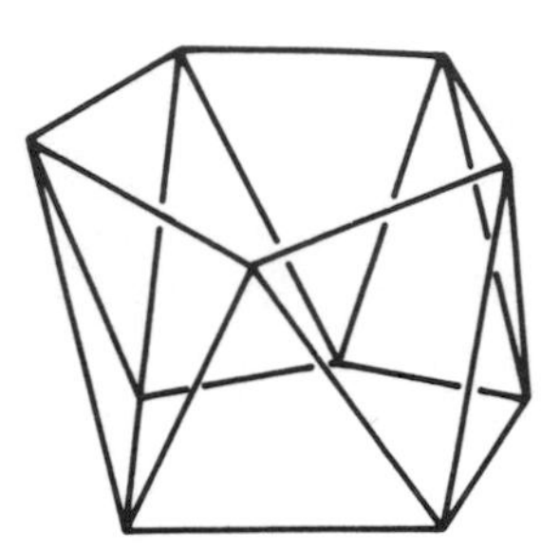

Figure A3.11 The pentagonal antiprism, of $D_{5d}$ symmetry

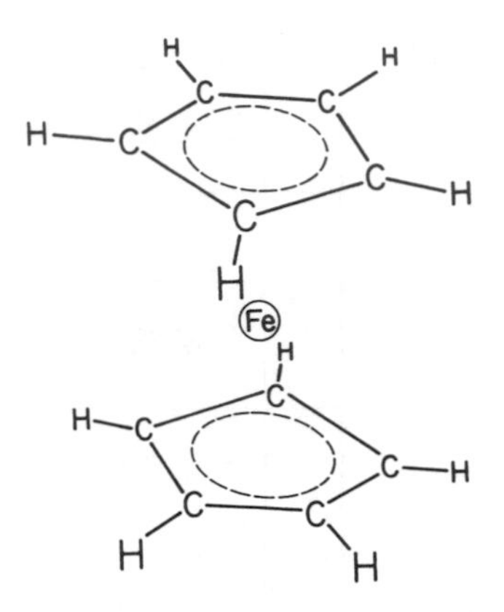


Figure A3.12 The staggered ferrocene molecule (Fe($C_5H_5$)$_2$, of $D_{5d}$ symmetry

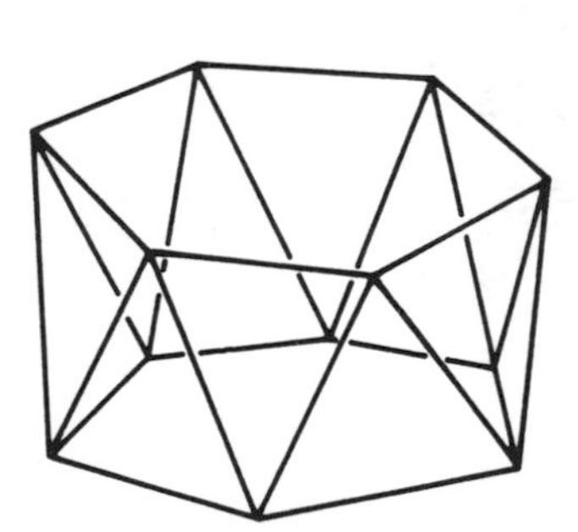

Figure A3.13 The hexagonal antiprism, of $D_{6d}$ symmetry

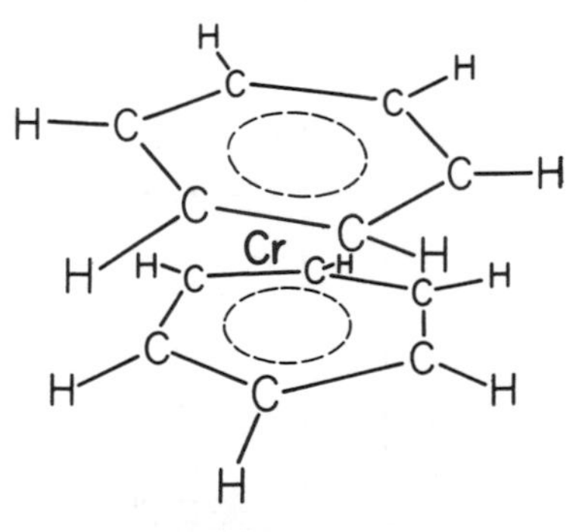


Figure A3.14 The staggered dibenzene chromium molecule, Cr($C_6H_6$)$_2$, of $D_{6d}$ symmetry

| $D_{6d}$ | E | $2S_{12}$ | $2C_6$ | $2S_4$ | $2C_3$ | $2S_{12}^5$ | $C_2$ | $6C_2'$ | $6\sigma_d$ | | |
|---|---|---|---|---|---|---|---|---|---|---|---|
| $A_1$ | 1 | 1 | 1 | 1 | 1 | 1 | 1 | 1 | 1 | | $x^2+y^2$; $z^2$ |
| $A_2$ | 1 | 1 | 1 | 1 | 1 | 1 | 1 | −1 | −1 | $R_z$ | |
| $B_1$ | 1 | −1 | 1 | −1 | 1 | −1 | 1 | 1 | −1 | | |
| $B_2$ | 1 | −1 | 1 | −1 | 1 | −1 | 1 | −1 | 1 | $T_z$ | z |
| $E_1$ | 2 | $\sqrt{3}$ | 1 | 0 | −1 | $-\sqrt{3}$ | −2 | 0 | 0 | $(T_x, T_y)$ | (x, y) |
| $E_2$ | 2 | 1 | −1 | −2 | −1 | 1 | 2 | 0 | 0 | | $[\frac{1}{\sqrt{2}}(x^2-y^2), xy]$ |
| $E_3$ | 2 | 0 | −2 | 0 | 2 | 0 | −2 | 0 | 0 | | |
| $E_4$ | 2 | −1 | −1 | 2 | −1 | −1 | 2 | 0 | 0 | | |
| $E_5$ | 2 | $-\sqrt{3}$ | 1 | 0 | −1 | $\sqrt{3}$ | −2 | 0 | 0 | $(R_x, R_y)$ | (yz, zx) |

*Note:* Many of the operations of the group may be taken to be derived from $S_{12}$ because

$$S_{12}^2 = C_6$$
$$S_{12}^3 = S_4$$
$$S_{12}^4 = C_3$$
$$S_{12}^5 = S_{12}^5$$
$$S_{12}^6 = C_2$$

*Examples:* Hexagonal antiprism (Figure A3.13) and staggered dibenzene chromium (Figure A3.14).

## A3.5 THE GROUPS $D_n$

These are groups of proper (i.e. pure) rotations corresponding to bodies in which there are $n$ $C_2$ axes perpendicular to a principal $C_n$ axis. To obtain solid figures of these geometries it is simplest to take a polyhedron shown for a $D_{nh}$ or $D_{nd}$ symmetry and to systematically introduce zig-zag edges such as used to derive figures for the groups O and T. Molecules of $D_{nh}$ or $D_{nd}$ symmetries drop to $D_n$ symmetry when the 'top' and 'bottom' parts of the molecule are given small, arbitrary, twists in opposite directions about the z axis.

$D_{nh}$ and $D_{nd}$ ($n$ odd) groups are direct products of $D_n$ with either $C_i$ or $C_s$. For problems in these groups it is often simplest to work in $D_n$ symmetry and move to the full group at a later stage.

| $D_2$ | E | $C_2(z)$ | $C_2(y)$ | $C_2(x)$ | | |
|---|---|---|---|---|---|---|
| A | 1 | 1 | 1 | 1 | | $x^2$; $y^2$; $z^2$ |
| $B_1$ | 1 | 1 | −1 | −1 | $T_z$; $R_z$ | z; xy |
| $B_2$ | 1 | −1 | 1 | −1 | $T_y$; $R_y$ | y; xz |
| $B_3$ | 1 | −1 | −1 | 1 | $T_x$; $R_x$ | x; yz |

*Notes:* 1. Because there are three mutually perpendicular $C_2$ axes the choice of x, y and z is arbitrary. Interchange of labels of these axes will lead to an interchange of the labels $B_1$, $B_2$ and $B_3$.
2. Because $x^2$ and $y^2$ transform, separately, as A, it follows that the function $x^2 - y^2$ also has A symmetry.

*Example:* Ethylene in which the two $CH_2$ groups have been made non-coplanar by a counter-rotation of these two units about the C—C axis (Figure A3.15).

| $D_3$ | E | $2C_3$ | $3C_2$ | | |
|---|---|---|---|---|---|
| $A_1$ | 1 | 1 | 1 | | $x^2+y^2$; $z^2$ |
| $A_2$ | 1 | 1 | −1 | $T_z$; $R_z$ | z |
| E | 2 | −1 | 0 | $(T_x, T_y)$ $(R_x, R_y)$ | (x, y); $[\frac{1}{\sqrt{2}}(x^2-y^2), xy]$ (xz, yz) |

*Example:* Ethane in which the two $CH_3$ units have been counter-rotated about the C—C axis so that the molecule is neither eclipsed or staggered (Figure A3.16).

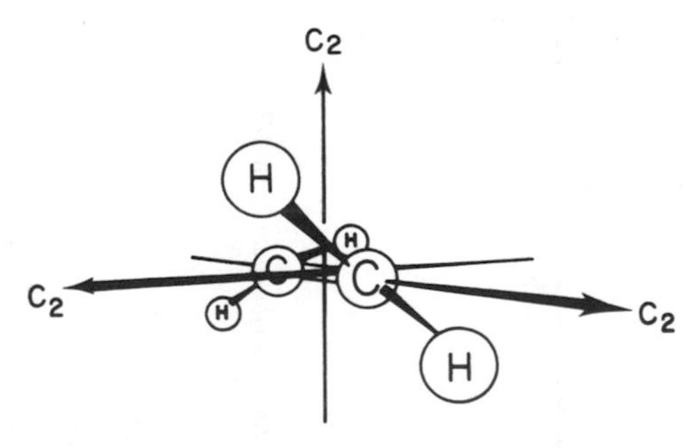


Figure A3.15 A slightly twisted ethylene molecule, of $D_2$ symmetry

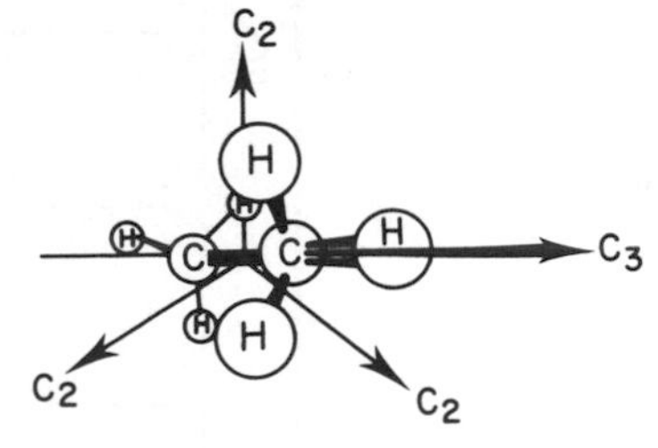


Figure A3.16 An ethane molecule which is neither eclipsed nor staggered, of $D_3$ symmetry

| $D_4$ | $E$ | $2C_4$ | $C_2(=C_4^2)$ | $2C_2'$ | $2C_2''$ | | |
|---|---|---|---|---|---|---|---|
| $A_1$ | 1 | 1 | 1 | 1 | 1 | | $x^2+y^2$; $z^2$ |
| $A_2$ | 1 | 1 | 1 | −1 | −1 | $T_z$; $R_z$ | z |
| $B_1$ | 1 | −1 | 1 | 1 | −1 | | $x^2-y^2$ |
| $B_2$ | 1 | −1 | 1 | −1 | 1 | | xy |
| $E$ | 2 | 0 | −2 | 0 | 0 | $(T_x, T_y)$; $(R_x, R_y)$ | (x, y); (xz, yz) |

*Notes:* 1. The x axis has been taken as coincident with one $C_2'$.
2. The choice between which set of two $C_2$ axes is called $2C_2'$ and which is called $2C_2''$ is arbitrary. If the choice opposite to that above is taken then the labels $B_1$ and $B_2$ have to be interchanged ($B_1$ has a character of 1 under the $C_2'$ operations).

*Example:* A twisted cube (Figure A3.17).

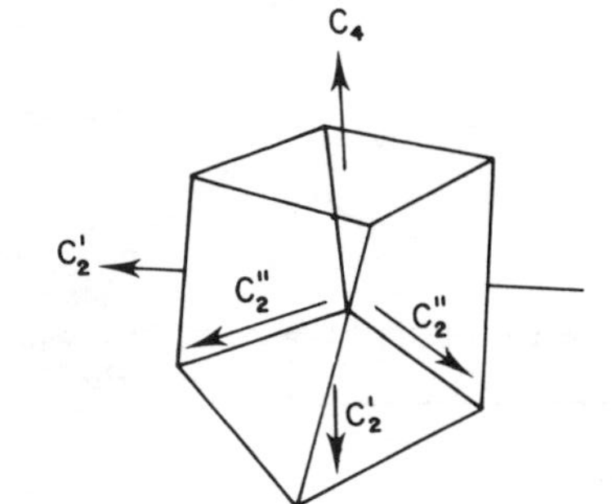


Figure A3.17 A slightly twisted cube, of $D_4$ symmetry

| $D_5$ | E | $2C_5$ | $2C_5^2$ | $5C_2$ | | |
|---|---|---|---|---|---|---|
| $A_1$ | 1 | 1 | 1 | 1 | | $x^2 + y^2$; $z^2$ |
| A | 1 | 1 | 1 | −1 | $T_z$; $R_z$ | z |
| $E_1$ | 2 | 2 cos 72° | 2 cos 144° | 0 | $(T_x, T_y)$; $(R_x, R_y)$ | (x, y); (xz, yz) |
| $E_2$ | 2 | 2 cos 144° | 2 cos 72° | 0 | | $[\frac{1}{\sqrt{2}}(x^2 - y^2), xy, yz]$ |

*Note:* Before working with this group refer to Note 2 under the $D_{5h}$ character table.
*Example:* Ferrocene when the carbon atoms in opposite rings are neither staggered nor eclipsed (Figure A3.18).

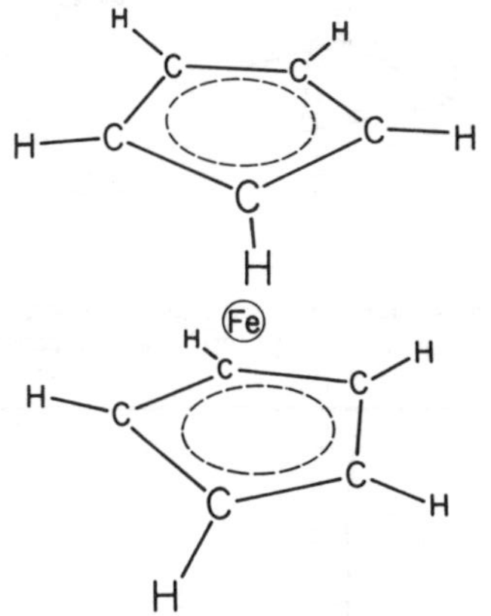

Figure A3.18 A ferrocene molecule, $Fe(C_5H_5)_2$, in which the two rings are neither staggered or eclipsed, of $D_5$ symmetry

| $D_6$ | E | $2C_6$ | $2C_3$ | $C_2$ | $3C_2'$ | $3C_2''$ | | |
|---|---|---|---|---|---|---|---|---|
| $A_1$ | 1 | 1 | 1 | 1 | 1 | 1 | | $x^2+y^2$; $z^2$ |
| $A_2$ | 1 | 1 | 1 | 1 | −1 | −1 | $T_z$; $R_z$ | z |
| $B_1$ | 1 | −1 | 1 | −1 | 1 | −1 | | |
| $B_2$ | 1 | −1 | 1 | −1 | −1 | 1 | | |
| $E_1$ | 2 | 1 | −1 | −2 | 0 | 0 | $(T_x, T_y)$; $(R_x, R_y)$ | (x, y); (xz, yz) |
| $E_2$ | 2 | −1 | −1 | 2 | 0 | 0 | | $[\frac{1}{\sqrt{2}}(x^2-y^2), xy]$ |

*Notes:* 1. The x axis has been taken as coincident with one $C_2'$.
2. The choice between which set of three $C_2$ axes is called $3C_2'$ and which is called $3C_2''$ is arbitrary. If the choice opposite to that above is taken then the labels $B_1$ and $B_2$ on functions will have to be interchanged ($B_1$ has a character of 1 under the $C_2'$ operations).

*Example:* Dibenzene chromium when the carbon atoms in opposite rings are neither staggered nor eclipsed (Figure A3.19).

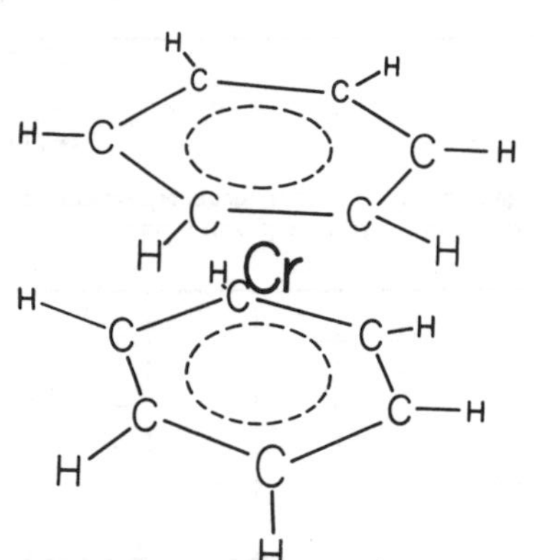


Figure A3.19 A dibenzene chromium molecule, $Cr(C_6H_6)_2$, in which the two rings are neither staggered nor eclipsed, of $D_6$ symmetry.

## A3.6 THE GROUPS $C_{nv}$

| $C_{2v}$ | E | $C_2$ | $\sigma_v$ | $\sigma_v'$ | | |
|---|---|---|---|---|---|---|
| $A_1$ | 1 | 1 | 1 | 1 | $T_z$ | $z; x^2; y^2; z^2$ |
| $A_2$ | 1 | 1 | −1 | −1 | $R_z$ | $xy$ |
| $B_1$ | 1 | −1 | 1 | −1 | $T_y; R_x$ | $y; yz$ |
| $B_2$ | 1 | −1 | −1 | 1 | $T_x; R_y$ | $x; xz$ |

*Notes:* 1. x is taken as lying in $\sigma_v'$.
2. Interchange of the labels $\sigma_v$ and $\sigma_v'$ or (equivalently) interchange of the choice of direction of x and y axes (one lies in each mirror plane) interchanges the labels $B_1$ and $B_2$ ($B_1$ has a character of 1 under the $\sigma_v$ operation).

*Example:* The water molecule (see Chapters 2 and 3).

| $C_{3v}$ | E | $2C_3$ | $3\sigma_v$ | | |
|---|---|---|---|---|---|
| $A_1$ | 1 | 1 | 1 | $T_z$ | $z; x^2 + y^2; z^2$ |
| $A_2$ | 1 | 1 | −1 | $R_z$ | |
| E | 2 | −1 | 0 | $(T_x, T_y); (R_x, R_y)$ | $(x, y); [\frac{1}{\sqrt{2}}(x^2 - y^2), xy]; (xz, yz)$ |

*Example:* The ammonia molecule (see Chapter 6).

| $C_{4v}$ | E | $2C_4$ | $C_2$ | $2\sigma_v$ | $2\sigma_v'$ | | |
|---|---|---|---|---|---|---|---|
| $A_1$ | 1 | 1 | 1 | 1 | 1 | $T_z$ | $z; x^2+y^2; z^2$ |
| $A_2$ | 1 | 1 | 1 | −1 | −1 | $R_z$ | |
| $B_1$ | 1 | −1 | 1 | 1 | −1 | | $x^2-y^2$ |
| $B_2$ | 1 | −1 | 1 | −1 | 1 | | $xy$ |
| E | 2 | 0 | −2 | 0 | 0 | $(T_x, T_y), (R_x, R_y)$ | $(x, y); (xz, yz)$ |

*Notes:* 1. x is taken as lying in one $\sigma_v$ plane.
2. Interchange of the labels $\sigma_v$ and $\sigma_v'$ (and the choice is arbitrary) leads to an interchange of the labels $B_1$ and $B_2$ ($B_1$ has a character of 1 under the $\sigma_v$ operations).

*Example:* The $BrF_5$ molecule (Chapter 5).

| $C_{5v}$ | E | $2C_5$ | $2C_5^2$ | $5\sigma_v$ | | |
|---|---|---|---|---|---|---|
| $A_1$ | 1 | 1 | 1 | 1 | $T_z$ | $z; x^2+y^2; z^2$ |
| $A_2$ | 1 | 1 | 1 | −1 | $R_z$ | |
| $E_1$ | 2 | 2 cos 72° | 2 cos 144° | 0 | $(T_x, T_y); (R_x, R_y)$ | $(x, y); (xz, yz)$ |
| $E_2$ | 2 | 2 cos 144° | 2 cos 72° | 0 | | $[\frac{1}{\sqrt{2}}(x^2-y^2), xy]$ |

*Note:* Before working with this group refer to Note 2 under the $D_{5h}$ character table.

*Example:* $\eta^5$-cyclopentadiene nickel monocarbonyl (Figure A3.20).

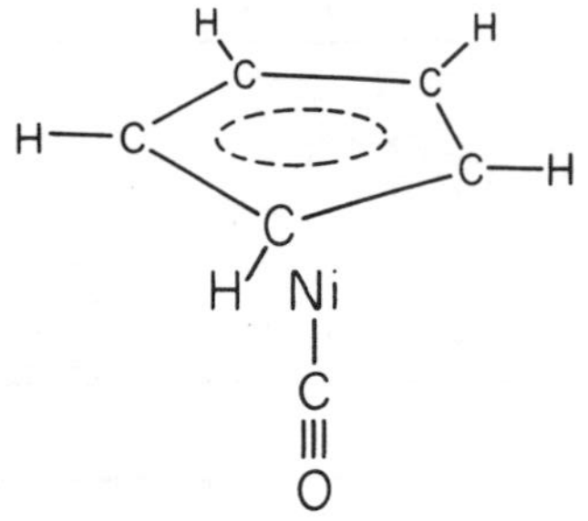

Figure A3.20 The $\eta^5$-cyclopentadiene nickel monocarbonyl molecule, $Ni(C_5H_5)CO$, of $C_{5v}$ symmetry

| $C_{6v}$ | E | $2C_6$ | $2C_3$ | $C_2$ | $3\sigma_v$ | $3\sigma_v'$ | | |
|---|---|---|---|---|---|---|---|---|
| $A_1$ | 1 | 1 | 1 | 1 | 1 | 1 | $T_z$ | $z; x^2+y^2; z^2$ |
| $A_2$ | 1 | 1 | 1 | 1 | −1 | −1 | $R_z$ | |
| $B_1$ | 1 | −1 | 1 | −1 | 1 | −1 | | |
| $B_2$ | 1 | −1 | 1 | −1 | −1 | 1 | | |
| $E_1$ | 2 | 1 | −1 | −2 | 0 | 0 | $(T_x, T_y); (R_x, R_y)$ | $(x, y); (xz, yz)$ |
| $E_2$ | 2 | −1 | −1 | 2 | 0 | 0 | | $[\frac{1}{\sqrt{2}}(x^2-y^2), xy]$ |

*Note:* Interchange of the labels $\sigma_v$ and $\sigma_v'$ (and the choice is arbitrary) leads to an interchange of the labels $B_1$ and $B_2$ on functions ($B_1$ has a character of 1 under the $\sigma_v$ operations).
*Example:* The compound $\eta^6$-hexamethylbenzene $\eta^6$-benzene chromium (Figure A3.21).

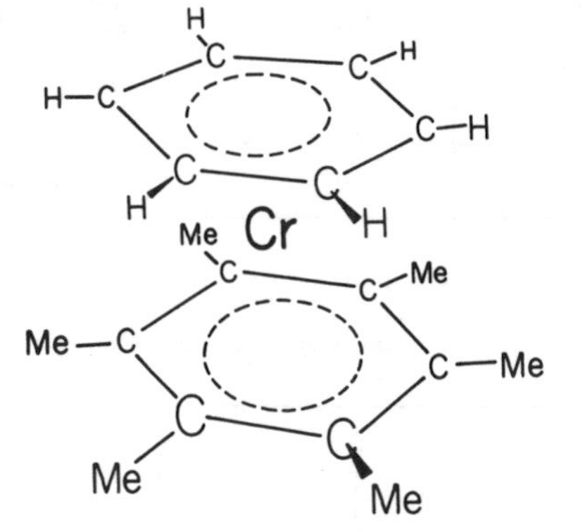


Figure A3.21 The molecule $\eta^6$-hexamethyl benzene $\eta^6$-benzene chromium, $Cr(C_6H_6)(C_6Me_6)$, of $C_{6v}$ symmetry

## A3.7 THE GROUPS $C_{nh}$

These groups have a derivation similar to that of the $D_{nh}$ groups — they are direct products of $C_n$ with either $C_i$ ($n$ even) or $C_s$ ($n$ odd). The only one which has been found to be of real chemical importance is $C_{2h}$; however, $C_{3h}$ is included to give an example of the $n$ odd case.

| $C_{2h}$ | E | $C_2$ | i | $\sigma_h$ | | |
|---|---|---|---|---|---|---|
| $A_g$ | 1 | 1 | 1 | 1 | $R_z$ | $x^2$; $y^2$; $z^2$; xy |
| $B_g$ | 1 | −1 | 1 | −1 | $R_x$; $R_y$ | xz; yz |
| $A_u$ | 1 | 1 | −1 | −1 | $T_z$ | z |
| $B_u$ | 1 | −1 | −1 | 1 | $T_x$; $T_y$ | x; y |

*Note:* This group is a direct product of the $C_2$ and $C_i$ groups, indicated in the character table by the dotted lines.

| $C_{3h}$ | $E$ | $C_3$ | $C_3^2$ | $\sigma_h$ | $S_3$ | $S_3^5$ | | |
|---|---|---|---|---|---|---|---|---|
| $A'$ | 1 | 1 | 1 | 1 | 1 | 1 | $R_z$ | $x^2 + y^2; z^2$ |
| $E'$ | $\begin{cases} 1 \\ 1 \end{cases}$ | $\begin{matrix} \varepsilon \\ \varepsilon^2 \end{matrix}$ | $\begin{matrix} \varepsilon^2 \\ \varepsilon \end{matrix}$ | $\begin{matrix} 1 \\ 1 \end{matrix}$ | $\begin{matrix} \varepsilon \\ \varepsilon^2 \end{matrix}$ | $\begin{matrix} \varepsilon^2 \\ \varepsilon \end{matrix}$ | $[(T_x + iT_y), (T_x - iT_y)]$ | $[(x + iy), (x - iy)]; [\frac{1}{\sqrt{2}}(x^2 - y^2), xy]$ |
| $A''$ | 1 | 1 | 1 | $-1$ | $-1$ | $-1$ | $T_z$ | $z$ |
| $E''$ | $\begin{cases} 1 \\ 1 \end{cases}$ | $\begin{matrix} \varepsilon \\ \varepsilon^2 \end{matrix}$ | $\begin{matrix} \varepsilon^2 \\ \varepsilon \end{matrix}$ | $\begin{matrix} -1 \\ -1 \end{matrix}$ | $\begin{matrix} -\varepsilon \\ -\varepsilon^2 \end{matrix}$ | $\begin{matrix} -\varepsilon^2 \\ -\varepsilon \end{matrix}$ | $[(R_x + iR_y), (R_x - iR_y)]$ | $(xz, yz)$ |

*Notes:* 1. See the notes on the $C_3$ group for the meaning of $\varepsilon$ and $\varepsilon^2$.
2. This group is a direct product of the $C_3$ and $C_s$ group, indicated in the character table by the dotted lines.

## A3.8 THE GROUPS $C_n$

These are cyclic groups with character tables that look rather strange when compared with most of those encountered earlier in this appendix. They only look strange when compared with other point groups. For many other groups — for instance, the translational groups encountered in theories of crystal structure — the appearance of complex numbers is the norm.

Chapter 11 gives detailed examples of working with the $C_4$ group.

| $C_2$ | E | $C_2$ | | |
|---|---|---|---|---|
| A | 1 | 1 | $T_z$; $R_z$ | z; $x^2$; $y^2$; $z^2$; xy |
| B | 1 | −1 | $T_x$; $T_y$; $R_x$; $R_y$ | x; y; yz; xz |

| $C_3$ | E | $C_3$ | $C_3^2$ | | |
|---|---|---|---|---|---|
| A | 1 | 1 | 1 | $T_z$; $R_z$ | $z$; $x^2+y^2$; $z^2$ |
| E | $\begin{cases}1\\1\end{cases}$ | $\begin{matrix}\varepsilon\\\varepsilon^*\end{matrix}$ | $\begin{matrix}\varepsilon^*\\\varepsilon\end{matrix}$ | $[(T_x+iT_y),\ (T_x-iT_y)]\ [(R_x+iR_y),\ (R_x-iR_y)]$ | $[(x+iy),\ (x-iy)]$; $[\frac{1}{\sqrt{2}}(x^2-y^2),\ xy]$; $(yz,\ xz)$ |

*Note:* In this and two of the next three tables we use the notation that

$$\varepsilon=\exp\left(\frac{2\pi i}{n}\right)=\cos\left(\frac{2\pi}{n}\right)+i\sin\left(\frac{2\pi}{n}\right)$$

$$\varepsilon^*=\exp\left(\frac{-2\pi i}{n}\right)=\cos\left(\frac{2\pi}{n}\right)-i\sin\left(\frac{2\pi}{n}\right)$$

so that here ($n=3$) we have

$$\varepsilon=\cos 120+i\sin 120=-\tfrac{1}{2}+i\frac{\sqrt{3}}{2}$$

$$\varepsilon^*=\cos 120-i\sin 120=-\tfrac{1}{2}-i\frac{\sqrt{3}}{2}$$

Note that $\varepsilon$ and $\varepsilon^*$ are complex conjugates.
In the case of $n=3$, $\varepsilon^2=\varepsilon^*$ and the $\varepsilon^2$ notation has been used in the T, $T_h$ and $C_{3h}$ character tables.

| $C_4$ | E | $C_4$ | $C_2$ | $C_4^3$ | | |
|---|---|---|---|---|---|---|
| A | 1 | 1 | 1 | 1 | $T_z$; $R_z$ | $z$; $x^2+y^2$; $z^2$ |
| B | 1 | −1 | 1 | −1 | | $x^2-y^2$; $xy$ |
| E | $\begin{cases}1\\1\end{cases}$ | $\begin{matrix}i\\-i\end{matrix}$ | $\begin{matrix}-1\\-1\end{matrix}$ | $\begin{matrix}-i\\i\end{matrix}$ | $[(T_x+iT_y),(T_x-iT_y)]\ [(R_x+iR_y),(R_x-iR_y)]$ | $[(x+iy),(x-iy)]$ |

*Note:* It is easy to show by substitution in the equations given above that for $n=4$ $\exp(2\pi i/n)=i$.

| $C_5$ | E | $C_5$ | $C_5^2$ | $C_5^3$ | $C_5^4$ | | |
|---|---|---|---|---|---|---|---|
| A | 1 | 1 | 1 | 1 | 1 | $T_z$; $R_z$ | $z$; $x^2+y^2$; $z^2$ |
| $E_1$ | 1 | $\varepsilon$ | $\varepsilon^2$ | $\varepsilon^{2*}$ | $\varepsilon^*$ | $[(T_x+iT_y), (T_x-iT_y)]$ $[(R_x+iR_y), (R_x-iR_y)]$ | $[(x+iy), (x-iy)]$; $(yz, xz)$ |
| | 1 | $\varepsilon^*$ | $\varepsilon^{2*}$ | $\varepsilon^2$ | $\varepsilon$ | | |
| $E_2$ | 1 | $\varepsilon^2$ | $\varepsilon^*$ | $\varepsilon$ | $\varepsilon^{2*}$ | | $[\frac{1}{\sqrt{2}}(x^2-y^2), xy]$ |
| | 1 | $\varepsilon^{2*}$ | $\varepsilon$ | $\varepsilon^*$ | $\varepsilon^2$ | | |

*Note:* For a definition of $\varepsilon$, etc., see the note under $C_3$.

| $C_6$ | E | $C_6$ | $C_3$ | $C_2$ | $C_3^2$ | $C_6^5$ | | |
|---|---|---|---|---|---|---|---|---|
| A | 1 | 1 | 1 | 1 | 1 | 1 | $T_z$; $R_z$ | |
| B | 1 | $-1$ | 1 | $-1$ | 1 | $-1$ | | |
| $E_1$ | 1 | $\varepsilon$ | $-\varepsilon^*$ | $-1$ | $-\varepsilon$ | $\varepsilon^*$ | $[(T_x+iT_y), (T_x-iT_y)]$ | $[(x+iy), (x-iy)]$ |
| | 1 | $\varepsilon^*$ | $-\varepsilon$ | $-1$ | $-\varepsilon^*$ | $\varepsilon$ | $[(R_x+iR_y), (R_x-iR_y)]$ | $(xz, yz)$ |
| $E_2$ | 1 | $-\varepsilon^*$ | $-\varepsilon$ | 1 | $-\varepsilon^*$ | $-\varepsilon$ | | $[\frac{1}{\sqrt{2}}(x^2-y^2), xy]$ |
| | 1 | $-\varepsilon$ | $-\varepsilon^*$ | 1 | $-\varepsilon$ | $-\varepsilon^*$ | | |

*Note:* For a definition of $\varepsilon$, etc., see the note under $C_3$.

## A3.9 THE GROUPS $S_n$ ($n$ EVEN) (INCLUDES $C_i$)

Another set of cyclic groups denoted $S_n$. These only exist for $n$ even because odd values of $n$ do not satisfy the requirement $(S_n)^n = E$. The $S_2$ group is usually labelled $C_i$ because the operations $S_2$ and i are identical (this is demonstrated in Figure 7.29).

| $C_i$ | E | i | | |
|---|---|---|---|---|
| $A_g$ | 1 | 1 | $R_x$; $R_y$; $R_z$ | $x^2$; $y^2$; $z^2$; xy; xz; yz |
| $A_u$ | 1 | −1 | $T_x$; $T_y$; $T_z$ | x; y; z |

*Note:* $C_i$ often forms a direct product with another group. In this case the 'g' and 'u' suffixes of the $C_i$ group are carried into the labels of the direct product group.

| $S_4$ | $E$ | $S_4$ | $C_2$ | $S_4^3$ | | |
|---|---|---|---|---|---|---|
| A | 1 | −1 | 1 | 1 | $R_z$ | $x^2+y^2$; $z^2$ |
| B | 1 | −1 | 1 | −1 | $T_z$ | $z$; $x^2-y^2$; $xy$ |
| E | $\left\{\begin{matrix}1\\1\end{matrix}\right.$ | $\begin{matrix}i\\-i\end{matrix}$ | $\begin{matrix}-1\\-1\end{matrix}$ | $\left.\begin{matrix}-i\\i\end{matrix}\right\}$ | $[(T_x+iT_y),(T_x-iT_y)]$; $[(R_x+iR_y),(R_x-iR_y)]$ | $[(x+iy),(x-iy)]$; $(xz, yz)$ |

*Note:* See the note under the $C_4$ group.

## A3.10 THE GROUP $C_s$ AND THE TRIVIAL GROUP $C_1$

The group $C_s$, like the group $C_i$, often participates in a direct product group. In the case of $C_s$ it is the superscript primes which carry over into the labels of the irreducible representations of the product group.

| $C_s$ | C | $\sigma$ | | |
|---|---|---|---|---|
| A′ | 1 | 1 | $T_x$; $T_y$; $R_z$ | x; y; $x^2$; $y^2$; $z^2$; xy |
| A″ | 1 | −1 | $T_z$; $R_x$; $R_y$ | z; yz; xz |

The groups $C_1$ is trivial because it is the symmetry of an object which has no symmetry! The only symmetry operation is the identity.

| $C_1$ | E |
|---|---|
| A | 1 |

In this group no bases are listed — all bases subtend the A irreducible representation!

## A3.11 THE INFINITESIMAL ROTATION (LINEAR) GROUPS $C_{\infty v}$ AND $D_{\infty h}$

Molecules in which all atoms lie on a common axis demand special attention because a rotation of any magnitude about this axis is a symmetry operation. The attack which proves profitable on this problem is to regard all such rotations to be (very large) multiples of an infinitesimally small rotation. That is, we have a $C_\infty$ axis and associated operations. The character table gives the character for the operation of rotation by an arbitrary angle $\phi$, denoted $C_\infty{}^\phi$. Not only is there an infinite number of operations based on $C_\infty$ but there is also an infinite number of $\sigma_v$ mirror planes. Fortunately, they all fall into a single class. If the linear molecule has no centre of symmetry then the appropriate group is $C_{\infty v}$. With a centre of symmetry the group is the direct product of $C_{\infty v}$ with $C_i$ and is denoted $D_{\infty h}$. Because the groups are infinite, the usual method of reducing a reducible representation will not work. However, reduction by inspection is usually possible. Appendix 5 discusses this problem in more detail. The alternative labels for irreducible representations for $C_{\infty v}$ and for $D_{\infty h}$ antedate the system used in this book. It is the $\Sigma$, $\Pi$, $\Delta$ system which is more commonly used.

The $D_{\infty h}$ point group is the direct product of $C_{\infty v}$ and $C_i$; we would therefore expect all terms in the irreducible representations carrying a ‘g’ suffix to appear with the same coefficients as their counterparts in $C_{\infty v}$. Similarly, the ‘u’ suffix representations should carry the coefficients with changed signs. Inspection of the character table of $D_{\infty h}$ reveals that these expectations do not appear to be

| $C_{\infty v}$ | E | $2C_\infty^\phi$ | ... | $\infty\sigma_v$ | | |
|---|---|---|---|---|---|---|
| $A_1 \equiv \Sigma^+$ | 1 | 1 | ... | 1 | $T_z$ | $z; x^2+y^2; z^2$ |
| $A_2 \equiv \Sigma^-$ | 1 | 1 | ... | −1 | $R_z$ | |
| $E_1 \equiv \Pi$ | 2 | $2 \cos \phi$ | ... | 0 | $(T_x, T_y); (R_x, R_y)$ | $(x, y); (xz, yz)$ |
| $E_2 \equiv \Delta$ | 2 | $2 \cos 2\phi$ | ... | 0 | | $[\frac{1}{\sqrt{2}}(x^2-y^2), xy]$ |
| $E_3 \equiv \Phi$ | 2 | $2 \cos 3\phi$ | ... | 0 | | |
| ... | ... | ... | ... | ... | | |

| $D_{\infty h}$ | E | $2C_\infty^\phi$ | ... | $\infty\sigma_v$ | i | $2S_\infty^\phi$ | ... | $\infty C_2$ | | |
|---|---|---|---|---|---|---|---|---|---|---|
| $A_{1g} \equiv \Sigma_g^+$ | 1 | 1 | ... | 1 | 1 | 1 | ... | 1 | | $x^2+y^2; z^2$ |
| $A_{2g} \equiv \Sigma_g^-$ | 1 | 1 | ... | −1 | 1 | 1 | ... | −1 | $R_z$ | |
| $E_{1g} \equiv \Pi_g$ | 2 | $2 \cos \phi$ | ... | 0 | 2 | $-2 \cos \phi$ | ... | 0 | $(R_x, R_y)$ | $(xz, yz)$ |
| $E_{2g} \equiv \Delta_g$ | 2 | $2 \cos 2\phi$ | ... | 0 | 2 | $2 \cos 2\phi$ | ... | 0 | | $[\frac{1}{\sqrt{2}}(x^2-y^2), xy]$ |
| ... | ... | ... | ... ... | ... | ... | ... | ... | | | |
| $A_{1u} \equiv \Sigma_u^+$ | 1 | 1 | ... | 1 | −1 | −1 | ... | −1 | | |
| $A_{2u} \equiv \Sigma_u^-$ | 1 | 1 | ... | −1 | −1 | −1 | ... | 1 | | |
| $E_{1u} \equiv \Pi_u$ | 2 | $2 \cos \phi$ | ... | 0 | −2 | $2 \cos \phi$ | ... | 0 | $(T_x, T_y)$ | $(x, y)$ |
| $E_{2u} \equiv \Delta_u$ | 2 | $2 \cos 2\phi$ | ... | 0 | −2 | $-2 \cos 2\phi$ | ... | 0 | | |
| ... | ... | ... | ... | ... | ... | ... | ... | ... | | |

obeyed. For instance, $E_1$ in $C_{\infty v}$ has the character 2 cos $\phi$ under $2C_\infty^\phi$ whereas $E_{1g}$ in $D_{\infty h}$ has a character of $-2$ cos $\phi$ under $2S_\infty^\phi$.

The reason for this apparent anomaly lies in the nature of $S_n$ operations. Consider $C_3$; the combination i.$C_3$ corresponds to an $S_6$ operation, not to an $S_3$ — it is i.$C_6$ which corresponds to an $S_3$. (See, for instance, Figure 7.5b; there is no $C_6$ axis in an octahedron, only $C_3$.) In general, corresponding to pure rotation operation involving a rotation of $\phi$ will be a rotation–inversion operation involving a rotation of $-(180-\phi)$. But cos $-(180-\phi) = -\cos\phi$, which explains the apparent anomalies in the $D_{\infty h}$ character table.

In previous character tables a prime (and double prime) notation has been used to indicate symmetry behaviour with respect to a mirror plane. These mirror planes were all of the $\sigma_h$ type. In $C_{\infty v}$ and $D_{\infty h}$, a +, − notation has been used but this is because the mirror planes are of the $\sigma_v$ type.

APPENDIX 4

# *The fluorine group orbitals of $\pi$ symmetry in $SF_6$*

It is inevitable that in the application of group theory to chemistry some short-cuts exist — and are exploited — which circumvent tedious or difficult mathematics. Thus, although at first sight it seems an advantage to have high symmetry this is sometimes not the case when carrying out a detailed calculation — for instance, there would be a considerable number of different interactions possible between two sets of triply degenerate orbitals in a bonding problem. In such a case it may help to pretend that the symmetry is lower than is in fact the case because the consequent reduced degeneracy forces a pairing between individual members of each set, thus reducing the number of interactions to be considered. Having paired the orbitals by this device, the low symmetry geometry can be forgotten and the correct point group used.

It is a similar trick which provides an alternative to the projection operator method of obtaining linear combinations of orbitals (Sections 4.6, 5.5 and 6.2) and which proves to be easier to use in high symmetry cases. It uses knowledge of the correct combinations in a lower symmetry case to obtain those of a higher symmetry molecule, the lower symmetry group being a subgroup of the higher. There is no unique path in this approach — different workers might choose different low symmetry groups. For a given choice of subgroup there may be several equally valid ways of proceeding. One develops a 'nose' for the method which is based on a mixture of experience and the ability to anticipate problems that will be encountered along each alternative path. Something of this 'nose' will be evident in the next section where we have tried to give the reasons for expecting a particular approach to be fruitful (or not, as the case may be).

In tackling the problem of generating the fluorine group orbitals of $\pi$ symmetry in $SF_6$ — of $O_h$ symmetry — we must first make a choice of lower symmetry group. It is usually sensible to choose the subgroup of the highest symmetry for which detailed results are available. In the present case this suggests that the $C_{4v}$ subgroup of $O_h$ be chosen because we have obtained some ligand group orbitals for a molecule of this symmetry — $BrF_5$ — in Chapter

5. It is true that in that chapter we only considered Br—F $\sigma$-bonding interactions but perhaps we can use these as a base from which to obtain the $\pi$ combinations. In Chapter 5 we explicitly recognized that the fluorine $\sigma$ orbitals would be mixtures of s and p atomic orbitals; for simplicity we there drew them as pure s orbitals. In Figure A4.1 we draw them again, but this time as pure p orbitals; the $C_{4v}$ symmetry labels are included. Suppose we tilt the p orbitals out of the plane, as in Figure A4.2. The symmetry labels of Figure A4.1 remain appropriate as do the linear combinations. This is really an indication that in $BrF_5$ there is no symmetry-dictated requirement that the Br—F $\sigma$-bonding orbitals have their maxima in the plane defined by the fluorine atoms. If we now complete the tilting processes we obtain Figure A4.3, which shows that we have obtained the $\pi$ orbital combinations starting from the $\sigma$. We were able to use this method because there is no operation in $C_{4v}$ which interchanges — and thus compares — the 'top' with the 'bottom' of each p orbital in Figure A4.3. We could not have used this trick in the $D_{4h}$ subgroup because the $\sigma_h$ mirror plane in that group gives this comparison and so distinguishes between $\sigma$ and $\pi$ orbitals. Nonetheless, the combinations shown in Figure A4.3 remain correct in $D_{4h}$ because $C_{4v}$ is also a subgroup of $D_{4h}$, but the symmetry labels would have to be changed.

The next step is that of recognizing that the twelve $p_\pi$ orbitals of the fluorine atoms in $SF_6$ are obtained if we correctly put together three planes of four atoms of the type shown in Figures A4.1 to A4.3. This step is shown in Figure A4.4.

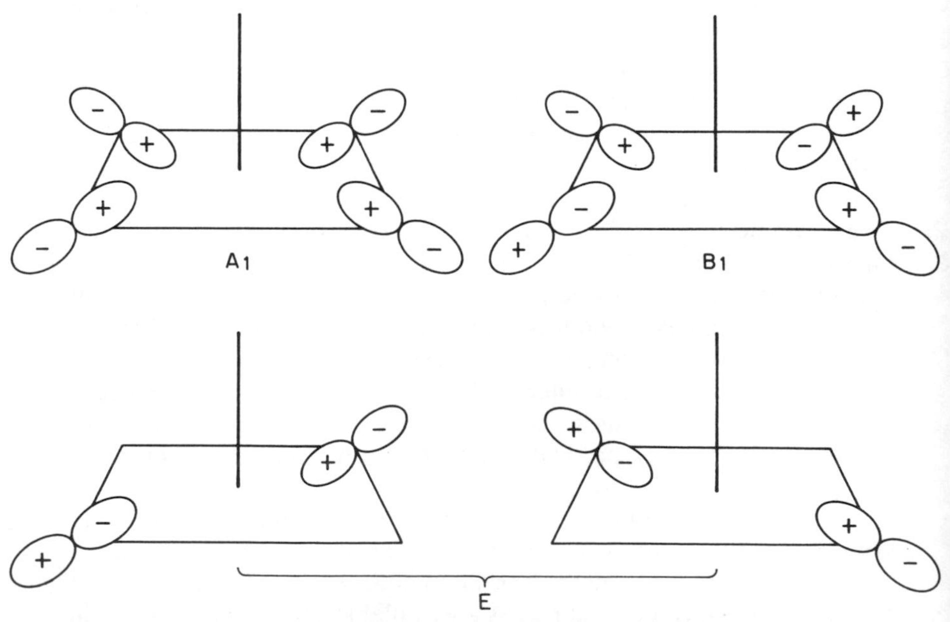


Figure A4.1 The pure p orbital representation of coplanar fluorine $\sigma$ orbital symmetry-adapted combinations in $BrF_5$

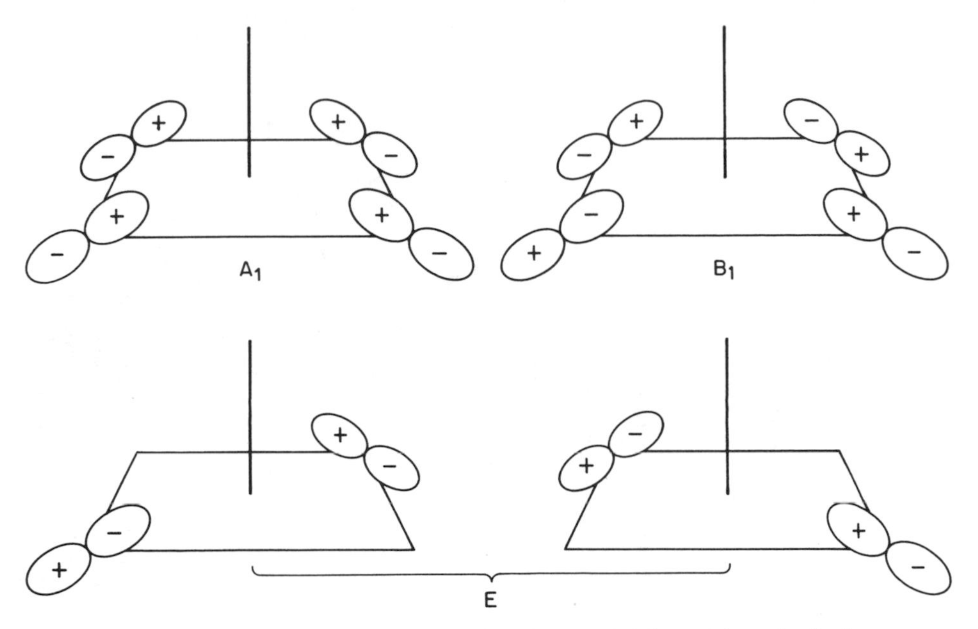


Figure A4.2 The same p orbitals as in Figure A4.1 but with each tilted out of the plane in the direction of the apical fluorine in $BrF_5$

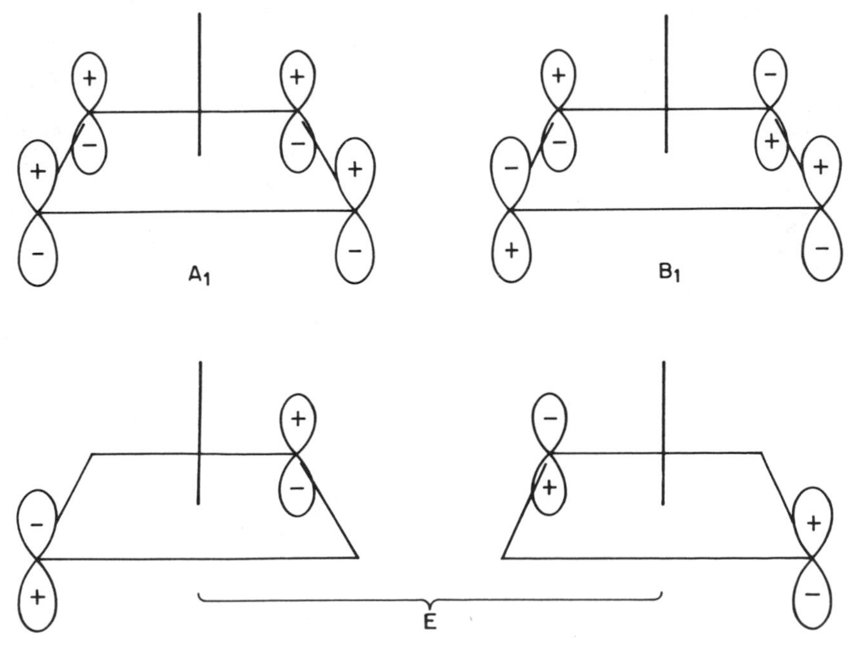


Figure A4.3 The same p orbital combinations as in Figure A4.1 but reoriented so as to be perpendicular to the original set

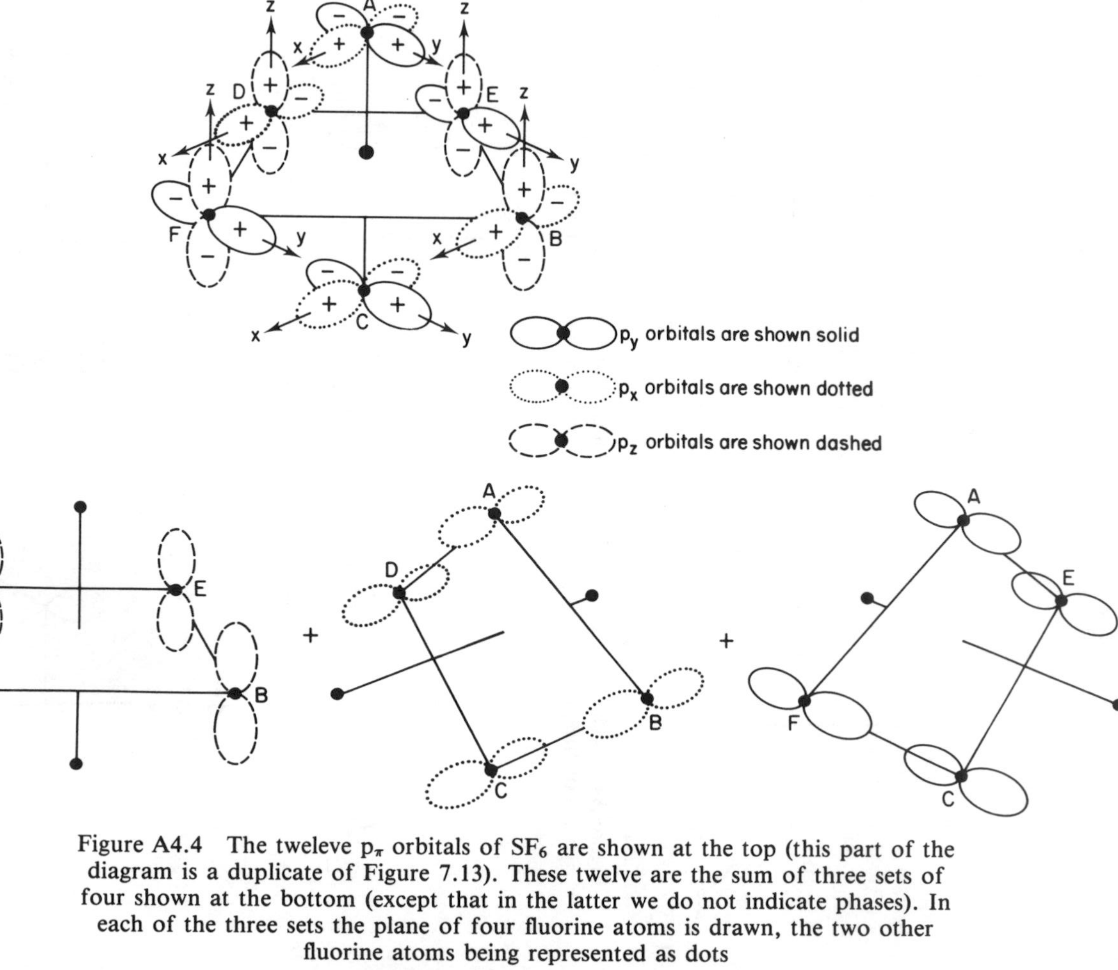


Figure A4.4 The tweleve $p_\pi$ orbitals of $SF_6$ are shown at the top (this part of the diagram is a duplicate of Figure 7.13). These twelve are the sum of three sets of four shown at the bottom (except that in the latter we do not indicate phases). In each of the three sets the plane of four fluorine atoms is drawn, the two other fluorine atoms being represented as dots

We recall that the twelve $\pi$ orbitals of Figure A4.4 transform as $T_{1g} + T_{1u} + T_{2g} + T_{2u}$; i.e. four different sets of triply degenerate orbitals. This triple degeneracy neatly matches the three planes, and associated sets of orbitals, shown in Figure A4.4. If we exploit this and group together the three orbitals which correspond to the $A_1$ combination in Figure A4.3:

$$\tfrac{1}{2}[p_z(B) + p_z(D) + p_z(E) + p_z(F)] \quad \Leftarrow$$
$$\tfrac{1}{2}[p_y(A) + p_y(C) + p_y(E) + p_y(F)]$$
$$\tfrac{1}{2}[p_x(A) + p_x(B) + p_x(C) + p_x(D)]$$

we obtain the $T_{1u}$ set of ligand $\pi$ orbitals in $O_h$ (as may be checked by considering their transformation as a set). The combination shown in Figure 7.15 is indicated by an arrow.

Similarly, the combinations corresponding to the $B_1$ in Figure A4.3 are

$$\tfrac{1}{2}[p_z(B) + p_z(D) - p_z(E) - p_z(F)] \quad \Leftarrow$$
$$\tfrac{1}{2}[p_y(A) + p_y(C) - p_y(F) - p_y(F)]$$
$$\tfrac{1}{2}[p_x(A) + p_x(B) - p_x(C) - p_x(D)]$$

which is the $T_{2u}$ set of ligand $\pi$ orbitals in $O_h$, the combination shown in Figure 7.17 being arrowed.

It is at this point that anticipation cautions us against plunging on and finishing the problem. There are two indicators that we should pause. Firstly, the next step would have involved three *pairs* of orbitals (the degenerate E set in Figure A4.3). We would obtain six orbitals, apparently all degenerate. However, we are looking for two sets of three ($T_{1g}$ and $T_{2g}$) and the two sets are not expected to be degenerate. Secondly, we have exercised some arbitrariness in the procedure that we have followed. In particular, when working with the planes shown in Figures A4.1 to A4.4 we have chosen to consider only $p_\pi$ orbitals which have their maximum amplitude perpendicular to this plane. We could equally have chosen to work with $p_\pi$ orbitals which 'lie in' the plane, as shown in Figure A4.5, although it is not immediately clear how we would have proceeded had we made this alternative choice.

Experience tells us that when six apparently degenerate orbitals are obtained this is because symmetry-distinct combinations have been mixed together. After all, this is always mathematically possible even if it is a step we would be unlikely to make from choice. Further, experience is that because a choice exists between 'perpendicular $p_\pi$ orbitals' and 'coplanar $p_\pi$ orbitals' we must expect each alternative to appear to an equal extent in the answer.

The way to extract symmetry-distinct combinations from sets in which they have been mixed together is to take suitable linear combinations of members of the mixed-up sets (a set of six orbitals in our case). These six (unnormalized) are:

$$\psi_1 = [p_z(E) - p_z(F)] \qquad \psi_2 = [p_z(B) - p_z(D)]$$
$$\psi_3 = [p_y(A) - p_y(C)] \qquad \psi_4 = [p_y(E) - p_y(F)]$$
$$\psi_5 = [p_x(A) - p_x(C)] \qquad \psi_6 = [p_x(B) - p_x(D)]$$

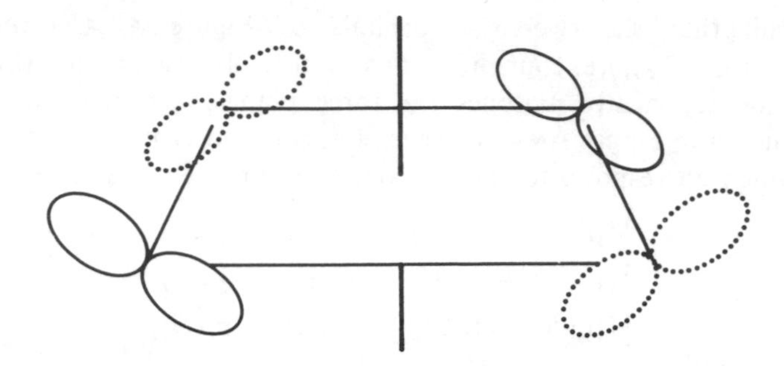

Figure A4.5 A set of four fluorine $p_\pi$ orbitals which lie in the plane of the fluorine atoms (cf. Figure A4.3 and the second part of Figure A4.4, where the sets of fluorine $p_\pi$ orbitals shown are all perpendicular to the plane of the four fluorine atoms)

Which orbitals should we combine together? It is here that the expectation of a 'coplanar' set of $p_\pi$ orbitals comes to our aid. We note that the set of 'coplanar' $p_\pi$ orbitals shown in Figure A4.5 contain contributions from

$$p_y(F), \quad p_x(B), \quad p_y(E) \quad \text{and} \quad p_x(D)$$

i.e. those orbitals contained in $\psi_4$ and $\psi_6$ in the list above. We expect, then, to have to combine these two. Becuse $\psi_4$ and $\psi_6$ are symmetry-equivalent (a $C_4$ rotation turns $\psi_4$ into $\psi_6$) we expect them to contribute equally to the combinations. The only way for this to occur is to combine them firstly with the same and then with opposite signs. We obtain (giving the final combinations in normalized form):

$$\psi_6 + \psi_4: \quad \tfrac{1}{2}[p_x(B) - p_x(D) + p_y(E) - p_y(F)]$$
$$\psi_6 - \psi_4: \quad \tfrac{1}{2}[p_x(B) - p_x(D) - p_y(E) + p_y(F)]$$

We proceed similarly with the pairs:

$$\psi_1 \text{ and } \psi_5$$
$$\psi_2 \text{ and } \psi_3$$

and thus obtain the complete sets:

$$
\begin{array}{ll}
T_{1g}: & \tfrac{1}{2}[p_x(A) - p_x(C) + p_z(E) - p_z(F)] \quad \Leftarrow \\
 & \tfrac{1}{2}[p_y(A) - p_y(C) - p_z(B) + p_z(D)] \\
 & \tfrac{1}{2}[p_x(B) - p_x(D) + p_y(E) - p_y(F)] \\
T_{2g}: & \tfrac{1}{2}[p_x(A) - p_x(C) - p_z(E) + p_z(F)] \quad \Leftarrow \\
 & \tfrac{1}{2}[p_y(A) - p_y(C) + p_z(B) - p_z(D)] \\
 & \tfrac{1}{2}[p_x(B) - p_x(D) - p_y(E) + p_y(F)]
\end{array}
$$

The combinations illustrated in Figures 7.14 ($T_{1g}$) and 7.16 ($T_{2g}$) are indicated by arrows.

How did we apportion the six combinations between the $T_{1g}$ and $T_{2g}$ sets? Formally, of course, the answer is 'by considering their transformations', but, fortunately, experience relieves us of the tedium of this step. The character table for the $O_h$ point group given in Appendix 3 shows that $T_{1g}$ functions have the characteristic of a rotation whilst $T_{2g}$ functions behave like products of coordinate axes. Figure A4.6 shows how these two observations may be used as a yardstick to discriminate between $T_{1g}$ and $T_{2g}$ functions.

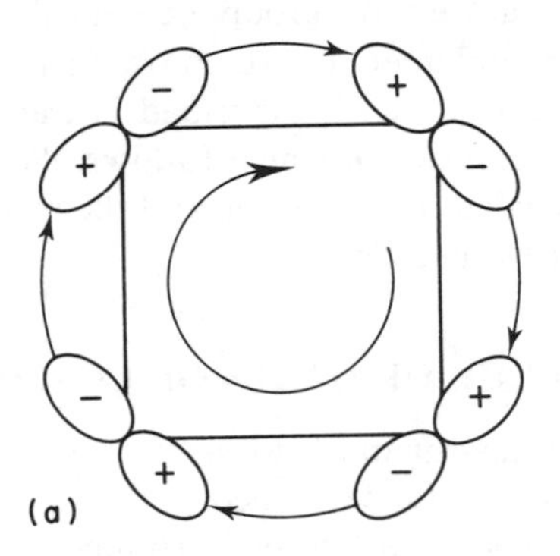


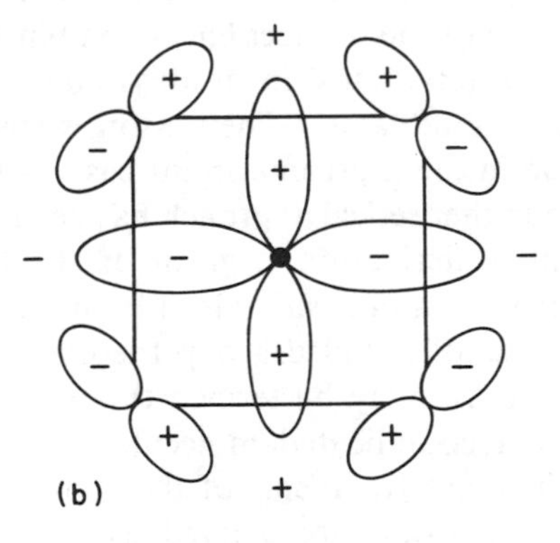


Figure A4.6 (a) A $T_{1g}$ function compared to a rotation (central arrow). Note the arrows drawn between lobes of adjacent $p_\pi$ orbitals. (b) Comparison of a typical $T_{2g}$ function (xy, $d_{xy}$, etc.) — at the centre — with the nodal pattern of the corresponding $T_{2g}$ combination of $p_\pi$ orbitals

This method appears to break down when there is no basis function listed against an irreducible representation in a character table — for instance in $O_h$ there is nothing listed against $A_{1u}$. This does not mean that no basis functions exist — they always do. Instead, they tend to be rather complicated, containing many nodes (thus, in Problem 4.2 we had to invoke the $f_{xyz}$ orbital). This itself is usually enough to identify functions transforming under such an irreducible representation — it seldom happens that one is interested in more than one of this type at a time.

In this appendix we have attempted to give some insight into the way that experienced practitioners tackle some group theoretical problems. The approach that we have used is complementary to more formal short-cut treatments which can be given, one of which is described in the reprint of an article in the *Journal of Chemical Education* which follows. The case of the fluorine $\pi$ orbitals in $SF_6$ is not included in this article and the reader may find it of value to extend the treatment to include it.

## LIGAND GROUP ORBITALS OF COMPLEX IONS*

Several articles which discuss ligand field theory reflect the growing interest in this refinement of simple crystal field theory.[1–5]

The reasons why covalency needs to be introduced into the latter theory are too well known to need elaboration here. Rather, we shall discuss a problem which arises in teaching the theory to undergraduate classes. As we have pointed out, the derivation of the 'correct linear combinations of lignd orbitals' in ligand field theory is a step almost invariably omitted in expositions of the subject suitable for undergraduates.[5] The reason is simple: the derivation is difficult. The derivation in the most important case — that of the octahedral complex — using a group theoretical approach has been discussed.[5] In the present article an alternative derivation is given of the form of ligand group orbitals (l.g.o's) which is more suitable for undergraduate tuition. In particular, no use is made of detailed group theory.

The method which we use may be termed the method of '*ascent in symmetry*'. The l.g.o's of a complicated molecule are derived from those of simpler 'molecules' which are fragments of the complicated one, i.e. vectors appropriate to any point group are derived as linear combination of the vectors of its subgroup, implicit use being made of the group correlation tables. The method is one of considerable power; e.g. one may obtain the l.g.o's for an icosahedral arrangement of equivalent $\sigma$-type orbitals in relatively simple form by this method. The standard group theoretical procedure is most unwieldy in this case, making it necessary to resort to a Schmidt orthogonalization procedure.

*Reprinted with minor alterations from an article by S. F. A. Kettle in the *Journal of Chemical Education*, **43** 652, December 1966. 

The method may conveniently be based on three axioms:

**Axiom 1** The l.g.o's of a complicated molecule are related to those of its fragments by the condition that only sets with non-zero overlap may interact.

**Axiom 2** Two equivalent orbitals are properly considered in in-phase and out-of-phase combinations. So, the correct combination of two localized orbitals $\sigma_1$ and $\sigma_2$ are, with neglect of overlap,*

$$\psi_s = \frac{1}{\sqrt{2}}(\sigma_1 + \sigma_2)$$

$$\psi_a = \frac{1}{\sqrt{2}}(\sigma_1 - \sigma_2)$$

**Axiom 3** If a set of l.g.o's is $d$-fold degenerate and is formed from $n$ equivalent orbitals then the sum of squares of coefficients with which each equivalent orbital appears in the set is $d/n$. This axiom may often be used in the simpler form that for every set of $n$ equivalent orbitals $(\sigma_1, \sigma_2, \ldots, \sigma_n)$ there is always a totally symmetric combination:

$$\frac{1}{\sqrt{n}}(\sigma_1 + \sigma_2 + \cdots + \sigma_n)$$

We now illustrate the use of these axioms by deriving the l.g.o's appropriate to five different stereochemistries. In all cases we include group theoretical labels although these are not essential to the argument.

**Example 1** The $\sigma$ l.g.o's of a planar $AB_3$ molecule ($D_{3h}$ symmetry).

Label the $\sigma$ orbitals $\sigma_1$, $\sigma_2$ and $\sigma_3$. Consider $\sigma_1$ and $\sigma_2$. From Axiom 2 the correct combinations, neglecting overlap, are:

$$\psi_s = \frac{1}{\sqrt{2}}(\sigma_2 + \sigma_2)$$

$$\psi_a = \frac{1}{\sqrt{2}}(\sigma_1 - \sigma_2)$$

By Axiom 1, of these only $\psi_s$ can interact with $\sigma_3$ (the nodal plane implicit in $\psi_a$ bisects $\sigma_3$). We have, then:

$$\psi_1 = \frac{1}{\sqrt{1+\lambda^2}}(\psi_s + \lambda\sigma_3)$$

*For consistency, we assign a positive phase to each localized orbital.

and

$$\psi_2 = \frac{1}{\sqrt{1+\lambda^2}}\,(\lambda\psi_3 - \sigma_3)$$

where the constant $\lambda$ has to be determined. Now, from Axiom 3 the first of these combinations must be, in expanded form, $\psi_1 = (1/\sqrt{3})(\sigma_1 + \sigma_2 + \sigma_3)$ so, by comparison of coefficients, $\lambda = 1/\sqrt{2}$ (we need only consider the positive root). It follows that $\psi_2 = (1/\sqrt{6})\,(\sigma_1 + \sigma_2 - 2\sigma_3)$. $\psi_1$ is of $A_1'$ symmetry and $\psi_a$ and $\psi_2$ together transform as $E'$.

**Example 2** The $\sigma$ l.g.o's of a planar $AB_4$ molecule ($D_{4h}$ symmetry).

Label the $\sigma$ orbitals cyclically $\sigma_1$, $\sigma_2$, $\sigma_3$ and $\sigma_4$. Consider the pairs $\sigma_1$ and $\sigma_3$; $\sigma_2$ and $\sigma_4$. By Axiom 2 we have the combinations:

$$\psi_1 = \frac{1}{\sqrt{2}}\,(\sigma_1 + \sigma_3) \qquad \psi_2 = \frac{1}{\sqrt{2}}\,(\sigma_1 - \sigma_3)$$

$$\psi_3 = \frac{1}{\sqrt{2}}\,(\sigma_2 + \sigma_4) \qquad \psi_4 = \frac{1}{\sqrt{2}}\,(\sigma_2 - \sigma_4)$$

The nodal plane implicit in $\psi_2$ contains atoms 2 and 4. Similarly, the $\psi_4$ nodal plane contains atoms 1 and 3. It follows, from Axiom 1, that only $\psi_1$ and $\psi_3$ interact. From Axiom 3, one combination is $\psi_5 = \frac{1}{2}(\sigma_1 + \sigma_2 + \sigma_3 + \sigma_4)$, i.e. $(1/\sqrt{2})\,(\psi_1 + \psi_3)$, so the other must be $\psi_6 = \frac{1}{2}(\sigma_1 - \sigma_2 + \sigma_3 - \sigma_4)$, i.e. $(1/\sqrt{2})\,(\psi_1 - \psi_3)$. $\psi_5$ is of $A_{1g}$ symmetry, $\psi_2$ and $\psi_4$ together transform under the $E_u$ irreducible representation and $\psi_6$ is of $B_{2g}$ symmetry.

**Example 3** The $\sigma$ l.g.o's of a tetrahedral $AB_4$ molecule ($T_d$ symmetry).

The derivation in this case is identical to that in Example 2. $\psi_5$ transforms as $A_1$ and $\psi_2$, $\psi_4$ and $\psi_6$ as $T_2$. In this $T_2$ set the Cartesian coordinates onto which $\psi_2$, $\psi_4$ and $\psi_6$ have a one-to-one mapping are not equivalently orientated. Two, those which map onto $\psi_2$ and $\psi_4$, pass through the edges of the cube corresponding to the tetrahedron, but the third passes through the mid-point of faces. The $T_2$ l.g.o. set which maps onto the usual choice of Cartesian axes for the tetrahedron is:

$$\frac{1}{\sqrt{2}}(\psi_2 + \psi_4) = \tfrac{1}{2}(\sigma_1 + \sigma_2 - \sigma_3 - \sigma_4)$$

$$\frac{1}{\sqrt{2}}(\psi_2 - \psi_4) = \tfrac{1}{2}(\sigma_1 - \sigma_2 - \sigma_3 + \sigma_4)$$

and

$$\psi_6 = \tfrac{1}{2}(\sigma_1 - \sigma_2 + \sigma_3 - \sigma_4)$$

**Example 4** The $\sigma$ l.g.o's of an octahedral $AB_6$ molecule ($O_h$ symmetry).

We isolate four ligand $\sigma$ orbitals in a plane and label them cyclically

$\sigma_1$, $\sigma_2$, $\sigma_3$, and $\sigma_4$. The correct combinations for this set are given in Example 2. Above and below this plane, respectively, lie the orbitals $\sigma_5$ and $\sigma_6$.

We consider the combinations:

$$\psi_1 = \tfrac{1}{2}(\sigma_1 + \sigma_2 + \sigma_3 + \sigma_4)$$

$$\psi_2 = \frac{1}{\sqrt{2}}(\sigma_1 - \sigma_3) \qquad \psi_5 = \frac{1}{\sqrt{2}}(\sigma_5 + \sigma_6)$$

$$\psi_3 = \frac{1}{\sqrt{2}}(\sigma_2 - \sigma_4) \qquad \psi_6 = \frac{1}{\sqrt{2}}(\sigma_5 - \sigma_6)$$

$$\psi_4 = \tfrac{1}{2}(\sigma_1 - \sigma_2 + \sigma_3 - \sigma_4)$$

Axiom 1, applied by the 'nodal plane' criterion, shows that in the two sets only $\psi_1$ and $\psi_5$ are non-orthogonal. The combination $(1/\sqrt{1+\lambda^2})$ $(\psi_1 + \lambda\psi_5)$ leads to (Axiom 3)

$$\psi_7 = \frac{1}{\sqrt{6}}(\sigma_1 + \sigma_2 + \sigma_3 + \sigma_4 + \sigma_5 + \sigma_6) = \sqrt{\frac{2}{3}}\,\psi_1 + \frac{1}{\sqrt{3}}\,\psi_5$$

It follows, by comparison of coefficients, that $\lambda = (1/\sqrt{2})$ so that the combination:

$$\frac{1}{\sqrt{1+\lambda^2}}(\lambda\psi_1 - \psi_5)$$

is

$$\psi_8 = \frac{1}{\sqrt{12}}(\sigma_1 + \sigma_2 + \sigma_3 + \sigma_4 - 2\sigma_5 - 2\sigma_6)$$

$\psi_7$ is of $A_{1g}$ symmetry, $\psi_4$ and $\psi_8$ transform as $E_g$ and $\psi_2$, $\psi_3$ and $\psi_6$ as $T_{1u}$.

**Example 5** The $\sigma$ l.g.o's of an $AB_8$ Archimedian antiprismatic molecule ($D_{4d}$ symmetry).

This example again uses the results of Example 2 by considering the allowed combinations between two square planar arrangements of ligand orbitals, rotated with respect to one another by 45°. In order to use Axiom 1 the nodal planes of the two sets must be brought into coincidence. This involves the rotation of coordinate axes as discussed in Exmaple 3. Label the ligand orbitals cyclically $\sigma_1, \ldots, \sigma_8$, those of one plane being $\sigma_1, \ldots, \sigma_4$ and those of the other $\sigma_5, \ldots, \sigma_8$. $\sigma_5$ is positioned so that viewed down the fourfold rotation axis it appears to lie between $\sigma_1$ and $\sigma_2$.

Appropriate combinations are:

$$\psi_1 = \tfrac{1}{2}(\sigma_1 + \sigma_2 + \sigma_3 + \sigma_4) \qquad \psi_5 = \tfrac{1}{2}(\sigma_5 + \sigma_6 + \sigma_7 + \sigma_8)$$

$$\psi_2 = \frac{1}{\sqrt{2}}(\sigma_1 - \sigma_3) \qquad \psi_6 = \tfrac{1}{2}(\sigma_5 - \sigma_6 - \sigma_7 + \sigma_8)$$

$$\psi_3 = \frac{1}{\sqrt{2}}(\sigma_2 - \sigma_4) \qquad \psi_7 = \tfrac{1}{2}(\sigma_5 + \sigma_6 - \sigma_7 - \sigma_8)$$

$$\psi_4 = \tfrac{1}{2}(\sigma_1 - \sigma_2 + \sigma_3 - \sigma_4) \qquad \psi_8 = \tfrac{1}{2}(\sigma_5 - \sigma_6 + \sigma_7 - \sigma_8)$$

Application of Axiom 1 shows that we must consider further combinations between the pairs:

$\psi_1$ and $\psi_5$

$\psi_2$ and $\psi_6$

$\psi_3$ and $\psi_7$

The first pair gives

$$\psi_9 = \frac{1}{\sqrt{8}}(\sigma_1 + \sigma_2 + \sigma_3 + \sigma_4 + \sigma_5 + \sigma_6 + \sigma_7 + \sigma_8)$$

$$\psi_{10} = \frac{1}{\sqrt{8}}(\sigma_1 + \sigma_2 + \sigma_3 + \sigma_4 - \sigma_5 - \sigma_6 - \sigma_7 - \sigma_8)$$

while use of Axiom 3, in its more detailed form, shows that the correct combinations of $\psi_2$ and $\psi_6$ are

$$\psi_{11} = \tfrac{1}{2}(\sigma_1 - \sigma_3) + \frac{1}{2\sqrt{2}}(\sigma_5 - \sigma_6 - \sigma_7 + \sigma_8)$$

$$\psi_{12} = \tfrac{1}{2}(\sigma_1 - \sigma_3) - \frac{1}{2\sqrt{2}}(\sigma_5 - \sigma_6 - \sigma_7 + \sigma_8)$$

and of $\psi_3$ and $\psi_7$ are

$$\psi_{13} = \tfrac{1}{2}(\sigma_2 - \sigma_4) + \frac{1}{2\sqrt{2}}(\sigma_5 + \sigma_6 - \sigma_7 - \sigma_8)$$

$$\psi_{14} = \tfrac{1}{2}(\sigma_2 - \sigma_4) - \frac{1}{2\sqrt{2}}(\sigma_5 + \sigma_6 - \sigma_7 - \sigma_8)$$

since it is evident that $\psi_{11}$ and $\psi_{13}$ must be degenerate as must also be $\psi_{12}$ and $\psi_{14}$. The symmetries of these combinations are

$\psi_4$ and $\psi_5 : E_2$

$\psi_9 : A_1$

$\psi_{10} : B_2$

$\psi_{11}$ and $\psi_{13} : E_1$

$\psi_{12}$ and $\psi_{14} : E_3$

**Example 6** As an example of the application of the method to combinations of ligand orbitals of diatomic $\pi$ symmetry we consider the l.g.o's of $\pi$ symmetry in a tetrahedral complex. Although the standard technique can be used to obtain the correct combinations, in practice the calculation is rather difficult.

We choose axes and orientations as shown in the diagram. One set of ligand $\pi$ orbitals (labelled $\alpha$) is 'coplanar' with the z axis. The other set (labelled $\beta$) lies in planes perpendicular to the z axis. If the x and y axes are chosen as shown some slight simplification results. Our basic combinations are

$$\psi_1 = \frac{1}{\sqrt{2}}(\alpha_1 + \alpha_2) \qquad \psi_2 = \frac{1}{\sqrt{2}}(\alpha_1 - \alpha_2)$$

$$\psi_3 = \frac{1}{\sqrt{2}}(\alpha_3 + \alpha_4) \qquad \psi_4 = \frac{1}{\sqrt{2}}(\alpha_3 - \alpha_4)$$

$$\psi_5 = \frac{1}{\sqrt{2}}(\beta_1 + \beta_2) \qquad \psi_6 = \frac{1}{\sqrt{2}}(\beta_1 - \beta_2)$$

$$\psi_7 = \frac{1}{\sqrt{2}}(\beta_3 + \beta_4) \qquad \psi_8 = \frac{1}{\sqrt{2}}(\beta_3 - \beta_4)$$

Following the usual procedure it is readily seen that

$$\psi_9 = \frac{1}{\sqrt{2}}(\psi_1 + \psi_3) = \tfrac{1}{2}(\alpha_1 + \alpha_2 + \alpha_3 + \alpha_4)$$

and

$$\psi_{10} = \frac{1}{\sqrt{2}}(\psi_1 + \psi_3) = \tfrac{1}{2}(\alpha_1 + \alpha_2 - \alpha_3 - \alpha_4)$$

are orthogonal to all other combinations. It follows that they must be members of degenerate sets for they contain no $\beta$ component (cf. Axiom 3). Consider $\psi_9$. This obviously transforms like the z axis and so will be a member of a triply degenerate set of which the other components transform as x and y. It follows that $d/n = \frac{3}{8}$. However, the coefficient of the $\alpha$'s, squared, in $\psi_9$ is $\frac{1}{4}$ so there must be an $\alpha$ component in the x and y transforming members. Evidently, these components are derived from $\psi_2$(y) and $\psi_4$(x), each of which must appear with a coefficient of $\frac{1}{2}$. ($\frac{3}{8} - \frac{1}{4} = \frac{1}{8} = [\frac{1}{2}(2/\sqrt{2})]^2$). Now, $\psi_6$ transforms like x and $\psi_8$ like y so we are evidently seeking combinations of $\psi_6$ with $\psi_4$ and of $\psi_8$ with $\psi_2$. The correct combinations are

$$\psi_{11} = \frac{\sqrt{3}}{2}\psi_6 - \tfrac{1}{2}\psi_4 = \frac{1}{2\sqrt{2}}[\sqrt{3}(\beta_1 - \beta_2) - \alpha_3 + \alpha_4]$$

and

$$\psi_{12} = \frac{\sqrt{3}}{2}\psi_8 - \tfrac{1}{2}\psi_2 = \frac{1}{2\sqrt{2}}[\sqrt{3}(\beta_3 - \beta_4) - \alpha_1 + \alpha_2]$$

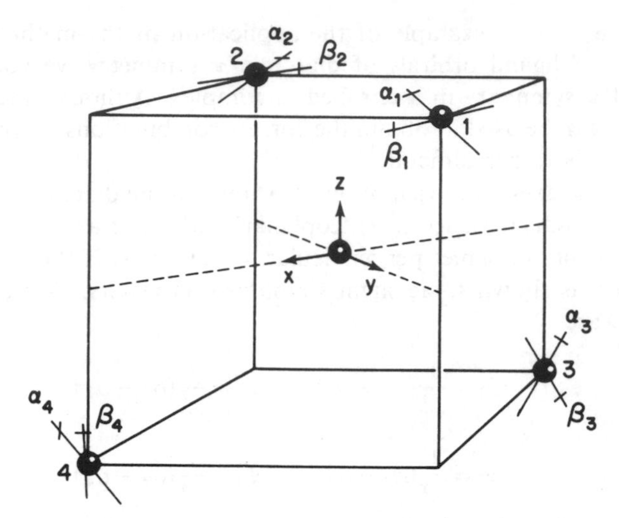


Figure A4.7 Cartesian axes and orientation of ligand $\pi$ orbitals in a tetrahedral complex ion

where we have been careful to make sure that the phases of $\psi_6$ and $\psi_4$ mapping onto x and of $\psi_8$ and $\psi_2$ onto y are identical.

Combinations of $\psi_6$ and $\psi_4$ and of $\psi_8$ and $\psi_2$ orthogonal to $\psi_{11}$ and $\psi_{12}$ are

$$\psi_{13} = \tfrac{1}{2}\psi_6 + \frac{\sqrt{3}}{2}\psi_4 = \frac{1}{2\sqrt{2}}\,[\beta_1 - \beta_2 + \sqrt{3}(\alpha_3 - \alpha_4)]$$

$$\psi_{14} = \tfrac{1}{2}\psi_8 + \frac{\sqrt{3}}{2}\psi_2 = \frac{1}{2\sqrt{2}}\,[\beta_3 - \beta_4 + \sqrt{3}(\alpha_1 - \alpha_2)]$$

We have only to deal with $\psi_5$ and $\psi_7$, which are not orthogonal; the correct combinations are

$$\psi_{16} = \tfrac{1}{2}(\beta_1 + \beta_2 + \beta_3 + \beta_4)$$
$$\psi_{17} = \tfrac{1}{2}(\beta_1 + \beta_2 - \beta_3 - \beta_4)$$

The symmetries of these combinations are

$$\psi_9,\ \psi_{11},\ \psi_{12} : \mathrm{T}_2$$
$$\psi_{13},\ \psi_{14},\ \psi_{16} : \mathrm{T}_1$$
$$\psi_{10},\ \psi_{17} : \mathrm{E}$$

## REFERENCES

1. W. Manch and W. C. Fernelius, *J. Chem. Educ.*, **38**, 192 (1961).
2. A. D. Liehr, *J. Chem. Educ.*, **39**, 135 (1962).
3. F. A. Cotton, *J. Chem. Educ.*, **41**, 466 (1964).
4. H. B. Gray, *J. Chem. Educ.*, **41**, 2 (1964).
5. S. F. A. Kettle, *J. Chem. Educ.*, **43**, 21 (1966).

# APPENDIX 5

# *The $C_{\infty v}$ and $D_{\infty h}$ groups*

The method developed in the text for the reduction of reducible representations does not work for infinite groups. This is because in the usual reduction formula (Equation A2.43)

$$a_i = \frac{1}{h} \sum_R \chi(R)\chi_i(R)$$

the quantity h is infinite, as is the set of R over which the summation has to be done. Instead, it is simplest to work with the equation from which this is derived (equation A2.41) and in which $a_i$ appears on the right-hand side:

$$\chi(R) = \sum_i a_i \chi_i(R) \tag{A5.1}$$

There have been several papers in the chemical education literature on the reduction of reducible representations of infinite point groups, most of them being based on equation (A5.1).[1-7] Comparison of them makes interesting reading; the method given in this appendix differs — just a little — from all of them.

The $C_{\infty v}$ character table is

| $C_{\infty v}$ | $E$ | $2C_\infty^\phi$ | ... | $\infty\sigma_v$ |
|---|---|---|---|---|
| $\Sigma^+$ | 1 | 1 | ... | 1 |
| $\Sigma^-$ | 1 | 1 | ... | −1 |
| $\Pi$ | 2 | $2\cos\phi$ | ... | 0 |
| $\Delta$ | 2 | $2\cos 2\phi$ | ... | 0 |
| $\Phi$ | 2 | $2\cos 3\phi$ | ... | 0 |
| ⋮ | ⋮ | ⋮ | ... | ⋮ |

A general reducible representation ($\Gamma_{red}$) may have contributions from all irreducible representations, i.e. be of the form

$$\Gamma_{red} = a_1\Sigma^+ + a_2\Sigma^- + a_3\Pi + a_4\Delta + a_5\Phi + \cdots$$

so the problem is to determine the coefficients $a_1$, $a_2$, $a_3$, ..., etc. Under the operation E the character of this reducible representation will be

$$\chi_1 = a_1 + a_2 + 2(a_3 + a_4 + a_5 + a_6 + \cdots) \tag{A5.2}$$

because $\Sigma^+$ and $\Sigma^-$ each have characters of unity under this operation and all other irreducible representations a character of 2. Similarly, under the operation $C_\infty{}^\phi$ the character of the reducible representation is

$$\chi_2{}^\phi = a_1 + a_2 + 2(a_3 \cos\phi + a_4 \cos 2\phi + a_5 \cos 3\phi + a_6 \cos 4\phi + \cdots)$$

Finally, under the operation $\sigma_v$ the reducible representation has a character

$$\chi_3 = a_1 - a_2 \tag{A5.3}$$

We have, then, a series of equations from which the unknown coefficients $a_1$, $a_2$, $a_3$, ... can be determined, often by inspection. The fact that there are apparently only three equations but possibly more than three unknowns is no problem because there are really as many equations as needed. This is because it is possible to extract subsidiary equations by equating coefficients of, for example, $\cos\phi$, of $\cos 2\phi$, and so on. Thus, if $\Gamma_{red}$ contains a term $n \cos\phi$ under the $C_\infty{}^\phi$ operation this can only arise from the $\Pi$ irreducible representation and we can immediately conclude that $a_3 = n/2$.

For some cases it may be simpler to allow $\phi$ to assume specific values; in such cases $\chi_2{}^\phi$ may be chosen to become selective of the terms within the brackets. Thus, for $\phi = 90^\circ$ $\cos\phi = \cos 3\phi = \cos 5\phi = \cdots = 0$, but $\cos 2\phi = \cos 6\phi = \ldots = -1$ and $\cos 4\phi = \cos 8\phi = 1$ giving

$$\chi_2{}^{90} = a_1 + a_2 + 2(-a_4 + a_6 \cdots) \tag{A5.4}$$

whilst for $\phi = 180^\circ$ $\cos\phi = \cos 3\phi = \cos 5\phi = \cdots = -1$ and $\cos 2\phi = \cos 4\phi = \cdots = 1$, giving

$$\chi_2{}^{180} = a_1 + a_2 + 2(-a_3 + a_4 - a_5 + a_6 \cdots) \tag{A5.5}$$

In this way we can generate as many simultaneous equations in the unknowns, the $a_n$'s, as are needed to solve the problem. Our second example illustrates the use of this method. In practice, if there are more than a few terms the method becomes cumbersome.

**Example 1**

*Problem:* What are the symmetry species of the vibrations of a linear X—Y—Z molecule?

*Solution:* Following the method developed in Chapter 9 we consider the transformation of Cartesian displacement coodinates associated with each atom of the molecule:

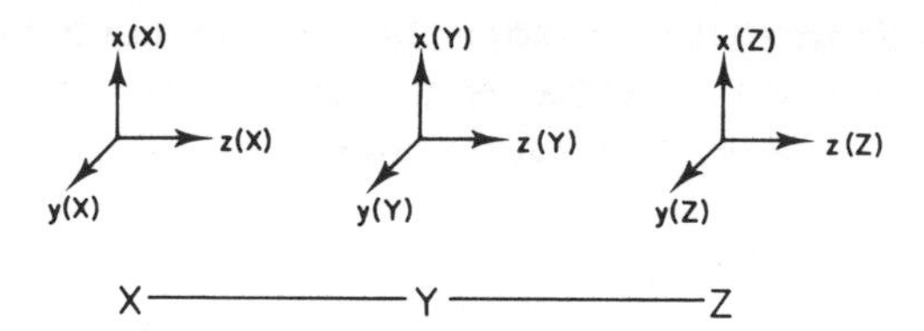


and obtain the reducible representation

| | E | $2\,C_\infty^\phi$ | $\infty\sigma_v$ |
|---|---|---|---|
| $\Gamma_{red}$: | 9 | $3+6\cos\phi$ | 3 |

(Under the $C_\infty^\phi$ rotations each of the z displacement coordinates remains itself; each of the x and y are rotated by $\phi$ and so the projection of the rotated displacement onto the original is $\cos\phi$ – see either equation (A2.6) or Section 6.1).

As indicated above, the $\cos\phi$ factor can at once be separated out from this reducible representation; given a character of $6\cos\phi$ under $2C_\infty^\phi$, the characters under E and $\infty\sigma_v$ follow:

| E | $2C_\infty^\phi$ | $\infty\sigma_v$ |
|---|---|---|
| 6 | $6\cos\theta$ | 0 |

which is $3\Pi$ (see the $C_{\infty v}$ character table above).

Subtracting these characters from the original reducible representation we are left with

| E | $2C_\infty^\phi$ | $\infty\sigma_v$ |
|---|---|---|
| 3 | 3 | 3 |

which is $3\,\Sigma^+$.

We conclude that the original reducible representation reduces into the components

$$3\Sigma^+ + 3\Pi$$

We use the character table for $C_{\infty v}$ given in Appendix 3 to determine the symmetry species of the molecular translations and rotations. However, because the molecule is linear the $3N-5$ rule operates and we do not include $R_z$ in the rotations. From Appendix 3 we have

Translations: $\Sigma^+ + \Pi$

Rotations: $\Pi$

We conclude that the vibrations of a linear X—Y—Z molecule transform as

$$2\Sigma^+ + \Pi$$

(The two $\Sigma^+$ vibrations arise from the X—Y and Y—Z stretching motions — the $\Pi$ from the X—Y—Z angle deformation.)

*Comment:* Had we in this development obtained a term in cos $2\phi$ under $2C_\infty{}^\phi$ we would first have extracted a $\Delta$ contribution, a term in cos $3\phi$ would have led to a $\Phi$ contribution, and so on.

**Example 2**

*Problem:* The central atom in a linear X—Y—Z molecule is a transition metal atom. Determine the symmetries of its d orbitals.

*Comment:* Given the $\phi$ angular dependence of the d orbitals — and this is explicit in their algebraic forms — we could use the first method above. However, many will prefer to use schematic diagrams of the angular dependence and the present method is well suited to this, more qualitative, description.

*Solution:* Take the coordinate axes of the central atom to be those given in Example 1. By drawing the individual d orbitals (and viewing them 'down' the z axis, as is appropriate for a rotation about this axis) or by inspection we can derive characters. We take specific values of $\phi$ of 90° and 180° to obtain the characters

| | E | $2C_\infty{}^{90}$ | $2C_\infty{}^{180}$ | $\infty\sigma_v$ |
|---|---|---|---|---|
| $d_{z^2}$: | 1 | 1 | 1 | 1 |
| $d_{xz}$, $d_{zy}$: | 2 | 0 | −2 | 0 |
| $d_{x^2-y^2}$, $d_{xy}$: | 2 | −2 | 2 | 0 |
| $\Gamma_{red}$: | 5 | −1 | 1 | 1 |

It is, of course, simplest not to work with $\Gamma_{red}$ but, separately, with each of the three representations of which it is the sum. However, for the purpose of the present example, we will tackle the more difficult problem of working with $\Gamma_{red}$ in a systematic manner.

We do this by the use of equations (A5.2) to (A5.4). In these equations we must make a decision about how many of $a_1$, $a_2$, $a_3$, . . . to include on the right-hand side. A wrong choice will either lead to a set of equations which are not internally consistent or, the final check, an answer which does lead to the regeneration of the original reducible representation. In the present example we will only include $a_1$, $a_2$, $a_3$ and $a_4$ on the right-hand side of these equations (we have four characters of the reducible representation and so can determine four unknowns); if we wished to include $a_5$ on the right-hand side we would have had to include another rotation, $C_\infty{}^{45}$, for instance.

The substitutions give

In (A5.2)(E):

$$5 = a_1 + a_2 + 2a_3 + 2a_4$$

In (A5.4)($2C_\infty^{90}$):

$$-1 = a_1 + a_2 \qquad - 2a_4$$

In (A5.5)($2C_\infty^{180}$):

$$1 = a_1 + a_2 - 2a_3 + 2a_4$$

In A5.3($\infty\sigma_v$):

$$1 = a_1 - a_2$$

By straightforward manipulation of these equations we deduce

$$a_1 = 1, \qquad a_2 = 0, \qquad a_3 = 1, \qquad a_4 = 1$$

i.e.$\Gamma_{red} = \Sigma^+ + \Pi + \Delta$.

Had we worked with the three simpler problems rather than the complicated one we would have identified $d_{z^2}$ as $\Sigma^+$, ($d_{xz}$, $d_{yz}$) as $\Pi$ and ($d_{x^2-y^2}$, $d_{xy}$) as $\Delta$.

**Problem A5.1** Show that if an attempt is made to solve the above problem using only $a_1$, $a_2$ and $a_3$ then the equations obtained are not internally consistent.

**Problem A5.2** Show that the set of p orbitals on an atom in a molecule of $C_{\infty v}$ symmetry transform as

$$\Sigma^+ + \Pi$$

Use this result to show that the configuration $p^1p^1$ (where two different sets of p orbitals are involved) on such an atom gives rise to states which transform as

$$2\Sigma^+ + \Sigma^- + 3\Pi$$

(*Hint:* consider the direct product $(\Sigma^+ + \Pi) \times (\Sigma^+ + \Pi)$).

In this appendix we have used $C_{\infty v}$ as an illustrative infinite point group; the methods described may be immediately extended to $D_{\infty h}$ by including characters appropriate to $S_n$ and $C_2$ operations.

## REFERENCES


1. L. Schäfer and S. J. Cyvin, *J. Chem. Ed.*, **48**, 295 (1971).
2. D. P. Strommen and E. P. Lippincott, *J. Chem. Ed.*, **49**, 341 (1972).
3. H. P. Fritzer, *Match*, **3**, 21 (1977).
4. J. M. Alvariño, *J. Chem. Ed.*, **55**, 307 (1978).
5. R. L. Flurry Jr., *J. Chem. Ed.*, **56**, 638 (1979).
6. D. P. Strommen, *J. Chem. Ed.*, **56**, 640 (1979).
7. J. M. Alvariño and A. Chamorro, *J. Chem. Ed.*, **57**, 785 (1980).

# Index